INFORMED ASSESSMENTS
An Introduction to Information, Entropy and Statistics

Alan Jessop

ELLIS HORWOOD
NEW YORK LONDON TORONTO SYDNEY TOKYO SINGAPORE

First published 1995 by
Ellis Horwood Limited
Campus 400, Maylands Avenue
Hemel Hempstead
Hertfordshire, HP2 7EZ
A division of
Simon & Schuster International Group

© Ellis Horwood 1995

All rights reserved. No part of this publication may be reproduced, stored in a retrieval system, or transmitted, in any form, or by any means, electronic, mechanical, photocopying, recording or otherwise, without prior permission, in writing, from the publisher.

Typeset by BeWare CRC

Printed and bound in Great Britain by
Redwood Books, Trowbridge, Wiltshire

Library of Congress Cataloging-in-Publication Data

Jessop, Alan.
   Informed assessments : an introduction to information, entropy, and statistics / Alan Jessop.
   p. cm.
   Includes bibliographical references and index.
   ISBN 0-13-109299-5
   1. Uncertainty (Information theory) 2. Entropy (Information theory) 3. Bayesian statistical decision theory. I. Title.
Q375.J47 1994
003-dc20
                                                    94-42323
                                                      CIP

British Library Cataloguing in Publication Data

A catalogue record for this book is available from the British Library

ISBN 0-13-109229-5

1 2 3 4 5   99 98 97 96 95

INFORMED ASSESSMENTS

Contents

Preface — xi

Acknowledgements — xiii

Chapter 1 Probability

    1.1   A little history — 1
    1.2   More than one event — 4
    1.3   Joint events — 7
    1.4   Distributions — 10
    1.5   Density functions — 13
    1.6   Summary measures — 18
    1.7   Expectation — 24
    1.8   Further reading — 25
    1.9   Examples — 26
    1.10 Exercises — 35

Chapter 2 Information

    2.1   Surprise — 39
    2.2   Searching — 42
    2.3   Signalling — 45
    2.4   Disorder — 49
    2.5   Drawing together — 50
    2.6   Redundancy — 51
    2.7   Further reading — 52
    2.8   Examples — 53
    2.9   Exercises — 56

Chapter 3 Unbiased estimates

    3.1   Jaynes' criterion — 58
    3.2   Finding unbiased estimates — 60
    3.3   Assessment when the mean is known — 63
    3.4   Assessment when mean and variance are given — 66
    3.5   Further reading — 69
    3.6   Examples — 70
    3.7   Exercises — 73

Chapter 4 More on unbiased estimates

4.1	Information change	75
4.2	Using an existing distribution	77
4.3	Using information about parts of the distribution	79
4.4	The answer!	82
4.5	Maximum entropy density functions	83
4.6	The negative exponential distribution	88
4.7	The Normal distribution	90
4.8	Bounded real variable: mean and variance given	92
4.9	Some practical considerations	93
4.10	Examples	95
4.11	Exercises	102

Chapter 5 Bayes' Rule

5.1	Background and basic statement	105
5.2	Bayes' Rule with a Binomial likelihood	109
5.3	Convenience priors: the Beta distribution	112
5.4	Poisson likelihood and Gamma prior	116
5.5	Normal likelihood and prior	120
5.6	Normal approximations	122
5.7	Fitting distributions	122
5.8	Bayesian learning	124
5.9	Further reading	127
5.10	Examples	128
5.11	Exercises	138

Chapter 6 Decision theory

6.1	Decisions	141
6.2	Utility	143
6.3	A decision rule	148
6.4	Loss	149
6.5	Further reading	150
6.6	Examples	151
6.7	Exercises	154

Chapter 7 Using probability estimates

7.1	Averages	158
7.2	Intervals	159
7.3	Estimating the population mean	163
7.4	Hypotheses	169
7.5	Discrimination	171
7.6	Further reading	177
7.7	Examples	177
7.8	Exercises	184

Chapter 8 Two dimensions

8.1	Structure in contingency tables	186
8.2	Correlation and independence	191
8.3	Testing for independence	193
8.4	Information transmission	196
8.5	Bayes' Rule	198
8.6	Further reading	200
8.7	Examples	200
8.8	Exercises	208

Chapter 9 Least biased estimates of joint distributions

9.1	Marginal distributions as constraints	212
9.2	Using a previous estimate	214
9.3	Joint distribution with given mean	216
9.4	Modifying a previous estimate with a new mean	218
9.5	Estimating marginal distributions	219
9.6	The structural comparison of tables	220
9.7	Modelling two-dimensional contingency tables	225
9.8	Modelling three-dimensional contingency tables	228
9.9	Further reading	236
9.10	Examples	236
9.11	Exercises	245

Chapter 10 Correlation and regression

10.1	Covariance and correlation	250
10.2	Fitting a linear model	255
10.3	Model performance	258
10.4	Entropy reduction and correlation	260

Contents

	10.5	Inference and forecast	261
	10.6	Examples	263
	10.7	Exercises	283

References 289

Answers to numerical exercises 305

Appendices

    A1.   The mathematics of entropy maximisation   311
              Finding a simple maximum
              Constrained optimisation
              Maximising entropy
              Minimising relative entropy
              Joint distributions
              Three-dimensional tables and the log-linear model

    A2.   Logarithms   319

    A3.   Numerical integration   322
              The mid-point rule
              Simpson's Rule
              Example

    A4.   The Binomial, Poisson and Normal distributions   326
              The Binomial distribution
              The Poisson distribution
              The Normal distribution

    A5.   Statistical tables   331

	A5.1	Binomial distribution	332
	A5.2	Poisson distribution	341
	A5.3	Normal distribution	346
	A5.4	Beta distribution	348
	A5.5	Gamma distribution	349
	A5.6	χ^2 distribution	350
	A5.7	The t distribution	353

    A6.   Measures of location and loss functions   355

	A7.	Conditions for a credible interval	357
	A8.	The regression model	358
Index			363

Preface

Information is the reduction of uncertainty.

Probability and statistics are concerned with the measurement of uncertainty.

These definitions give the motivation for this book.

Information theory was originally developed for the study of the transmission of telegraphic messages but soon found wider applications in areas such as psychology and philosophy as well as in mathematics and physics. The common thread was that information provided a fundamental viewpoint in the understanding of the behaviour of many apparently disparate systems. The application of a statistical measure of information exposed the underlying similarities.

It subsequently became clear that information provided a methodology that was useful in various areas of statistics itself. The development of the entropy maximising formalism following the work of Jaynes has provided a unifying base.

All of this has been known for a number of years and yet, after an initial flurry (notably by Tribus, 1969), no introductory texts of general interest have been published, the material usually being aimed at those with some mathematical expertise and particular specialist interest. This book provides a simple introduction that makes no heavy mathematical demands of the reader, though some mathematical material is presented in the appendices to satisfy the curious. As in all introductions a fair amount has to be taken on trust.

Chapter 1 gives an introduction to those aspects of probability theory that are helpful in describing uncertainty, and so will be useful in what follows.

Chapter 2 introduces the basic ideas of information and entropy from a number of standpoints, drawing out those aspects which are common.

Chapters 3 and 4 show how the maximum entropy methodology can be used to obtain unbiased estimates of uncertain quantities.

Chapter 5 describes the incorporation of data via Bayes' Rule as a way of improving estimates and learning from observation.

In *Chapters 6 and 7* the use of probabilistic estimates in the solution of decision problems is described and some common statistical applications are discussed.

The extension of maximum entropy methods to relationships between variables presented in tabular form is given in *Chapters 8 and 9*, while *Chapter 10* covers the relationship between two variables when this is presented as a straight line graph.

You will need to do calculations. Some are quite simple and can be done using a calculator. Others require a bit more work and for these I have assumed that you know how to use a spreadsheet and have access to one. If this represents uncharted water for you now is the time to take the plunge.

Finally, a word about references. Information touches on a number of topics and disciplines and so I have tried to give plenty of references to help you to follow your interests. Some are mathematically a bit tough but many are not. You will have to try them and see how you get on. Wherever possible I have given books rather than papers since they are likely to provide a more useful next step for the general reader. I have also given the original references where possible. Some are quite old now, but still worth reading.

Acknowledgements

The lines on page 48 are from the poem "Not Waving but Drowning" which is to be found in *The Collected Poems of Stevie Smith* published by Penguin Twentieth Century Classics. Permission for their reproduction here has been given by James MacGibbon, 8 Quay Street, Manningtree, Essex CO11 1AU.

The tables of the Normal, t and χ^2 distributions to be found in the appendices are taken from *Tables for Statisticians* by White, J., Yeats, A. and Skipworth, G., published by Stanley Thornes, and are reproduced here with the permission of the publisher.

1

Probability

This book is concerned with making estimates of the chance that an event will happen or that a statement is true.

We may be concerned with the chance that the sales of a new product will not reach some target, the chance that a batch of some manufactured goods is below specification, the chance that a train will arrive late, and so on.

Estimating these chances inevitably involves some speculation on our part. On what basis do we speculate? How do we describe the chances?

1.1   A LITTLE HISTORY

We shall quantify chances by using probabilities. A probability is a number between 0 and 1. A probability of 0 means that an event will certainly *not* happen or a statement is certainly *not* true and a probability of 1 means that an event certainly *will* happen or a statement certainly *is* true.

In effect just about the only examples of the extreme cases are logical: anything assigned a probability of 1 is a tautology while the only thing that could be assigned a probability of 0 is a logical contradiction.

Intermediate probability values represent degrees of uncertainty.

These ideas are quite familiar. If a friend and I discuss a cricket match and decide that "the odds of the local team winning are even" then we have decided that the probability of a home win is 0.5. If we think that the odds that it will rain tomorrow are 3:1 then the probability of rain is 0.75.

We can use a shorthand for writing assessments such as these:

prob(rain tomorrow) = 0.75

or p(rain tomorrow) = 0.75

But where do these assessments come from?

There are two ways to think about probability. The first is as a measure of belief, an encoding of the degree to which I believe a statement or proposition to be true. Why I have this degree of belief does not matter for the moment, only that the assigned probability value reflects that belief accurately, so far as I can judge.

Probability assignments are personal and subjective. I can assess the probability of rain tomorrow as 0.75 based on hunch, reading the tea leaves, extensive viewing of weather forecasts or whatever.

Alternatively, I can count. By examining past meteorological records I could find the proportion of days with tomorrow's date on which rain fell in, say, the last hundred years and take this proportion as my probability assessment. This assessment is apparently neither personal nor subjective. Given this procedure any two people will arrive at the same probability value. There is an appeal here to empirical objectivity. However, you and I may differ on the number of years over which records should be taken. You may choose a more complex procedure that takes into account whether it rained on the previous day. Perhaps there is some personal preference involved after all.

There are then two broad views: that probability is an expression of the degree to which one believes in something and that probability is in some sense based on data, on experimentation. These two approaches have characterised debates about probability for hundreds of years. In the seventeenth century one philosophical concern was for discovering the properties of games of chance while another was speculating on the existence of God. Pascal, Fermat and others pursued these ideas and introduced the concept of probability.

Hacking (1975) gives a lively account of all this and notes that different words have been proposed to describe the different concepts:

	assessment based on	
author	evidence	belief
Poisson and Cournot	*chance*	*probabilité*
Condorcet	*facilité*	*motif de croire*
Russell	probability	credibility

Braithwaite (in his foreword to Keynes, 1973) also quotes the names "statistical probability" and "long-run frequency" for assessments based on evidence and "degree of confirmation" and "acceptability" when discussing belief.

We now use the word probability for both concepts even though, as Hacking notes, there are fundamental differences:

> "The limit of increasing probability of opinion might be certain belief, but it is not knowledge: not because it lacks some missing ingredient, but because in general the objects of opinion are not the kinds of propositions that can be the objects of knowledge."

In these debates the idea of evidence referred to any sensory data as opposed to opinion. The notion of using relative frequency of occurrence as a probability value, as in the rain example above, was put forward in 1843 by Ellis and Cournot, working independently. The view was also championed by Venn. This *frequentist* stance became the dominant popular definition and is still presented, quite uncritically, as such in many texts.

In 1921 Keynes reasserted the earlier approach using the idea of the "probable degree of rational belief" (Keynes, 1973). Indeed, Keynes pointed out that probability does not always need to be quantified; we say that "A is more likely than B" without assigning probability values to A or B. Keynes and Jeffreys (1939) were leading members of the *logical* school pursuing an axiomatic approach to probability as a measure of degree of belief, but one rooted in *objective* observable facts.

For Ramsey (1926) the degree of belief was measured by the extent to which we are prepared to act upon it. This appeal to a decision oriented framework was also employed by Savage (1954, 1962) and de Finetti (1937, 1972, 1974) (see also DeGroot, 1970, and Good, 1983) and was closely related to ideas of game theory and utility (von Neumann and Morgenstern, 1947). Probabilities were seen by this *subjectivist* school as personal probabilities and the assessment of the probabilities was intimately related to the decisions which rested, in part, upon them. The objective of the theory was not to *describe* how decisions were made but rather to *prescribe* how they should be made. Key words here are *rational* and *coherent*. We shall return to the relationship between assessed probabilities and decision making in Chapter 6.

In the post-war period decision analysts working on business problems (notably Schlaiffer, 1959, 1961, and Raiffa and Schlaiffer, 1961) realised that many business decisions were inevitably based on subjective assessments of uncertainty because time pressures prevented the collection of data, or the data were not available, or the objects of concern were apparently unique future events. This led to a revival of interest in the work of Bayes (which is covered in Chapter 5) and to a view of probability which is now generally termed Bayesian and which is accepted in this book.

We start with a subjective probability assessment and this is revised in the light of subsequent evidence. Bayes' Rule provides the means of making the revision. There is no restriction on the object of interest, although it is often the value of some parameter, such as an average.

We may, for instance, have an opinion about the proportion of defectives on a particular production line. We may then go and find the number of defectives in a sample of a hundred units. Given this new evidence our opinion about the proportion of defectives will change. Collecting larger samples will result in revised opinions which are more and more precise: a tendency towards knowledge.

Alternatively, we may speculate on the price of crude oil next year. Collecting evidence in the form of the opinions of experts will give decreasingly vague revised opinions: a tendency towards belief, for here we can only have knowledge in retrospect.

Whatever the object of our speculation, the calculations with probabilities are the same. For convenience we shall generally refer to the things about which probability statements are being made as *events* whether they lead to knowledge or to belief.

Here are some basic rules for manipulating probabilities.

1.2 MORE THAN ONE EVENT

We have seen that probability values must lie between 0 and 1, so that for some event A

$$0 \leq \text{prob}(A) \leq 1 \qquad (1.1)$$

If you are unfamiliar with symbols such as "$\leq$" here is a brief dictionary:

    $<$ "less than"
    $\leq$ "less than or equal to"
    $>$ "greater than"
    $\geq$ "greater than or equal to"

Probability values can be pictured as being areas on a square with sides of unit length. The area of the whole square is 1 and the areas of any part of the square can represent probabilities. Figure 1.1 shows prob(rain tomorrow) = 0.75. Since it must either rain or not the remainder of the square shows prob(no rain tomorrow) and this must have a value of 1 − 0.75 = 0.25.

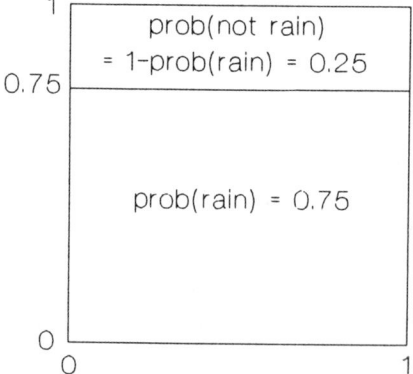

Figure 1.1 The probability square

In general, if either one or the other event *must* occur they are said to be *complementary* and so

$$\text{prob}(A) + \text{prob}(\text{not } A) = 1$$

or   prob(not A) = 1 − prob(A)   (1.2)

and "not A" is called the complement of "A".

This, and the other ideas in this chapter, are easily extended to more than two events.
When two events are such that if one occurs the other cannot they are said to be *mutually exclusive*. If the mutually exclusive events are A and B then

prob(A and B) = 0   (1.3)

If we now think about how much it will rain we can describe a set of events based upon the number of hours for which rain will fall. Call the number of hours X so that, for instance,

prob($X = 0$)   = 0.25
prob($0 < X \leq 2$) = 0.45
prob($X > 2$)   = 0.30

which is shown in Figure 1.2.

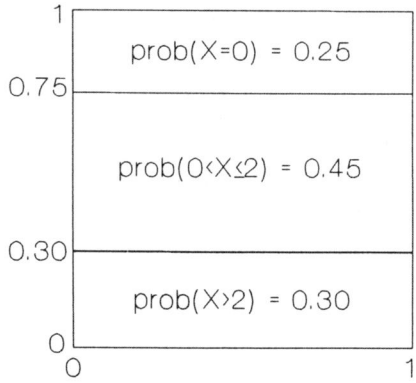

Figure 1.2

If we now want to know the probability either that there will be no rain or that it will rain for no longer than 2 hours we can see from the diagram that this is the sum of the two lower rectangles:

prob($X \leq 2$) = prob[($X = 0$) or ($0 < X \leq 2$)]
             = prob($X = 0$) + prob($0 < X \leq 2$)
             = 0.25 + 0.45
             = 0.70

Generalising, if A and B are any two mutually exclusive events then

$$\text{prob}(A \text{ or } B) = \text{prob}(A) + \text{prob}(B) \tag{1.4}$$

Continuing our obsession with the weather suppose that we look 2 days ahead and assess the probability of rain on the day after tomorrow as 0.6. Now apply (1.4) to get the probability of rain on either day:

$$\text{prob}[(\text{rain tomorrow}) \text{ or } (\text{rain next day})] = 0.75 + 0.6$$
$$= 1.35$$

Something has gone wrong. Probability values cannot exceed 1.0. Figure 1.3 shows that we have double counted the probability of rain on both days. The events we are dealing with are *not* mutually exclusive since rain tomorrow does not mean that the following day will be dry. Equation (1.4) needs to be modified to deal with *any* two events:

$$\text{prob}(A \text{ or } B) = \text{prob}(A) + \text{prob}(B) - \text{prob}(A \text{ and } B) \tag{1.5}$$

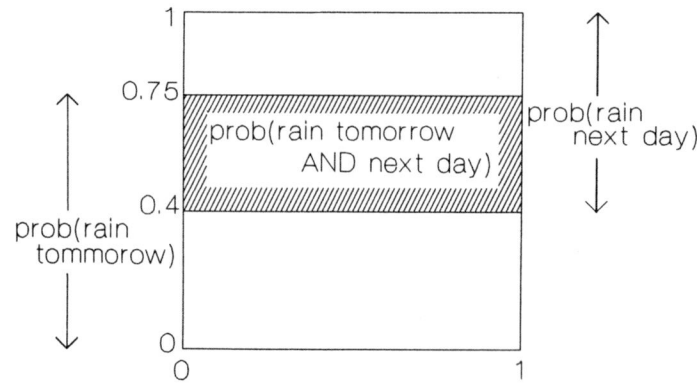

Figure 1.3 Non-mutually exclusive events

Using (1.3) gives (1.4) as a special case.

1.3 JOINT EVENTS

Sometimes we have a number of ways of thinking about the same object. We can, for example, classify people by age, sex or height. These classifications intersect to give categories such as "tall old men" and "short young women".

As well as classifying days as wet or dry (= not wet) we may also classify them as being bright or dull (= not bright) to give the four classes shown in Table 1.1.

		rain category	
		wet	dry
sunshine category	bright	wet and bright	dry and bright
	dull	dull and wet	dull and dry

Table 1.1
Four types of weather

When trying to forecast what sort of day tomorrow may be we now have four probabilities to assess. We know that some combinations are less likely than others. Our language often reflects this and so we might say "wet *but* bright" rather than "wet *and* bright".

How might we go about making the probability assessments? We have already decided that the probability of rain tomorrow is 0.75. This assessment was made without reference to the expected amount of sunshine, and so these probabilities are shown in Table 1.2 written along the bottom edge (or margin) of the table. They are called *marginal probabilities*.

	wet	dry
bright		
dull		
	0.75	0.25

Table 1.2
Marginal probabilities

There are other marginal probabilities which describe, without reference to the amount of rain expected, the chance that it will be a bright day. We have not thought about these probabilities yet.

Because we believe that it is more likely to be bright when it is dry it seems natural for us now to say that *if* it is dry *then* the probability of it being bright is, say, 0.6. We write this as

prob(bright | dry) = 0.6

where the symbol " | " is read as "given".

We may similarly believe that on a wet day it is more likely to be dull than bright and so assess

$$\text{prob}(\text{dull}\,|\,\text{wet}) = 0.8$$

For obvious reasons these are called *conditional* probabilities and they are shown in Table 1.3. Because the categories are both mutually exclusive and exhaustive the sum of each pair of conditional probabilities is 1.0 and this is shown as the column sum.

	wet	dry
bright	0.2	0.6
dull	0.8	0.4
sum:	1.0	1.0

Table 1.3
Conditional probabilities

But we want the column sums to be 0.75 and 0.25 as in Table 1.2 and so scale the columns so that they give these required sums. The two values in the "wet" column are $0.2 \times 0.75 = 0.15$ and $0.8 \times 0.75 = 0.60$.

The complete set of probabilities is shown in Table 1.4.

	wet	dry	
bright	0.15	0.15	0.30
dull	0.60	0.10	0.70
	0.75	0.25	

Table 1.4
Joint and marginal probabilities

The first value, 0.15, is the probability of tomorrow being bright and wet, prob(bright and wet). This is called a *joint* probability since it refers to the joint occurrence of two events. Remember that the value was found by multiplication of the conditional and joint probabilities

$$\text{prob}(\text{bright and wet}) = \text{prob}(\text{bright}\,|\,\text{wet})\text{prob}(\text{wet})$$

For any two events, A and B, their joint probability is

$$\text{prob}(A \text{ and } B) = \text{prob}(A\,|\,B)\text{prob}(B) \qquad (1.6)$$

We could alternatively have used the sunshine category to condition our forecasts and thought of prob(wet | bright) and so on. In other words we could have constructed

$$\text{prob}(A \text{ and } B) = \text{prob}(B | A)\text{prob}(A)$$

Putting this with (1.6):

$$\text{prob}(A \text{ and } B) = \text{prob}(A | B)\text{prob}(B) = \text{prob}(B | A)\text{prob}(A) \qquad (1.7)$$

Notice from Table 1.4 that by summing the rows we obtain the marginal distribution for the sunshine category showing a probability of 0.7 of a dull day tomorrow.

Conditional probabilities are important because they model the relationship between two or more descriptors or events. Whether we think it will be bright or dull is conditional upon whether it might rain. This means that we will make different assessments (forecasts) of the chance that it will be bright given knowledge, or assumption, about rainfall. From Table 1.3

$$\text{prob(bright)} = 0.2 \quad \text{if we assume rain}$$
$$= 0.6 \quad \text{if we assume no rain}$$

while from Table 1.4

$$\text{prob(bright)} = 0.3 \quad \text{if we make no assumption about rain}$$

Conditional probabilities are a way of making use of knowledge of one event to make a better forecast of the chance of occurrence of another. We would expect these conditional assessments to be more accurate or, at worst, no less accurate.

Suppose, on the other hand, that we are concerned to assess the chance of my beating my friend at our weekly game of squash tomorrow. He has won about 60% of our matches so far. I have no reason, empirical or otherwise, to believe that the results of our contests are affected by rainfall so knowing the weather tomorrow will not help me forecast the result of our match. Table 1.5 gives the situation.

	wet	dry	
I win	0.4	0.4	0.4
I lose	0.6	0.6	0.6
	1.0	1.0	

Table 1.5
Conditional and marginal probabilities for independent events

Here the probability of my winning is the same whether I assume that it will rain, or will not rain, or whether I make no assumption about rainfall at all. The conditional probabilities are equal to each other and to the marginal distribution. The events are said to be *statistically independent* of each other. In equation (1.6) prob($A \mid B$) = prob(A) and so, for independent events,

$$\text{prob}(A \text{ and } B) = \text{prob}(A)\text{prob}(B) \tag{1.8}$$

This means that if the marginal probabilities are known then a table of joint probabilities is easily found by simple multiplication. Using the previous probability assessments:

prob(I win at squash and it rains) = prob(I win)prob(wet)
= 0.4 × 0.75
= 0.3

Table 1.6 shows the complete set of joint and marginal probabilities.

	wet	dry	
I win	0.30	0.10	0.40
I lose	0.45	0.15	0.60
	0.75	0.25	

Table 1.6

1.4 DISTRIBUTIONS

When we have a probability assessment for each possible manifestation or value of an event we call that set of probabilities a *probability distribution*.

In the examples above one of the events in which we were interested was whether it would rain or not. This was described by a *variable* called "rain category". A variable is an object that can take different *values*. In this case the values were the simple descriptors "dry" and "wet". There is no sense of order or magnitude associated with these values; they are just names of states of the weather. A variable characterised by these sorts of values is called *nominal*.

The assessed probability distribution was

prob(wet) = 0.75
prob(dry) = 0.25

Figure 1.4 shows a graphical representation.

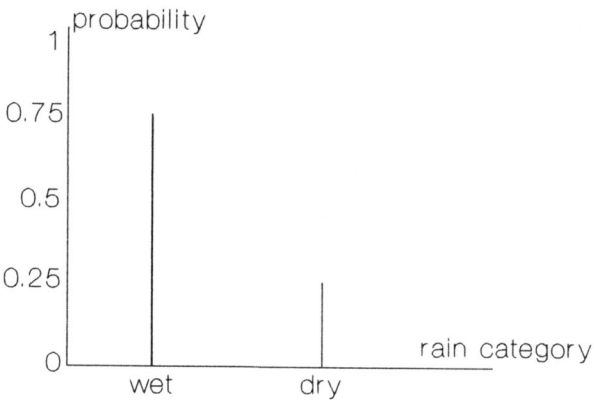

Figure 1.4 Probability distribution for a nominal variable

It is common to wish to deal with variables whose values are whole numbers (*integers*), usually because we are concerned with counting in some way. We could, for example, construct a probability distribution for the number of Sunday newspapers delivered to a household. If N is the number delivered we may obtain

$$\begin{aligned}
&\text{prob}(N = 0) = 0.05 & &\text{prob}(N = 3) = 0.04 \\
&\text{prob}(N = 1) = 0.40 & &\text{prob}(N = 4) = 0.01 \\
&\text{prob}(N = 2) = 0.50 & &\text{prob}(N \geq 5) = 0
\end{aligned} \quad (1.9)$$

This distribution is shown in Figure 1.5.

From this distribution we can, by simple addition, construct the probabilities of N not exceeding given values:

$$\begin{aligned}
\text{prob}(N \leq 0) &= 0.05 \\
\text{prob}(N \leq 1) &= 0.05 + 0.40 = 0.45 \\
\text{prob}(N \leq 2) &= 0.45 + 0.50 = 0.95 \\
\text{prob}(N \leq 3) &= 0.95 + 0.04 = 0.99 \\
\text{prob}(N \leq 4) &= 0.99 + 0.01 = 1.00
\end{aligned}$$

This new distribution is called the *cumulative probability distribution* and is shown in Figure 1.6.

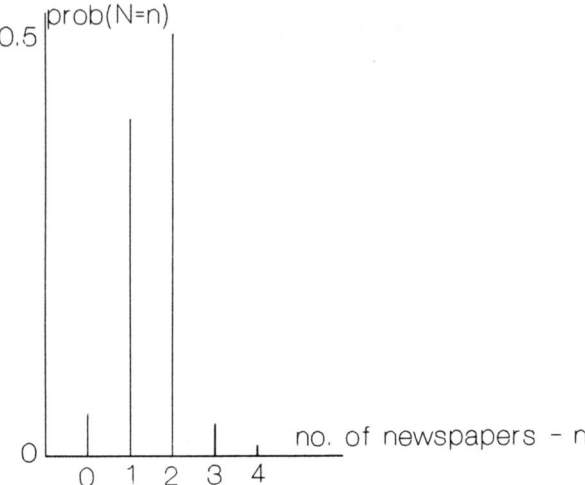

Figure 1.5 Probability distribution

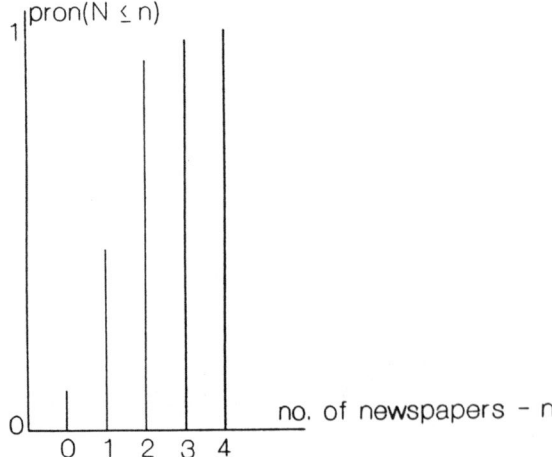

Figure 1.6 Cumulative probability distribution

Finding probabilities using this distribution is quite easy, for instead of adding a number of values a simple subtraction is all that is needed. To find the probability that N is 1 or 2 or 3,

$$\text{prob}(1 \leq N \leq 3) = \text{prob}(N \leq 3) - \text{prob}(N \leq 0)$$
$$= 0.99 - 0.05$$
$$= 0.94$$

1.5 DENSITY FUNCTIONS

In the previous section the variables could only take a certain number of *discrete* values. This was emphasised in the diagrams by drawing the probability values as distinct vertical lines with gaps between them indicating that intermediate values of the variables were not defined.

Suppose now that we are to make a probability assessment of the average number of Sunday newspapers taken by households in this country. What will the diagram of that assessment look like?

Quickly looking at (1.9) it is likely that the average, call it X, will be between 1 and 2 but it could take any particular value, 1.1587263548 for example. The number of figures given after the decimal point could be infinite. Values such as this are called *real* values. How can we draw a picture of this distribution?

Suppose that we decided to tabulate values of the average in units of 0.01 so that we had, by some means, probability assessments for values 1.00, 1.01, 1.02, 1.03 and so on. The result would look something like Figure 1.7. If we now tabulated at a smaller interval the vertical lines would be more numerous and closer together. Eventually we could tabulate at an infinitely small interval in which case the lines would be touching and would form one continuous solid figure such as that shown in Figure 1.8.

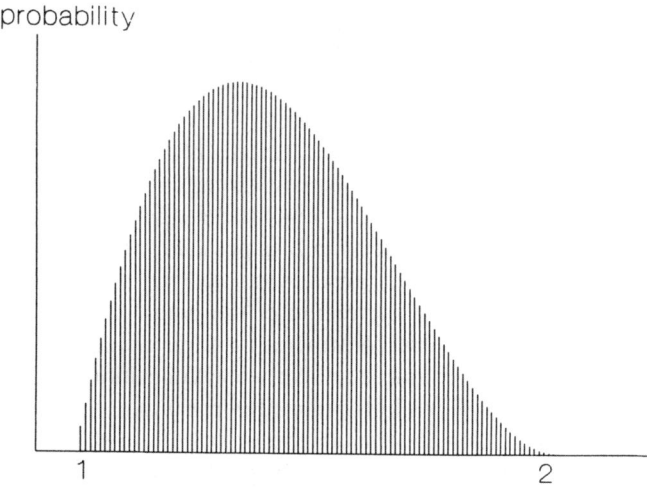

Figure 1.7

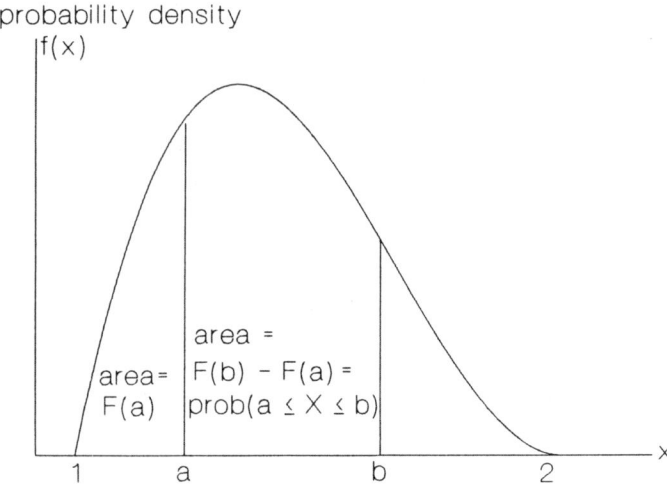

Figure 1.8   Probability density function (pdf)

What is the probability that X will have a value somewhere between the values a and b? It is just the sum of the infinite number of bars which comprise the figure between these limits, in other words the area under the curve between $X = a$ and $X = b$.

This is important. It is the *area* under this diagram that is equal to probability. This means that the vertical axis is *probability density*.

Think of a hollow tube that has been filled with all sorts of material (sawdust, pieces of metal, polystyrene chips) and sealed. We wish to say something about how the weight of this object varies along the length of the tube. We cannot find the weight of a slice of the tube taken at *exactly* 12 cm from one end since a slice must have some width in order for it to be a tangible disc that can be weighed. What we could do, however, is to take a slice as thin as possible, weigh it, and then divide the weight by the width of the slice to give a value in g/cm. This is the rate at which weight increases with length in the region of the point 12 cm from one end of the tube. By repetition we could then construct a diagram similar to Figure 1.8 showing how this density of weight per unit length varies with position on the tube. In similar fashion probability density shows how probability per unit of the variable changes.

Just as we constructed the cumulative probability distribution in the previous section we can here see that the area under the probability density curve to the left of a given value is the probability of that value not being exceeded.

The formula or rule that defines the probability density curve is called the *probability density function* or *pdf* for short. If the variable is X then the pdf $f(x)$ represents the value of probability density at $X = x$. (The notation $f()$ simply means a function called, in this case, f). The cumulative probability is denoted by $F(x)$ and for real valued variables is called the *distribution function*.

Sec. 1.5] Density functions

In summary, for a real valued variable, X,

$f(x)$ is the pdf and gives the value of the probability density at $X = x$

$F(x)$ is prob$(X \leq x)$ and is the area under the pdf curve to the left of $X = x$

and   prob$(a \leq X \leq b) = F(b) - F(a)$, $b > a$

Figure 1.9 shows the distribution function that corresponds to the density function of Figure 1.8. The graph of the density function gives a better picture of the shape of the distribution; a better feel for how probability varies with different values of X. On the other hand, it is the distribution function that we need to find probabilities. Distribution functions are typically S-shaped, as shown here.

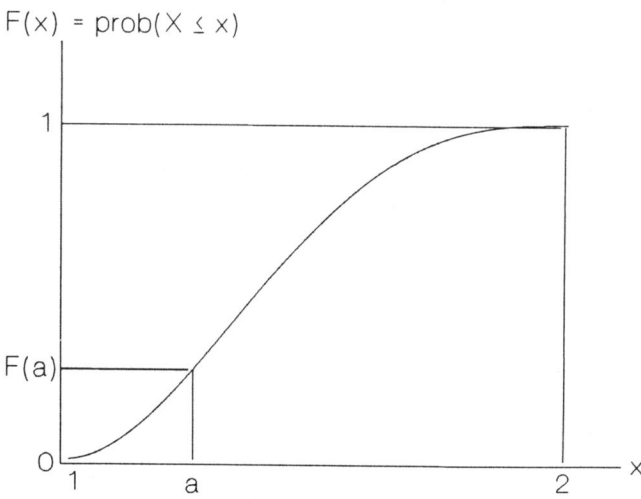

Figure 1.9 Distribution function

For illustration suppose that I wish to assess the probability of an overrun on the cost of a project. I do not believe that the final cost will be less than the estimated cost but neither do I believe that the overrun will be more than 10%. The contractor has quite good cost controls but the project is a tricky one. After some thought I decide on the triangular density function shown in Figure 1.10. (We shall see in later chapters how justifiable this shape is.) Let X be the percentage overrun.

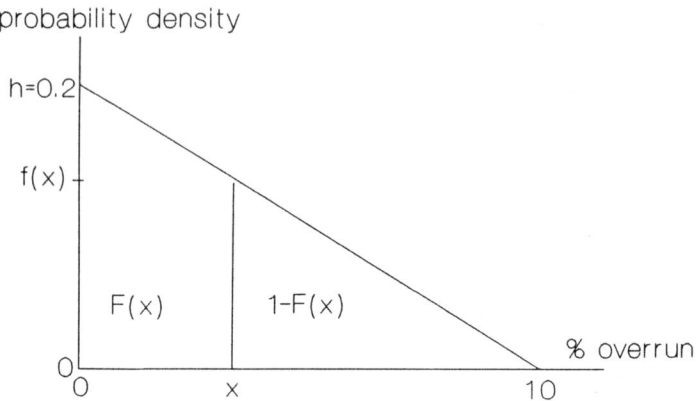

Figure 1.10

Since I assess that X must be between 0 and 10, then prob$(0 \leq X \leq 10) = 1.0$ and so this is the area under the triangle. By simple geometry the height of the triangle, h, can be found from

$$\frac{h \times 10}{2} = 1.0$$

so $h = 0.2$.

The value of the density function at $X = x$ is, by simple proportion,

$$\frac{f(x)}{h} = \frac{10-x}{10}$$

so $f(x) = \dfrac{10-x}{50}$

To find the value of the distribution function consider the triangle to the right of $X = x$ in Figure 1.10.

$$1 - F(x) = \frac{f(x)(10-x)}{2}$$

$$= \frac{(10-x)^2}{100}$$

so

$$F(x) = 1 - \frac{(10-x)^2}{100} = 1 - \frac{(100 - 20x + x^2)}{100} = \frac{20x - x^2}{100}$$

(we have used the standard result $(a + b)^2 = a^2 + b^2 + 2ab$)

$F(x)$ is shown in Figure 1.11. (Notice that because of the rather special form of the density function in this example, this distribution function does not follow the usual S-shape.)

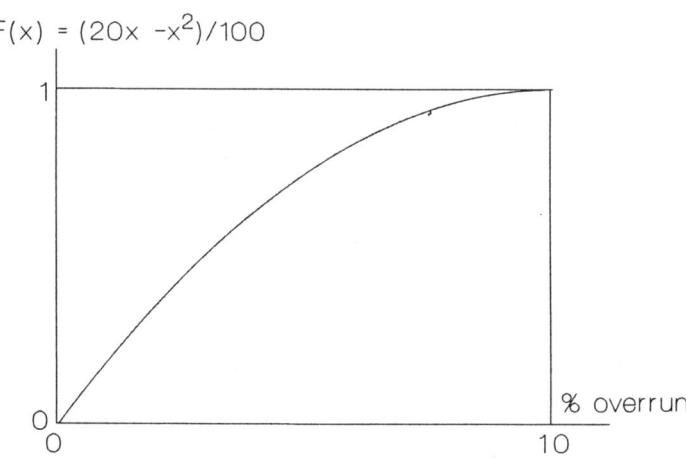

Figure 1.11

Suppose now we want the probability that the overrun will be between 2% and 6%.

$$\text{prob}(2 \leq X \leq 6) = F(6) - F(2)$$
$$= \frac{120 - 36}{100} - \frac{40 - 4}{100}$$
$$= 0.84 - 0.36 = 0.48$$

1.6 SUMMARY MEASURES

A probability distribution or density function, whether displayed as a table or a graph, contains all that we need to know about a probability assessment. Sometimes, however, it is useful to be able to describe the characteristics of an assessment by a couple of summary figures. The two characteristics most often described are those of *average* and *spread*, ideas which are self-explanatory.

As an example consider the distribution (1.9), also shown in Figure 1.5. This showed an estimate of the number of Sunday newspapers, N, delivered to a household. What is the average value that best characterises this distribution?

An obvious reply is that the average value is 2 since this is the value to which the greatest probability has been assigned; no other value is thought to be more likely. This notion of an average is called the *mode*.

Alternatively, we could look for the value that was in some sense "in the middle of" the distribution, the value that split the range of values into two equiprobable halves. This is called the *median* and in this example also has the value 2. It can be read from the cumulative probability table as the value of N that corresponds to a cumulative probability of 0.5. This is shown in the last column of Table 1.7 and later in Figure 1.13(a).

n	prob(n)	$n$prob(n)	cumulative probability	
0	0.05	0.00	0.05	
1	0.40	0.40	0.45	
2	0.50	1.00	0.95	← median
3	0.04	0.12	0.99	
4	0.01	0.04	1.00	
	1.00	1.56		

Table 1.7

The final idea of an average is that which we would be most likely to calculate if left to our own devices. Given a list of values we would sum them and then divide by the number of values in the list. The result is called the *mean*.

$$\text{mean} = \sum_i n_i \text{prob}(n_i) \tag{1.10}$$

where n_i is the i'th possible value of N, so, in this case, $n_1 = 0$, $n_2 = 1$, etc.

The Σ sign (Greek capital "sigma") means "sum". What is to be summed is all the terms $n_i$prob(n_i) where i takes the values 1,2,3 and so on.

The mean is also denoted by the symbol μ ("mu"). You can see from Table 1.7 that $\mu = 1.56$. Notice that in the table I have not used the subscript i since its omission does not result in any ambiguity.

Some distributions are symmetrical and some are not. Distributions that are not symmetrical are said to be *skewed*. Figure 1.12 shows distributions of different shapes. Notice how the relationship between the three measures of the average varies with skew.

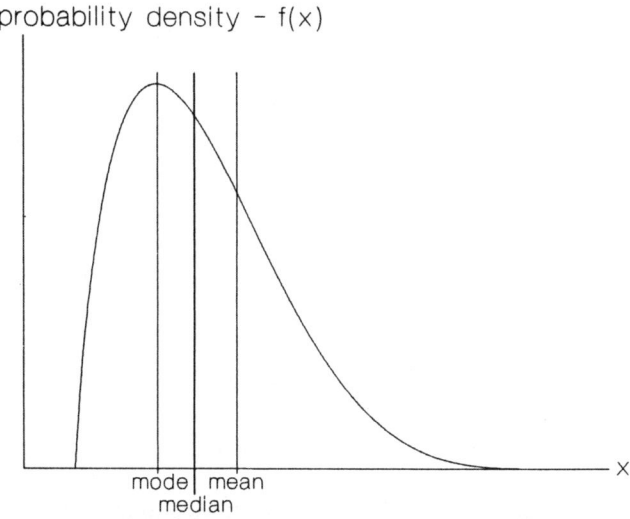

Figure 1.12(a)   Right or positive skew

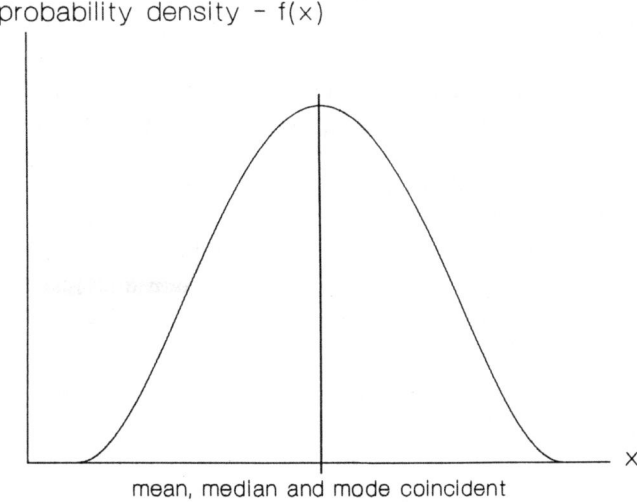

Figure 1.12(b)   Symmetric distribution: no skew

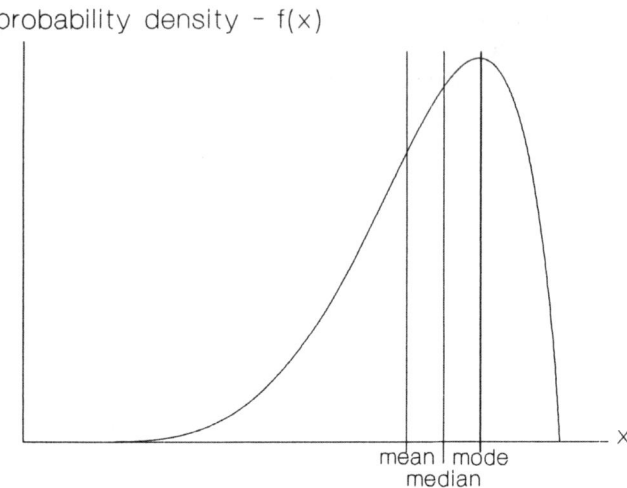

Figure 1.12(c) Left or negative skew

We turn now to ideas of spread and look for a simple descriptive figure that will convey this feature of a distribution.

It is easiest just to look at the *range* of values that the variable may take. In the current example the range is from 0 to 4, written as (0,4), a range length of 5. This is a rather coarse summary, being of limited use because it is insensitive to changes in the distribution. Suppose, for instance, that an alternative probability assessment gave all five possible values the probability 0.2. Clearly, the spread of the distribution has increased yet the range is unchanged.

One way of reducing this problem is to consider only the range of the central part of the distribution, so overcoming the effect of the attenuation in the extremes (the *tails*). The definition of the extremes is arbitrary but it is common to take the central 50% of the distribution. The values which divide a distribution into equiprobable quarters are called *quartiles*; the second quartile is, of course, the median. The range defined by the first and third quartiles is called the *inter-quartile range* and is shown in Figure 1.13 for both discrete and real variables. For the Sunday newspaper distribution the inter-quartile range is (1,2), a length of 1.

Summary measures

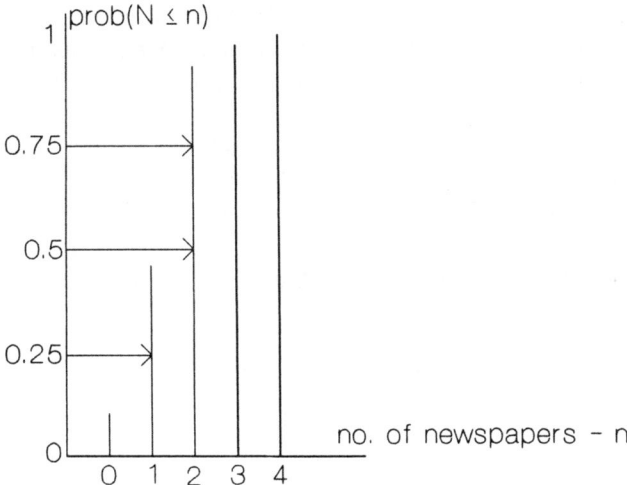

Figure 1.13(a)   Quartiles from Figure 1.6

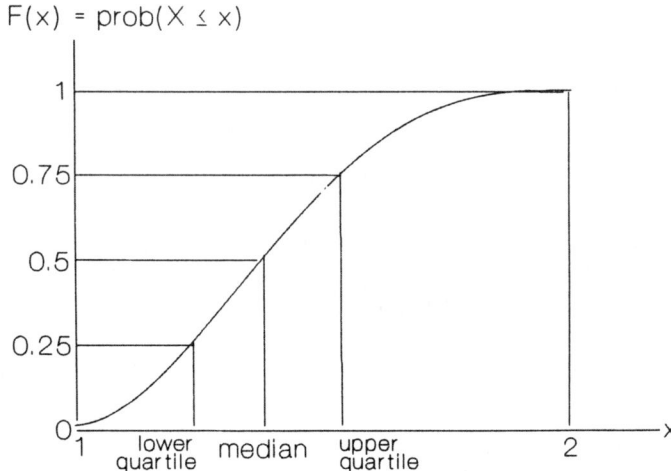

Figure 1.13(b)   Quartiles from Figure 1.9

A frequently used alternative is to take the central 80% of the distribution by giving the *inter-decile range*. You must judge for yourself which range best reflects what you want to say about spread.

Alternative measures of spread are found by thinking about how far, on average, a value is from the mean of the distribution. Simply taking the mean value of $(n - \mu)$ is no good since this will always be identically zero by definition of μ (Table 1.8, column 4).

It is sensible, therefore, to take the mean value of deviations from μ measured in absolute terms, i.e. by disregarding whether a value is higher or lower than the mean. The result is called the *mean absolute deviation*, or *MAD* for short.

$$\text{mean absolute deviation} = MAD = \sum_i \text{prob}(n_i)|n_i - \mu|$$

where the two vertical bars indicate the absolute value so that, for example, $|-2| = |2| = 2$.

The last column of Table 1.8 shows $MAD = 0.6$.

n	prob(n)	$(n - \mu)$	prob(n)($n - \mu$)	$|n - \mu|$	prob(n) $|n - \mu|$
0	0.05	−1.56	−0.08	1.56	0.08
1	0.40	−0.56	−0.22	0.56	0.22
2	0.50	0.44	0.22	0.44	0.22
3	0.04	1.44	0.06	1.44	0.06
4	0.01	2.44	0.02	2.44	0.02
			0.00		0.60

Table 1.8

Alternatively we could take the squares of the deviations, which also ensures positive values. The mean of these squared deviations is called the *variance* and is shown as σ^2 (σ is lower case "sigma") so that

$$\text{variance} = \sigma^2 = \sum_i \text{prob}(n_i)(n_i - \mu)^2 \tag{1.11}$$

Table 1.9 shows that $\sigma^2 = 0.49$.

n	prob(n)	$(n - \mu)$	$(n - \mu)^2$	prob(n)($n - \mu)^2$
0	0.05	−1.56	2.43	0.12
1	0.40	−0.56	0.31	0.13
2	0.50	0.44	0.19	0.10
3	0.04	1.44	2.07	0.08
4	0.01	2.44	5.95	0.06
	1.00			0.49

Table 1.9

Variance has units which are the square of those in which the variable is measured. If N is measured in newspapers then σ^2 is measured in (newspapers)^2. This is not generally what is wanted and so we use the square root of variance which is called the *standard deviation*, σ:

$$\text{standard deviation} = \sigma = \sqrt{\sigma^2}$$

In the current example, therefore, $\sigma = \sqrt{0.49} = 0.7$. Alternative notations for variance and standard deviation are var(X) and sd(X).

The significance of a given amount of variation is often related to the average value. For example, we would be more concerned about fluctuations of a few grams in bags of grain nominally containing 1 kg than in sacks containing 100 kg. To reflect this we sometimes use the ratio of standard deviation to mean, which is called the *coefficient of variation*. For the distribution describing the number of newspapers delivered

$$\text{coefficient of variation} = \frac{\sigma}{\mu} \tag{1.12}$$

$$= \frac{0.7}{1.56} = 0.45$$

Which of these summary measures should you use? We shall see later that the mean and variance are both descriptors and parameters of some useful standard distributions and they will be of value in that context. They are not, however, particularly good summaries for describing the characteristics of a distribution to third parties. Not many people would find "variance = 32.6" an intuitively appealing, or even comprehensible, report.

There is some evidence that if I ask you how long, on average, you spend travelling to work each day that you would recall the time that you most commonly spent on the journey. On these grounds the mode has a good claim as a useful average. However, the mode could be quite eccentrically placed. Suppose that we had a fairly flat distribution with a small peak very close to one of the extremes. This would undoubtedly be the mode, but would it conform to your idea of an average?

For myself, I find the concepts of median and inter-quartile or inter-decile ranges intuitively simple to comprehend. They are, for me, essentially pictorial in a descriptive sense but also allow me to make simple probability statements for they are, after all, points on a distribution function. Figure 1.14 shows a *box plot* that depicts these key features of a distribution.

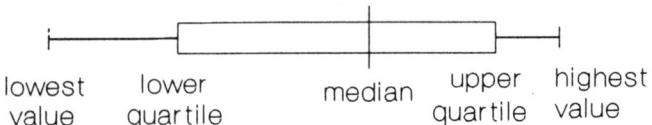

Figure 1.14 Box plot

Remember though that the measures described are only summaries and that a clearly presented graph or table is inherently a more powerful way to present data. Do not feel that you must impress by quoting numerous summary statistics whose significance may at best be opaque to the reader. Often a clear and simple diagram together with a brief commentary is much more effective. Avoid the flashy confusions of most computer graphics packages.

1.7 EXPECTATION

The *expected value* of a function $f(x)$ is simply

$$E[f(x)] = \Sigma f(x)\text{prob}(x) \qquad (1.13)$$

so that the mean is just the expected value of X:

$$\mu = E[x] = \Sigma x\text{prob}(x) \qquad (1.14)$$

(the subscript i may be omitted, as it has been here, if there is no danger of confusion as a result).

Similarly

$$\text{MAD} = E[\,|x - \mu|\,] = \Sigma\,|x - \mu|\,\text{prob}(x) \qquad (1.15)$$

Variance is the expected value of the squared deviations:

$$\sigma^2 = E[(x - \mu)^2] = \Sigma (x - \mu)^2 \text{prob}(x) \qquad (1.16)$$

$$= \Sigma (x^2 + \mu^2 - 2x\mu)\text{prob}(x)$$

$$= \Sigma x^2 \text{prob}(x) + \mu^2 \Sigma\,\text{prob}(x) - 2\mu\Sigma\,x\text{prob}(x)$$

$$= E[x^2] + \mu^2 - 2\mu^2$$

so $\quad \sigma^2 = E[x^2] - \mu^2$ (1.17)

This formula provides an alternative way of calculating variance. Using the data from Tables 1.7 to 1.9:

n	prob(n)	$n$prob(n)	n^2prob(n)
0	0.05	0.00	0.00
1	0.40	0.40	0.40
2	0.50	1.00	2.00
3	0.04	0.12	0.36
4	0.01	0.04	0.16
	1.00	1.56	2.92

Table 1.10

$\mu = E[N] = 1.56$

$\sigma^2 = E[N^2] - \mu^2 = 2.92 - 1.56^2 = 0.49$

as before.

1.8 FURTHER READING

Good brief summaries of the views of the main thinkers about probability are given in Barnett (1973, Ch. 3), de Finetti (1972, Ch. 9), Hartigan (1983, Ch. 1) and Kyburg (1961). Some important contributions are collected in Kyburg and Smokler (1964). If you are interested in the history of statistics generally you might find Owen (1976) and Stigler (1986) worth reading.

Ehrenberg (1982) provides a good guide to sensible and effective ways to present and understand data. Tukey (1977) will also be of interest.

1.9 EXAMPLES

Example 1.1

Suppose that the results of a survey into which of two chocolate bars, *Fatso* or *Lite*, were preferred were tabulated as proportions and according to the sex of the respondent. Taking the relative frequencies as probabilities this gives:

		Lite	*Fatso*	
sex	female	0.2	0.4	Joint
	male	0.1	0.3	probabilities

This is the joint probability distribution. What other probability statements can we make?

Solution

Summing rows and columns gives the marginal probabilities. These show, for instance, that the probability that a respondent is female is 0.6, disregarding her preference in chocolate bars.

		Lite	*Fatso*		
sex	female	0.2	0.4	0.6	Marginal
	male	0.1	0.3	0.4	probabilities
		0.3	0.7		

If we wish to know the preference for chocolate bars given the sex of the respondent then we find the conditional probability distributions by dividing the first row by 0.6 and the second by 0.4 to give:

		Lite	*Fatso*		
sex	female	0.33	0.67	1.00	Conditional
	male	0.25	0.75	1.00	probabilities

The probability that a man will prefer the *Lite* bar is 0.25. The other conditional distributions are:

		Lite	*Fatso*	
sex	female	0.67	0.57	Conditional
	male	0.33	0.43	probabilities
		1.00	1.00	

Example 1.2

In a group of 50 people 30 went abroad for their holidays last year; 10 live in a city and 5 of the city dwellers stayed at home for their holidays.

What is the probability that someone chosen at random from this group and known to live in the country took his or her holidays abroad?

Solution

Start by writing the data given in an ordered form.

	abroad	home	
city		5	10
country			
	30		50

The missing figures are easily found by simple subtraction. For example, the number of country dwellers is $50 - 10 = 40$. Ensure that you can obtain the complete distribution:

	abroad	home	
city	5	5	10
country	25	15	40
	30	20	50

Finally,

$$\text{prob}(\text{abroad} \mid \text{country}) = 25/40 = 0.63.$$

Example 1.3

How well do the data in Example 1.2 support or refute the proposition that there is no relationship between where people live and where they take their holidays?

Solution

The general strategy here is to calculate what we would expect to find if the proposition is true and then to compare with what we did find.

Assuming the two variables to be statistically independent we have, from (1.8), that the joint probabilities are the product of the marginal probabilities. For example,

$$\text{prob}(\text{live in city and holiday abroad}) = \frac{10}{50} \times \frac{30}{50}$$

We wish to compare like with like, not probability with frequency, and so calculate the expected frequency of living in a city and holidaying abroad as

$$\frac{10}{50} \times \frac{30}{50} \times 50 = \frac{300}{50} = 6$$

So if the proposition were true that place of residence and holiday destination are independent we would expect to observe

	abroad	home		
city	6	4	10	Expected
country	24	16	40	frequencies
	30	20		

which is almost the same as what was observed.

But not quite. There are only two possible explanations for the differences: that the proposition of statistical independence is wrong or that the underlying behaviour is indeed truly independent but that sampling fluctuations inevitably mean that exact correspondence is not to be expected. The smaller the difference between what was expected and what was observed the more we would favour the latter explanation. Large differences would lead us to believe that the proposition was ill-founded. But what is large and what is small in this context? What do you think? We shall return to this problem in Chapter 8.

Example 1.4

A factory owns three generators, A, B and C. It is thought that the chances of them suffering a major failure in a year are 4%, 2% and 3% respectively. It is thought to be inconceivable that a generator would suffer more than one such failure in any year.

What is the probability that exactly one generator fails in a year?

Solution

Remember here that one failure occurs if either A or B or C fails, the other two remaining in service. Because of this "or" relationship we need to add probabilities according to (1.4).

The probability that A is the only failure while B and C do not fail is

prob(only A fails) = prob((A fails) and (B does not fail)
    and (C does not fail))

Since we may reasonably assume failures to be independent we use (1.8) to give

prob(only A fails) = 0.04 × (1 − 0.02) × (1 − 0.03)

Finally we have

prob(one failure) = prob(only A fails) +
prob(only B fails) +
prob(only C fails)

$= (0.04 \times 0.98 \times 0.97) +$
$(0.96 \times 0.02 \times 0.97) +$
$(0.96 \times 0.98 \times 0.03)$

$= 0.0380 + 0.0186 + 0.0282$

$= 0.0848$

Example 1.5

The density function in Figure 1.15 shows the estimated time to complete a project. Construct the distribution function. What is the probability that the project will be completed in less than 32 days?

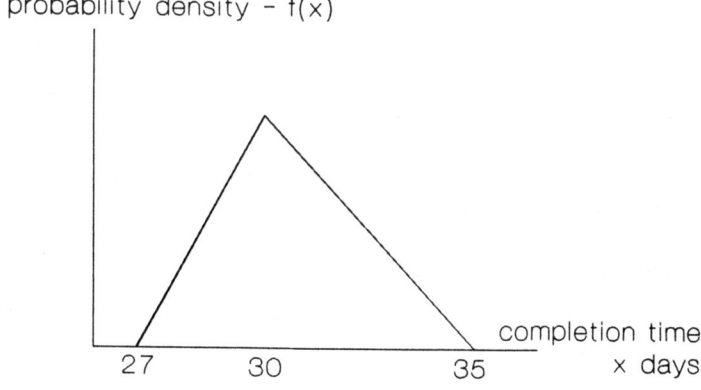

Figure 1.15

Solution

Refer to Figure 1.16.

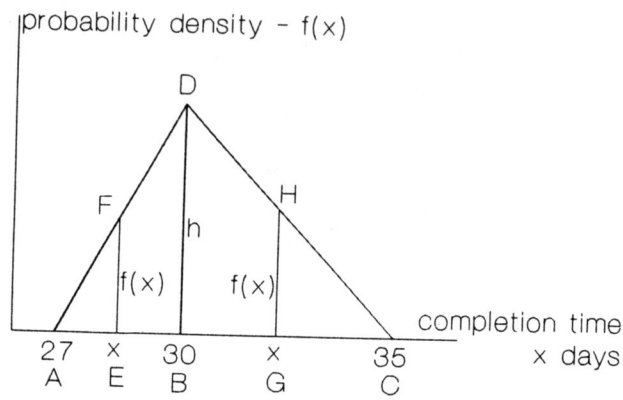

Figure 1.16

The area of a triangle is (½ × base × height). For a density function this area is 1.0 and so

$$0.5 \times (35 - 27) \times h = 1$$

giving

$$h = \tfrac{1}{4}$$

Get the formula for the density function, $f(x)$, and the distribution function, $F(x)$, in two parts. First, for $x \leq 30$.

From the similar triangles ABD and AEF

$$\frac{f(x)}{(x-27)} = \frac{h}{(30-27)}$$

so

$$f(x) = \frac{h}{3}(x-27) = \frac{(x-27)}{12}$$

The distribution function $F(x) = \text{prob}(X \leq x)$ is the area of triangle AEF:

$$F(x) = \frac{f(x)(x-27)}{2} = \frac{(x-27)^2}{24}$$

Similar calculations hold for $x > 30$. From similar triangles BCD and CGH:

$$\frac{f(x)}{(35-x)} = \frac{h}{(35-30)}$$

so

$$f(x) = \frac{h}{5}(35-x) = \frac{(35-x)}{20}$$

$F(x)$ is found by subtracting the area of triangle CGH from the area of the whole.

$$F(x) = 1 - \frac{f(x)(35-x)}{2} = 1 - \frac{(35-x)^2}{40}$$

The distribution function is shown in Figure 1.17.

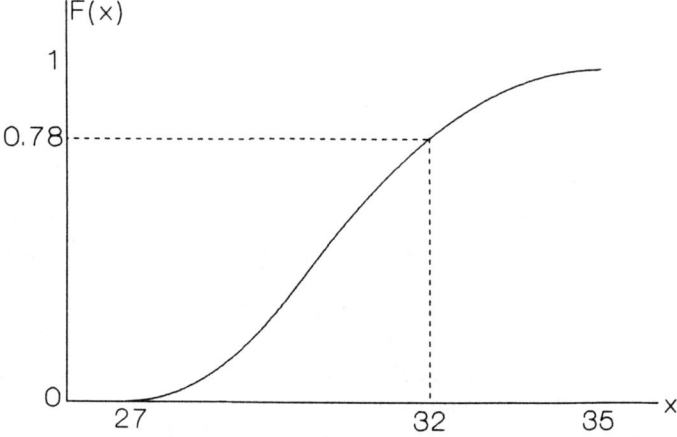

Figure 1.17

Finally, the probability that the project will be completed in less than 32 days is

$$F(32) = 1 - \frac{(35-32)^2}{40} = 0.78$$

Example 1.6

The attendance records of 20 workers in a factory were inspected to find the number of days lost by each due to sickness during the previous six months. Here are the results:

0, 3, 2, 2, 6, 0, 1, 0, 0, 1, 0, 2, 0, 1, 1, 0, 4, 0, 0, 4

Find the average number of days lost and the standard deviation.

Solution

We begin by counting how many times a particular value occurs in the data to obtain the frequency. The first two columns of the table below show the result.

days lost – x	frequency – f	fx	fx^2
0	9	0	0
1	4	4	4
2	3	6	12
3	1	3	9
4	2	8	32
5	0	0	0
6	1	6	36
	20	27	93

It is immediately clear that the mode of the data is 0.

The median value is that which divides the lowest ten values from the highest and so is the mean of the tenth and eleventh ordered values. The first nine values are 0 and the next four, which include the tenth and eleventh, are 1, so the median is 1.

To obtain the mean we wish to use (1.10) or (1.14). The probabilities in this case are just the relative frequencies

$$\text{prob}(x_i) = f_i \Big/ \sum_i f_i$$

where f_i is the frequency of x_i. For instance,

Sec. 1.9] Examples

$$\text{prob}(x_3) = \text{prob}(X = 2) = 3/20 = 0.15$$

If, as here, there is no chance of confusion we may omit the subscripts to give

$$\text{prob}(x) = f/\Sigma f \qquad (1.18)$$

But rather than form all the probabilities we can substitute the formula for relative frequency into (1.10) to give

$$\text{mean} = \Sigma\, x\text{prob}(x) = \Sigma\, x(f/\Sigma f) = \Sigma\, fx/\Sigma f$$

and so in this case

$$\text{mean} = 27/20 = 1.35$$

Substituting $\text{prob}(x) = f/\Sigma f$ into (1.17) gives

$$\text{variance} = \Sigma\, fx^2/\Sigma f - (\text{mean})^2 \qquad (1.19)$$

$$= 93/20 - 1.35^2 = 2.83$$

so standard deviation $= 2.83^{0.5} = 1.68$

(Remember that $\sqrt{2.83} = 2.83^{0.5}$).

Example 1.7

The following table shows the duration of the journey to work of 50 office workers. Find the mean and standard deviation of the times.

journey time (min)	no. of commuters
0– 30	4
30– 60	16
60– 90	19
90–120	11

Solution

This is fairly typical of the data which are often found in reports. A survey has been conducted and the results grouped. We have no access to the original data and so must use the table as given. We may sometimes have made such a summary table ourselves for ease of presentation and calculation.

First, a word about the description of the classes. In which class is a commuter whose journey took exactly 30 minutes? The first or second? We cannot know from what is written. To resolve the ambiguity the classes should be defined as

0 < time ≤ 30
30 < time ≤ 60

and so on. The 30 minute journey is now unambiguously placed in the first class. It may be that people have been asked to recall (rather than record) their journey times and do so only to the nearest 5 minutes and so we would have

0–30
35–60

and similarly for the rest. Tables of the sort shown are not uncommon. Be aware of the need for precise definition of class boundaries and avoid ambiguity when making your own tables. In what follows we shall assume that journey times have been recorded precisely and so show the data as the first two columns in the table below.

journey time t (min)	frequency f	class mid-point x	fx	fx^2
$0 < t \leq 30$	4	15	60	900
$30 < t \leq 60$	16	45	720	32400
$60 < t \leq 90$	19	75	1425	106875
$90 < t \leq 120$	11	105	1155	121275
	50		3360	261450

To proceed with calculations we need to know just what the journey times were, but we cannot because the data have been aggregated into classes and so the original values are lost to us. We therefore have to make some assumptions.

Consider the 16 commuters who travelled between 30 and 60 minutes. Although we do not know the times travelled by each one it seems reasonable to assume that for computational purposes they may be taken as 16 journeys each taking $(30 + 60)/2 = 45$ minutes. The sum of the 16 journey times is approximated as $(16 \times 45) = 720$ minutes. By taking these class mid-points we may carry out the calculations as in the previous example to give

$$\text{mean} = 3360/50 = 67.2$$
$$\text{standard deviation} = (261450/50 - 67.2^2)^{0.5} = 713.16^{0.5} = 26.71$$

One final point on classes. Sometimes you may encounter open ended classes such as "over 90" or "under 50". To get a mid-point in these cases you will just have to make whatever seems the most reasonable assumption.

1.10 EXERCISES

1.1 Here is the estimated distribution of ages of women in 1991. Calculate the mean age and the variance of the distribution.

age	number (thousands)
16–19	1015
20–24	1559
25–34	3012
35–44	2957
45–54	2343
55–59	786
over 59	513

(Source: Employment Department, 1992)

1.2 Listed below are the playing times, in minutes, of the first 50 compact discs reviewed in the October 1992 edition of *Jazz Journal International*. Group the data into classes and then find the mean and standard deviation of the distribution of playing times.

54.08	69.28	74.13	42.70	36.08	65.05	43.62
63.42	43.53	67.15	43.37	55.50	72.02	51.70
33.18	67.35	46.92	48.42	55.00	43.13	58.05
61.13	69.08	62.87	63.02	79.53	58.38	62.93
58.70	72.68	42.77	79.70	66.17	59.00	58.60
63.20	65.50	70.93	50.30	53.15	67.83	73.70
47.22	28.83	69.32	41.42	65.95	63.02	72.73
55.13						

1.3 Some characteristics of calls to the fire service of London, Ontario, are given below. Each of the 13 regions of the city is described by its area (km^2) and call rate (calls/km^2/year).

Find the mean number of calls per region and the standard deviation of the number of calls per region.

What is the probability that a call originates in either region 5 or region 12?

region	area	call rate
1	9.29	3.34
2	13.94	17.44
3	13.94	6.46
4	13.94	3.23
5	27.87	22.71
6	13.94	3.16
7	20.90	2.39
8	13.94	4.59
9	13.94	11.33
10	9.29	73.08
11	20.90	10.95
12	9.29	7.32
13	18.58	5.59

(Source: Wijeratne and Wirasinghe, 1985)

1.4 Using the data from Example 1.4 calculate the probability distribution describing the chances of zero, one, two or three failures in a year.

1.5 A component has to pass through six stages during its manufacture. At each stage the probability that the process is completed satisfactorily, with no errors or faults, is 0.98. What is the probability that a finished component will be fault free?

1.6 Two tests are available to a doctor each of which indicates whether or not a patient is suffering from a particular condition. The principles on which the tests are based are dissimilar. For one test the proportion of occasions on which the presence of the condition is correctly indicated is 80% and for the other test the figure is 70%. The doctor always uses both tests and only declares the patient unaffected if they both indicate this to be the case. What is the probability that a patient suffering from the condition is correctly diagnosed?

1.7 I have to undertake a rather complicated train journey. I leave Hollyville on the 08.53 which is due to arrive at Williamstown at 10.14 from where I catch the 10.26 to Everly. This is scheduled to arrive at 12.00 and so it should be a fairly simple matter to catch the 12.13 to Cline Junction which should arrive a quarter of an hour before the bus departs to take me to my final destination at Wangford.

Each train that I catch starts and ends at the stations that I will use: I do not have to catch any through trains.

It is known that in the system as a whole no trains complete their journey before the scheduled time and none are more than 20 minutes late. The distribution of arrival times follows the right triangular distribution of Figure 1.10. Trains (and buses) start their journeys on time. Changing trains does not involve changing platforms.

What is the probability that I will miss the bus?

1.8 In the braking system of an aircraft a sub-assembly consists of three components A, B and C connected in series such that the sub-assembly will fail if any of the three components fails.

In order to reach an acceptable level of reliability a number of sub-assemblies will be connected in parallel so that the braking system will remain viable provided at least one sub-assembly is working. The situation is shown in Figure 1.18.

The probabilities that components A, B and C will perform without fault during a specified operational period are 92%, 90% and 87% respectively. How many sub-assemblies are required if the braking system must be fault free with a probability of at least 99%?

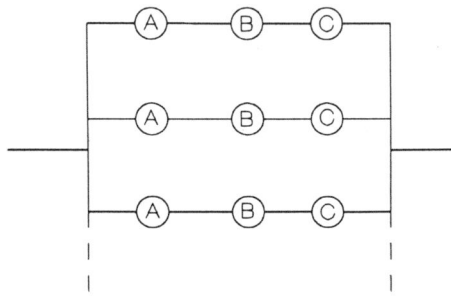

Figure 1.18

1.9 A survey of American college students in the 1950's determined the proportion drinking alcohol to be

religion	men	women
Jewish	94	94
Catholic	90	78
Protestant	77	60
Mormon	54	23

Percentage of students of different religions who drank alcohol
(Source: McCarthy, 1959, p220)

If in a group of students, 70% of whom were men, it was known that 10% were Mormons, 20% were Jewish and 40% were Catholics, what would have been the probability that a student chosen at random did not drink?

1.10 Incoming telephone calls are distributed uniformly throughout the peak hour 09.00 to 10.00. If, during that period, exactly two calls are received in a particular 1 minute interval what is the probability that they arrive within 15 seconds of each other?

[Hint: draw a graph with the times of the calls as the axes.]

2

Information

If probability is a measure of uncertainty then in some sense it must reflect the amount of *information* that would be needed to reduce that uncertainty. There are a number of ways that we can think about what we mean by information. In this chapter one measure will be introduced, though from four different starting points. It will enable us to quantify the *amount* of information associated with a probabilistic estimate.

2.1 SURPRISE

Suppose that my local football team, stuck firmly at the bottom of their division, were playing away and that their opponents were the runaway leaders of the division. I would not be surprised to hear on the radio that they had lost 3–0. If, on the other hand, I had heard that my team had *won* 0–1 then I would most certainly be surprised. What of the information carried in those two alternative radio announcements?

The two announcements could be considered as message transmissions that in some respects were equivalent: they might contain the same number of words, take the same amount of time to say, and so on. Yet I react very differently to them so surely they carry different amounts of information.

I react differently because one message was more or less what I had expected to hear while the other very much was not. The information contained in the two messages depends upon the degree to which my *a priori* expectations are confirmed and *not* on the set of symbols that comprise the transmitted message. If my friend had been at the match then for him the radio message would have been redundant, containing no information, since he already knew the score.

The smaller my *a priori* probability of the result the more surprised I am. This concept of *surprise* or *surprisal* can be modelled quite simply as

$$\text{surprisal} = -\log(\text{probability})$$

as shown in Figure 2.1. (If you need to brush up on logarithms turn to Appendix A2.)

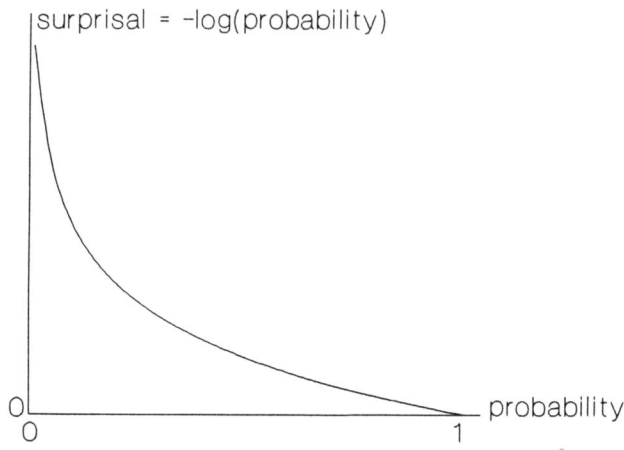

Figure 2.1

The base of the logarithms is generally arbitrary but when logarithms to the base 2 are used the units of measurement are called *bits* (for *bi*nary dig*its*) and when the base e is used they are called *nats* (for *na*tural uni*ts*). Unless stated otherwise the latter will be used in this book. Note that 1 nat = 0.693 bits.

The more surprised I would be to receive a message the greater the *information potential* of that message for me. Watanabe (1969) calls this *ignorance*. As more information is received ignorance or, synonymously, information potential is reduced.

There are, of course, a number of possible outcomes to the football match. The simplest situation is one in which there are just two alternatives: win and don't win, for example. If my *a priori* probability (called, more briefly, my *prior* probability) of a win is p then the probability of any other result is $(1-p)$. The expected surprise, the information potential, is then

$$H(p, 1-p) = -p\log(p) - (1-p)\log(1-p) \qquad (2.1)$$

where $H(\ )$ is the information potential of the probability distribution shown in the parentheses.

Figure 2.2 shows this function. Note that the maximum information potential corresponds to $p = (1-p) = 0.5$, the situation in which I am most unsure of the result.

When there are more than two possible outcomes with prior probabilities $p_1, p_2 \ldots$ then

$$H(p_1, p_2, \ldots, p_n) = -\Sigma\, p_i \log(p_i) \qquad (2.2)$$

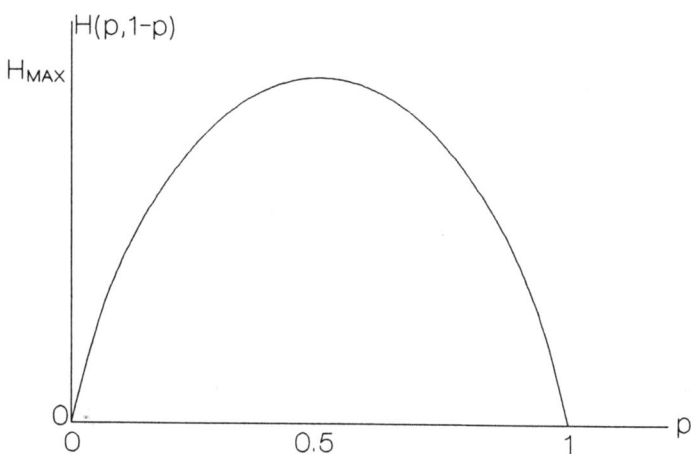

Figure 2.2 Information potential with two possibilities

It is worth emphasising that this measure, *H*, is the mean or expected surprisal (or expected information potential) of a probability distribution and it is this, rather than the surprisal associated with a particular event, that will be the concern of this book.

The maximum value of *H* is when the probability distribution is uniform and all probabilities are $1/n$, where *n* is the number of possible outcomes.

$$H_{MAX} = -\Sigma (1/n)\log(1/n) = -\log(1/n) = \log(n)$$

These uniform distributions are sometimes called *non-informative* because the uniformity is assumed to be the result of an inability or unwillingness to believe that any outcome is more likely than any other. Uniform distributions therefore correspond to maximum ignorance. (We shall see another view of this in the next chapter.)

Now suppose that in advance of the football match I believe that my team will score no more than one goal and their opponents no more than three. After some consideration I assign probabilities to each of the eight possible outcomes as shown in the second column of Table 2.1. Using natural logarithms for the calculation gives $H = 1.83$. Notice how the degree of surprisal varies from the least expected to the most expected result.

score	probability p	surprisal $-\ln(p)$	$p[-\ln(p)]$
3–0	0.25	1.39	0.35
3–1	0.20	1.61	0.32
2–0	0.20	1.61	0.32
2–1	0.15	1.90	0.28
1–0	0.10	2.30	0.23
1–1	0.06	2.81	0.17
0–0	0.03	3.51	0.11
0–1	0.01	4.61	0.05
	1.00		1.83

Table 2.1

If I had been unable to express a view on the outcome and so settled for a uniform distribution then, as shown in Table 2.2, H would have been 2.08.

score	probability p	surprisal $-\ln(p)$	$p[-\ln(p)]$
3–0	0.125	2.08	0.26
3–1	0.125	2.08	0.26
2–0	0.125	2.08	0.26
2–1	0.125	2.08	0.26
1–0	0.125	2.08	0.26
1–1	0.125	2.08	0.26
0–0	0.125	2.08	0.26
0–1	0.125	2.08	0.26
	1.000		2.08

Table 2.2

2.2 SEARCHING

I am thinking of one of the first eight letters of the alphabet. Guess which.

How difficult is this task? One answer is that the difficulty is measured by how many questions are likely to be needed to identify the letter. Assume that I am quite stupid (or just plain awkward) and so will only give you yes/no answers to your questions. How many questions will you have to ask?

You could simply go down the list asking "Is it A?" and then "Is it B?" and so on. If I am thinking of A you will only have to ask one question but if I am thinking of F you will have to ask six. If I am thinking of either G or H you will ask seven questions, since my seventh answer will indicate G if "yes" and H if "no".

You can only assume that I am equally likely to be thinking of any letter and so, using the data in Table 2.3, the expected number of questions is

$\Sigma \, np_n = 4.375$

letter	no. of questions n	prob. p_n
A	1	1/8
B	2	1/8
C	3	1/8
D	4	1/8
E	5	1/8
F	6	1/8
G	7	1/8
H	7	1/8

Table 2.3

There is another method of interrogation which will, on average, be quicker. You could first ask "Does the letter precede E in the alphabet?" My answer would tell you which group of four letters contained the one you were seeking. You could then repeat the strategy to determine which pair of letters contained the target. One more question will reveal the letter that I have in mind.

This method of repeatedly halving the (ordered) list will always terminate after a fixed number of questions, in this case three, whether the target is A or H or any other letter. The number of these halving questions you will have to ask given a list of length n is

$\log_2(n)$

which in this case is

$\log_2(8) = 3$

This procedure is called a *logarithmic search* and is shown diagrammatically in Figure 2.3. An important consequence of this strategy is that a doubling of the size of the problem only requires one extra question:

$\log_2(16) = 4$

$\log_2(32) = 5 = 1 + \log_2(16)$

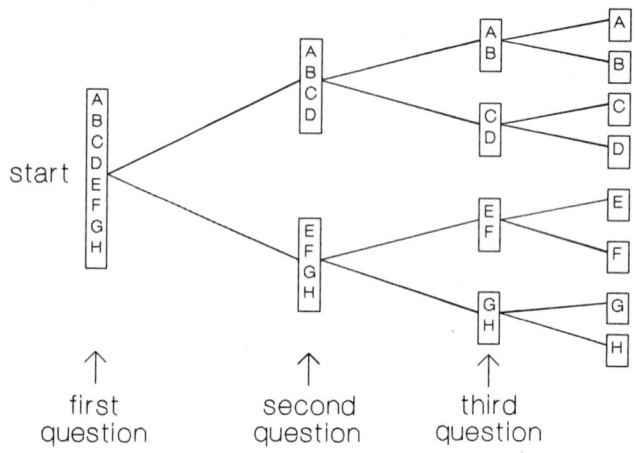

Figure 2.3   Search process with eight items

This is a measure of how difficult it is to identify the letter of which I am thinking.

Suppose now that there are N letters in this problem and that my answer to your first question will determine which of two groups contains the target. The difference is that the two groups are of unequal sizes n_1 and n_2 ($n_1 + n_2 = N$), as shown in Figure 2.4. What now is the expected number of questions? We can look at this in two ways. First, and directly, the answer is simply $\log_2(N)$.

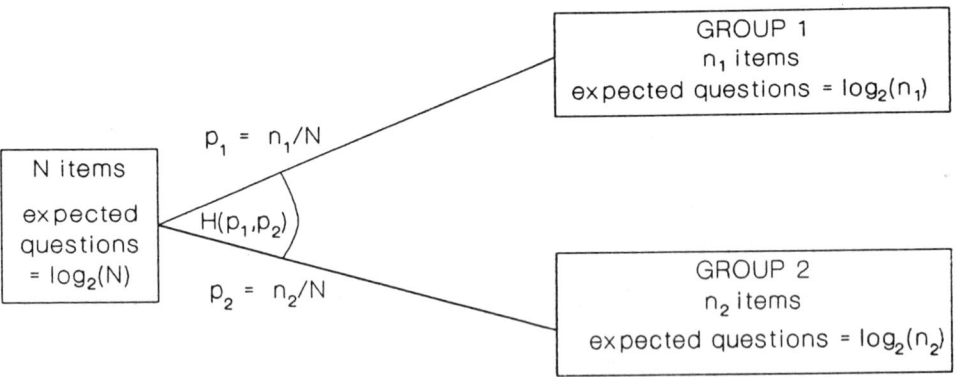

Figure 2.4   Information potential: two unequal groups

Second, think of the problem in two stages. After I have answered your first question the number of questions remaining will be either $\log_2(n_1)$ or $\log_2(n_2)$ and these with probabilities $p_1 = n_1/N$ and $p_2 = n_2/N$ so that the expected number of questions remaining is

$$p_1\log_2(n_1) + p_2\log_2(n_2)$$

But this is *after* the first question. Let the information potential of that first question be $H(p_1,p_2)$ so that the expected number of questions of the whole interrogation is

$$H(p_1,p_2) + p_1\log_2(n_1) + p_2\log_2(n_2)$$

We have already seen that, in total, the expected number of questions is $\log_2(N)$. Since $p_1 + p_2 = 1$, and allowing logarithms to any base, we have

$$\log(N) = (p_1 + p_2)\log(N) = H(p_1,p_2) + p_1\log(n_1) + p_2\log(n_2)$$

so

$$H(p_1,p_2) = -p_1(\log(n_1) - \log(N)) - p_2(\log(n_2) - \log(N))$$

$$= -p_1\log(p_1) - p_2\log(p_2)$$

For any number of subdivisions

$$H(p_1, p_2 ...) = -\Sigma\, p_i\log(p_i)$$

which is the same result as obtained above in equation (2.2). It is not unreasonable to think that the amount of effort needed to answer a question is in some sense equivalent to our *a priori* ignorance of the answer and so to the information potential.

2.3 SIGNALLING

Claude Shannon was a mathematician who worked on problems in signal transmission within communication systems such as that shown in Figure 2.5. Notice the differentiation between "message" and "signal". The latter is a purely technical issue and was the one that concerned Shannon. What the receiving system might make of the received signal is contingent on semantics and context as well as the efficiency of the decoding scheme. A particular signal may not only be an incorrect or inadequate coding of the message by the sender but also be misinterpreted by the receiver, as is famously shown in Stevie Smith's poem:

*Nobody heard him, the dead man,
But still he lay moaning:
I was much further out than you thought
And not waving but drowning.*

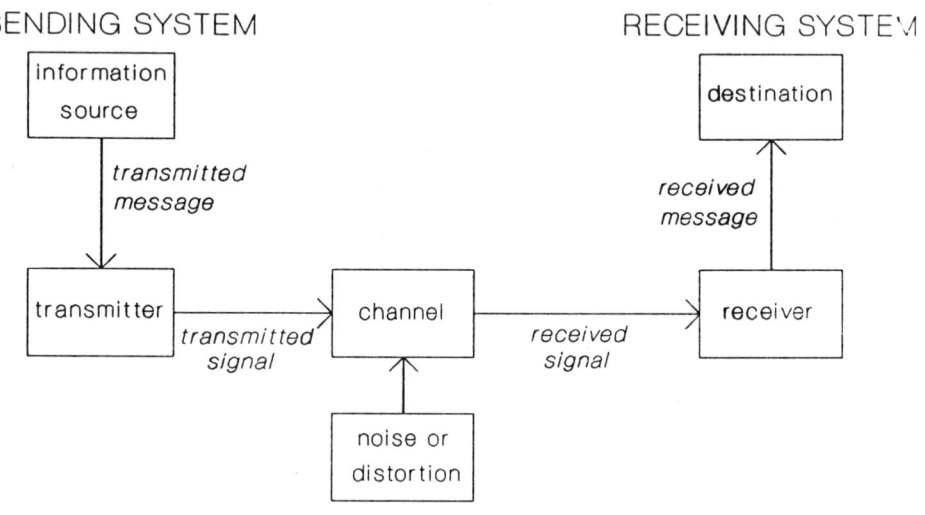

Figure 2.5   A communication system

Shannon was not concerned with the *meaning* of the information carried by a channel but only with the *amount*, and for this he had to be able to measure the information content of any signal.

(Note that Shannon used *information* in the same sense that in the previous section we used *information potential*. This just reflects the different origins of the measures. We shall clear up the question of what to call the quantity H towards the end of this chapter.)

A signal is a string of the symbols permitted by whatever code has been chosen to express the message. Not all symbols are equally likely. For example, Table 2.4 shows the probability of occurrence of symbols in written English as given by Zernike (1972).

Because of the rules of the language not all symbols are equally likely to occur in a message (English was not designed as an optimum code) and this structure must be taken into account when considering the information content of a signal. In English, for example, *e* is by far the most common letter and you would find it quite difficult to construct even a short piece of prose without using it several times.

symbol	space	e	t	a	o	n	i	r	s
prob.	.176	.105	.075	.064	.062	.060	.058	.056	.054
symbol	h	d	l	c	f	u	m	p	y
prob.	.046	.033	.032	.024	.023	.022	.021	.018	.014
symbol	w	g	b	v	k	x	q	j	z
prob.	.012	.011	.010	.009	.005	.004	.003	.002	.001

Table 2.4
Probability of occurrence of symbols in English

Remarkably, this did not prevent one Ernest Vincent Wright, in what must count as a fine display of ingenious whimsy, writing the 267 page novel *Gadsby* without once using the letter *e* (Cover and Thomas, 1991, p133).

The situation is even more complicated, of course, because some sequences of symbols are more likely to occur than others. For instance, the letter *q* in English is quite likely to be followed by *u* while the probability that it is followed by *t* is extremely small.

The receiver may have certain expectations. For instance, I may tell you that either Sue or Ann will be joining us for dinner and that I will let you know later who it is. On receipt of my subsequent badly transmitted fax (the signal) you just have to decide which of two three character strings I have sent, a much simpler problem than decoding a word made from just any three characters. We need to exercise some care here for if the expectations are not well founded then they will be quite misleading rather than helpful.

The simplest situation is when the *n* permissible symbols have probability of occurrence $p_1, p_2, p_3, \ldots p_n$. Shannon wanted an information measure, $H(p_1, p_2, p_3, \ldots, p_n)$, to have the following three reasonable properties:

1. H should be a smooth function of the p_i's.

2. If all symbols are equally likely (all $p_i = 1/n$) then H should increase with n.

3. Think of choosing a symbol from three possibilities. Now consider what happens if the choice is made in two stages. The original value of H should be the weighted sum of the H values of the disaggregated choices. Figure 2.6 shows such a situation in which we require

    $$H(0.5, 0.3, 0.2) = H(0.5, 0.5) + 0.5 H(0.6, 0.4)$$

    where the multiplier 0.5 in the last term reflects the fact that the second choice is only made with probability 0.5. (Note that this idea was used, without formality, in the preceding section.)

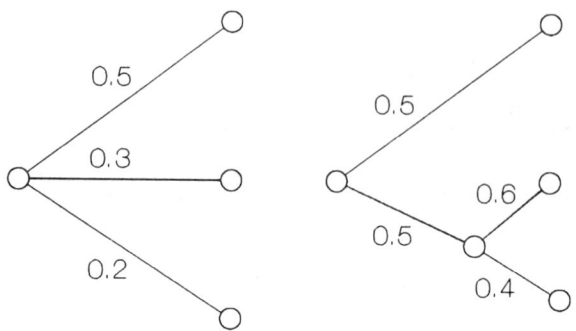

Figure 2.6 Two equivalent choice problems

Shannon showed (Shannon and Weaver, 1949) that the only measure which satisfies these three criteria is

$$H(p_1, p_2 ...) = -k\Sigma\, p_i \log(p_i)$$

Khinchin (1957) achieved the same with three different requirements:

1. H should be at a maximum when all probabilities are equal.

2. $H(p_1, p_2, p_3, ...) = H(p_1, p_2, p_3, ..., 0)$, i.e. H is unchanged by the inclusion of an impossible event.

3. $H(A.B) = H(B|A) + H(A)$

   the information potential of conditional and joint distributions are dealt with in Chapter 7.

You may feel more comfortable with one set of requirements than the other.

The scaling factor k depends only on the base of the logarithms, which is usually arbitrary, and so can be taken as 1 to give, once again,

$$H(p_1, p_2 ...) = -\Sigma\, p_i \log(p_i)$$

Remember that the richer the choice of symbol available to the sending system the greater the task faced by the receiving system.

Let us consider again the information in written English. If all 27 symbols were equally likely $H = \ln(27) = 3.296$. Using the distribution in Table 2.4 gives $H = 2.852$. The difference in the average information per symbol, 0.444 or 13.5%, is a measure of *redundancy* in the language. Further discussion may be found in, for example, Brillouin (1962) and also below.

2.4 DISORDER

Gases and other bodies contain heat, which is the sum of the kinetic energies of the constituent molecules. It is generally impossible (practically or theoretically or both) to know the state of these molecules in terms of their position, velocity or other properties but we may know the state of the whole system in terms of some macro level descriptor such as temperature. In 1872 Boltzmann proposed a probabilistic measure of the thermodynamic concept of *entropy* as

$$\text{entropy} = -k \Sigma\, p_i \log(p_i)$$

where p_i is the probability of the i'th microstate. A microstate is one particular realisation of the possible molecular states.

The degree to which the molecules in a gas are not all in the same state may be thought of as a measure of *disorder* and in 1882 Helmholtz suggested that entropy be used as an appropriate measure of this molecular disorder.

It seems reasonable to believe that the macrostate compatible with the largest number of microstates provides the best guess of the state of the system being considered. Consider, for example, Table 2.1. For the purpose of the football pools macrostates may be whether the game was drawn or won at home and so on, with the actual scores constituting the microstates:

MACROSTATE	MICROSTATE
The pools result	The score
score draw	1–1
no score draw	0–0
home win	3–0, 3–1, 2–0, 2–1, 1–0
away win	0–1

Table 2.5

Entropy has been presented here as a physical property that just happens to have the same functional form as the previously found information measure. There is, however, a suggestive demonstration of the relationship between the two ideas via a famous mind experiment proposed by James Clerk Maxwell in 1871. Imagine a sealed container divided into two chambers by an internal diaphragm in which is set an arbitrarily small opening operated by a frictionless door so arranged that no work is required to open and close it. Let both chambers contain gas at equal temperatures. A creature, called Maxwell's Demon, operates the door so that slowly moving molecules are allowed to pass from the left chamber to the right and fast moving molecules pass in the opposite direction. After a while the effect of this separation of high and low energy molecules will be that one chamber will be at a higher temperature than the other in defiance of the Second Law of Thermodynamics which states that it is not possible to raise the temperature in one part of a gas by using heat from another part without doing some work. The paradox was resolved by noting that the Demon had to possess *information* to enable it to identify the molecules and so allow their passage, or not. Brillouin (1962) provides further discussion, as do many other authors. (See, for example, Zurek, 1990).

Although this paradox has subsequently been resolved by the use of quantum theory the suggestive use of information served as a useful heuristic device (Denbigh and Denbigh, 1985) and it is in that spirit of provocation that it is repeated here.

2.5 DRAWING TOGETHER

In this chapter the same measure has been presented from four different viewpoints. This may strike you as overkill but the pervasiveness of $H = -\Sigma\, p_i \log(p_i)$ is worth emphasising. The first three viewpoints are closely related. A number of writers on information theory include the example of searching as part of the general discussion (e.g. Goldman, 1953) while the concept of surprise, implicit in Shannon's information measure, has found most favour with psychologists and philosophers.

It is really the entropy interpretation that stands out as different. Shannon noted the correspondence between his result and Boltzmann's well known formulation and talked of his measure giving the entropy of a probability distribution, though this use of the word entropy seems to have been suggested to him as a sort of marketing device (Denbigh and Denbigh, 1985, p104). Several writers (Jaynes, 1957, and Olmsted, 1988, among them) have pointed out that just because the formulae are similar does not mean that the underlying concepts are directly comparable. It has been suggested that using different names would help clear up any potential confusion. For example, Peters (1975) suggests that *spread* be used to describe the general property of a distribution and that it only be called entropy if it refers to the spread of microphysical state properties. In other circumstances he suggests *intropy* for *in*formation *t*heoretical ent*ropy*.

Such ideas would no doubt be useful but they have not been taken up. The idea that H measures the entropy of a quite general probability distribution has persisted and is common terminology: it will be used here.

Returning to the word *information* do not forget that H measures information *potential* or how much we *don't* know. If we have perfect knowledge (or belief) we have no need of further information.

Remember also the Stevie Smith poem. We will be talking about quantities of information and not with the semantic accuracy of the messages passed. Goldman (1953) suggested using two probabilities, one to measure the probability of a particular *message* in the language being used for communication and the other to measure the probability of the *event* described by the message. Values of H based on the first would then be measures of *language* information while the second would give measures of *semantic* information.

Jamison (1970) extends the argument,

> "Two alternative notions of semantic information are *reduction in uncertainty* and *change in belief.* Reduction in uncertainty is, clearly, a special case of change in belief."

while Hintikka (1970) remarks that the relationship between semantic information and statistical information theory is not always clear.

The various uses of the word *information* are, as we have seen, intriguing and suggestive but may also lead to some confusion. In what follows, therefore, we shall assume that

information potential = entropy = $H = -\Sigma p_i \log p_i$

We shall try to avoid the possibly confusing use of *information* alone, preferring to speak of a reduction in entropy, or in ignorance or in information potential.

2.6 REDUNDANCY

Since the units in which entropy is measured are arbitrary it will sometimes be helpful to consider the entropy of a distribution with reference to the maximum entropy possible and so we define

relative entropy = H/H_{max}

where

$H_{max} = 1/n$

For the English language (Table 2.4) the relative entropy is $2.852/3.296 = 0.865$.

A related measure is the redundancy of the distribution:

redundancy = 1 − (relative entropy)
= $1 - (H/H_{max})$

For English this is $1 - 0.865 = 0.135$, as before.

Think about what this means. If there were no redundancy and $H = H_{max}$ then in written English any letter would be equally likely to follow any other letter. The number of words that it would be possible to encode would be maximised but something would be lost. Although I cannot speak or read any language other than English I can nonetheless fairly reliably distinguish a page of written Italian from a page of French. I can do this because of the different patterns of characters: the letter frequencies are different as are the probabilities of letter pairs and the like. These different patterns are possible because each language exhibits some degree of redundancy.

For this reason Moles (1966) equates redundancy and structure saying that structure "expresses the ascendancy of the intelligible over the perceptible". Without redundancy we would have no styles and no crosswords. In this context entropy measures complexity while redundancy measures intelligibility.

The same ideas may be applied to music, for instance. Whitney Balliett, jazz critic of the *New Yorker*, once called jazz "the sound of surprise". The linear improvisations of mainstream jazz make it easy to "guess the next note". This music also has a readily accessible melodic structure and yet enough variety to be interesting. The chord based improvisations of bebop permit more realisations of the same theme and, to a degree, redundancy has decreased. The process is extended in free form improvisations. You can trace similar movements in other sorts of music. Why you or I prefer one type of music on this continuum to any other is, of course, a topic of lively discussion but must surely have something to do with interpreting a high degree of redundancy as banal rather than, favourably, whistleable, or vice versa. At the other extreme I might find that the presence of only a small amount of redundancy leads to incoherence while you might interpret it as permitting a satisfying intellectual richness. Our individual tastes and the conditioning effect of the cultures in which we live play some large part.

Thinking of the "guess the next note" (or letter or whatever) view of redundancy you may wish to consider

$$\text{surprisal} = -\log(\text{probability})$$

as a measure of originality.

Rather similar considerations lead Watanabe (1969, pp50–52), in discussing bureaucratic and administrative structures, to equate redundancy with organisation. He suggests that

$$\text{organisation} = (\text{sum of entropies of parts}) - (\text{entropy of the whole})$$

Quastler (1955b) has also written on entropy and structure.

These thoughts, which I have chosen to end this chapter, emphasise again the powerfully suggestive applications of the entropy concept. In the next chapter we return to statistical matters.

2.7 FURTHER READING

The concept of surprisal has most often been used by those concerned with problems of cognition and knowledge acquisition. Examples are provided by Attneave (1959), Dretske (1981) and Watanabe (1969).

In addition to that provided by Khinchin, Domotor (1970) refers to other axiomatic structures that may be used to derive H. This same measure of information was introduced by Wiener (1948) at about the same time as Shannon but independently of him.

There are many introductory texts on information theory and which you like depends on your taste and particular requirements. You might try, among others, Goldman (1953), Feinstein (1958), Jones (1979) and Mansuripur (1987). I particularly liked the emphasis of Cover and Thomas (1991). Cherry (1966) gives an interesting, broader, discussion, a little dated in some respects but still worth reading. Kullback (1959) is notable for linking information theory and statistics. Quastler (1955a) and Attneave (1959) discuss applications in psychology.

In similar vein there are many books on thermodynamics (e.g. Fast, 1962) and you will probably require only a small part of any of them to explain further the difference between the purely thermodynamic concept of entropy and that derived from statistical mechanics. The Second Law and entropy have often been seen, because of their emphasis on irreversible processes, as constituting a time directed prescription absent in many other theoretical frameworks. There are now several good popular books in which these interesting issues are discussed. Have a look at Coveney and Highfield (1990).

2.8 EXAMPLES

Example 2.1

It is the morning of Saturday 27 November 1993. At 2 p.m. the English rugby union side will play New Zealand's All Blacks at Twickenham. The visitors have not lost a match on this tour and last week comprehensively beat Scotland. England have only beaten the All Blacks three times this century. Swept up in a fervour of patriotic optimism I rate the probability of an English win as high as 30% and the probability of a draw as 10%. What is the entropy of this match?

Solution

Listing the outcomes from England's viewpoint gives the following calculation:

outcome	probability (p)	surprisal $-\ln(p)$	$-p\ln(p)$
win	0.3	1.204	0.361
draw	0.1	2.303	0.230
lose	0.6	0.511	0.307
	1.0		0.898

The entropy of the match is 0.898.

Example 2.2

It is 4 p.m. on Saturday 27 November 1993. England have just beaten the All Blacks 15–9. What is the information in this result?

Solution

We have to be clear what we mean when we use this elusive word *information*. It is clear that because we now know who won the information potential, the entropy, has been reduced to zero: we are no longer ignorant of the result. On this view the information in the result was 0.898. But this value is independent of who won. Remember that entropy measures the information potential in the situation before the signal (the result) is received.

We were (I was, certainly) surprised by the result and this is measured by the surprisal

$$-\ln(0.30) = 1.204$$

This measures the effect of the receipt of this *particular* signal and so better measures the effect of knowing that England won. This in no way contradicts the observation that the information potential has been reduced from 0.898 to zero, but any result would have done that.

Example 2.3

A number of boreholes are to be drilled to assess ground condition. The result will be expressed as a grade, A to D. I believe that the proportion of boreholes in each category will be

A	1/8
B	1/2
C	1/4
D	1/8

The results will be stored on a rather old fashioned data logger as a string of zeros and ones, rather like Morse code without the gaps. I may set the code. What ought it to be?

Solution

Here is one possible code:

Examples

A	0
B	1
C	10
D	11

The trouble with this is not in the coding but in the decoding. Does the sequence 10 mean one borehole of grade C or two, a B then an A? A little thought will give this as the most compact workable code:

A	0
B	10
C	110
D	111

Check for yourself that any sequence of zeros and ones may now be unambiguously decoded. Having constructed the words in the code how best ought they to be matched with the grade letters? It seems sensible to wish to minimise the expected length of the coded message for all the results. The number of symbols expected to be required in the above assignment is

grade	code	length	probability	(probability × length)
A	0	1	1/8	1/8
B	10	2	1/2	1
C	110	3	1/4	3/4
D	111	3	1/8	3/8
				18/8 = 2.25

On average we will need 2.25 symbols per borehole. The zeros and ones are usually called binary digits, or *bits*, and so we may say that this code requires 2.25 bits per borehole.

It may already have struck you that we haven't been too bright in the assignment of codes to grades and that it would make a lot of sense to assign the shortest code to the most likely grade, and so on. Here is the result:

grade	code	length	probability	(probability × length)
A	110	3	1/8	3/8
B	0	1	1/2	1/2
C	10	2	1/4	1/2
D	111	3	1/8	3/8
				14/8 = 1.75

We may now store results at the rate of 1.75 bits per borehole, a saving of 22% on the amount of storage required. This is the optimum code.

Before we leave this example let us calculate the entropy of the distribution of proportions. Departing from our usual practice we shall take logarithms to the base 2 and so express the result in bits.

$$H(1/2, 1/4, 1/8, 1/8) = -(1/2)\log_2(1/2) - (1/4)\log_2(1/4)$$
$$- (1/8)\log_2(1/8) - (1/8)\log_2(1/8)$$
$$= 1/2 + 2/4 + 3/8 + 3/8$$
$$= 14/8 = 1.75 \text{ bits}$$

which is exactly the same as the mean number of bits per borehole required for storage. This is, of course, no accident. The entropy provides a lower bound for the mean length of codeword. Sometimes, as here, a code may be devised to realise this lower bound exactly.

Coding is discussed more fully in just about any text on information theory, including those given in this chapter.

2.9 EXERCISES

2.1 After you next watch the television news ask yourself "how much information did it contain?"

2.2 Two dice are thrown and the score is the sum of the values shown on each. What is the entropy of this set of 11 outcomes? If you are told that the result is 10 how much information is there in the message?

2.3 In a certain community 25% of all girls are blondes and 75% of all blondes have blue eyes. Also 50% of all girls in the community have blue eyes. If you know that a girl has blue eyes, how much additional information do you get by being informed that she is blonde?

(Source: Goldman, 1953)

2.4 The following story is recounted by Perutz (1993).

The work that resulted in the first atomic bomb contained much that was of interest to the scientific community and, in the early period, papers were published in academic journals (without mention of the bomb, of course). In 1940 Louis Turner, a Princeton physicist, wrote a paper suggesting that uranium 238 might be transformed into uranium 239 (plutonium) which was superior to uranium 235 as a fissionable material for use in the bomb. Fearing that publication of this paper might give information to the Germans President Roosevelt's Defense Research Committee censored all papers on uranium and so Turner's work was not published.

George Flerov was a physicist serving in the Russian army and while on leave in 1942 he went to the library of the physics institute in Voronezh. He noted that publication of papers from America on this topical issue had stopped and so wrote to Stalin telling him that the Americans were working on a bomb. Stalin immediately ordered that work begin on a Russian bomb.

The censorship imposed to thwart the Germans may be thought to have led directly to the post-war arms race.

Analyse this event from the points of view of the American government and of Flerov. What were the signal, the message and the information? (If you are a Sherlock Holmes fan you will doubtless be reminded of *The Adventure of Silver Blaze* and the dog that did not bark.)

2.5 (a) I am about to toss a coin and you are interested in whether it will be a head or a tail. What is the information potential of this next toss?

(b) How would your guess have been changed if I had told you that I had already tossed the coin twice and obtained a head on both occasions?

(c) I am about to write a letter of the alphabet and you are interested in what it will be. What is the information potential?

(d) I have already written the two letters E and X. What is the information potential of the third letter?

2.6 Here are some pieces of text. Fill in the gaps, shown as *. What conclusions do you make about redundancy?

(a) *he c*t *at *n *he ***.
(b) I do n*t mu** l*ke pu**les.
(c) N** s*cl*p* d*uw we v*poas* br*h*kwe* * as*.
(d) L*ing is in**ed a* ac*ur*ed *ice.
(e) My *ife is ca*le* Syl**a a*d my da***t*r is *ate.
(f) *n *erit* *e ment*r **t *n m*u*it **ce.

3

Unbiased Estimates

We have seen how probability can be used as a tool in assessing the amount of information but have so far treated the assessment of probabilities quite informally. The question that must now be addressed is "where do our probability assessments come from?"

This chapter describes a methodology that answers this question and applies it to some discrete probability distributions.

3.1 JAYNES' CRITERION

I have just emptied all the coins that I have in my pocket onto my desk. There are 11 coins worth, in all, £2.07. If I now select one coin at random what is the probability that it will be a 10p piece?

How might you set about this problem? You could assume that we both accumulated coins by roughly the same process and so look at the proportion of the coins that you have that are 10p pieces and make that the answer.

Since the number of coins is quite small you could list the different combinations of 11 coins that add up to £2.07 and look at the occurrence of 10p pieces. There are 12376 combinations of which only 30 sum to £2.07. But can you find them? And what would you do if there were 110 coins or 11000? It would be useful to have available a method which looked directly at the relative frequencies of the different coins, and so worked equally well however many coins were involved, without having to enumerate all valid combinations. We wish to find the best estimate of the macrostate without having to list all the microstates, so even if you have managed to find the 30 combinations of 11 coins what follows will be of more general use.

You can probably think of other ways to approach the problem. Whatever you do it is likely that the method you choose will rest upon some analogy, assumption or piece of information additional to that provided in the previous paragraph.

This importation of perhaps sensible but ultimately hard to justify additional matter into the assessment can be thought of as introducing *bias*. An estimate is biased if, faced with the same problem, another reader of this book could arrive at a different assessment because of the introduction of some different line of auxiliary yet inexplicit reasoning. If the bias that you introduce is effective you will be called an expert, or perhaps just plain lucky.

The problem of bias is often used as an argument against the subjectivity inherent in a personal view of probability. Given the view of man as a "selective, step-wise information processing system with limited capacity ... ill-suited to the task of assessing probability distributions" (Hogarth, 1975) the difficulties may seem to be insurmountable.

Edwin Jaynes, a physicist, wished to remove this bias and suggested that we should make assessments that are as ignorant or uncertain as possible subject only to the data that we have available or, in his words, "maximally noncommittal with regard to missing information" (Jaynes, 1957). This means that the probability distribution to choose is the one with the highest uncertainty (i.e. entropy) consistent with what we know or assume, *a priori,* about the distribution. This principle, and the methodology to which it gives rise, is called *entropy maximisation.*

The elicitation of personal subjective probabilities is not uncontroversial. In addition to the problems of bias it has often been pointed out that there is an intimate relationship between assessed probabilities and the purposes for which they are needed: whether I am deciding to buy a raffle ticket or issue an earthquake warning, for instance. Probability assessments are unlikely to remain unaffected by the scale of the consequences of error and for this reason some believe it wrong to treat them separately and so propose that probabilities and the utilities of the outcomes be developed together. (See Chapter 6 for more on utility.)

Jaynes' views follow those of Ramsey (1926) and, particularly, Jeffreys (1939), that "statistical analysis should make use, not of anybody's personal opinions, but rather of the factual data on which those opinions are based" (Jaynes, 1968).

In this book I take a more catholic line. It seems problematic, to put it no higher, whether unaided humans can reliably and satisfactorily specify probability distributions or, perhaps, probabilities at all. It is also clear that if useful data are available then they should be used. But data may not always be available and on these occasions some expression of opinion would seem to be reasonable. This opinion will be based on some recollection or extrapolation to be sure, though perhaps not in a way that is easy to model.

There is the obvious objection which underlies Jaynes' point: that bias has just been shifted, not removed. Perhaps so, but remember that it is not recommended that opinion replace data, if they are available, rather that for some assessments they may not be, in which case a methodology which is both *structured* and *explicit* is required for the successful incorporation of opinion. You must make up your own mind on this. If you wish to restrict your considerations to those cases for which data are available then please do so. If, on the other hand, you wish to incorporate judiciously some opinion too then the same methods may be used. A textbook is no place for propaganda: it's your choice.

The maximum entropy formalism will be used to derive probability distributions given either *a priori* data or opinion about some characteristics of the distribution. As we shall see in Chapter 5 these distributions may be readily modified in the light of data subsequently acquired.

3.2 FINDING UNBIASED ESTIMATES

We have to find values for the probabilities p_i that maximise the entropy H subject only to what we know or assume *a priori*. We shall restrict our *a priori* conditions to those that can be expressed as simple formulae. For instance, we may require the mean of the distribution to be 3.6 and so our requirement is

$$\Sigma\, p_i x_i = 3.6$$

Such requirements impose *constraints* on the probability distribution. We therefore need to solve constrained maximisation problems of the type

$$\left. \begin{array}{ll} \text{maximise} & H = -\Sigma\, p_i \ln(p_i) \\ \text{subject to} & g(p_i) = k_1 \\ & h(p_i) = k_2 \\ & \text{etc.} \end{array} \right\} \qquad (3.1)$$

where the constraint equations are broadly of the type exemplified by $\Sigma\, p_i x_i = 3.6$. Notice that these constraints are shown as functions of the p_i's rather than the x_i's since we are determining values of the probabilities and not of the variable X.

It can be shown (see Appendix A1) that optimal values for the probabilities are given by

$$p_i = \exp[-1 + \alpha g'(p_i) + \beta h'(p_i) + \ldots] \qquad (3.2)$$

The parameters α, β, etc, are called *multipliers*.

$g'(p_i)$ and $h'(p_i)$ are called the *differentials* of the functions $g(p_i)$ and $h(p_i)$. They measure the rate at which the value of the function changes as p_i changes: if p_i changes by 1.0 then $g(p_i)$ changes by $g'(p_i)$.

Here are some common differentials:

Finding unbiased estimates

function $f(p)$	rate of change $f'(p)$
p	1
c	0
cp	c

c is a constant

Table 3.1

To see how all this works suppose that we have no *a priori* requirements and so no constraints at all. From (3.2) we just have

$$p_i = \exp(-1) = 0.368$$

There is something wrong here for this answer says that the probabilities, however many there are, all have the same value and that that value is 0.368. But we know from Chapter 1 that for a probability distribution

$$\Sigma p_i = 1.0 \qquad (3.3)$$

It is our failure to use this as part of the specification of our problem that has resulted in getting the wrong answer. We therefore augment the maximisation problem by adding the constraint

$$g(p_i) = \Sigma p_i = 1.0$$

Remember that we want an expression for a particular probability p_i and so all other probabilities are, for this purpose, taken as constants.

$$g(p_i) = \Sigma p_i = p_1 + p_2 + \ldots + p_i + \ldots$$

From Table 3.1

$$g'(p_i) = 0 + 0 + \ldots + 1 + \ldots$$

so $\quad g'(p_i) = 1.0$

and (3.2) becomes

$$p_i = \exp[-1 + \alpha g'(p_i)] = \exp[-1 + \alpha] = k$$

This tells us that all the probabilities have the same value, k. How is it to be found?

An appealing aspect of the maximum entropy methodology is that to see how to determine the value of a multiplier you just need to refer to the constraint with which it is associated. In this case α, and so k, comes from the constraint that says the sum of the probabilities must be 1.0, so choose k to ensure this.

If there are n probability values in the distribution then

$$\Sigma p_i = \Sigma k = nk = 1.0$$

so $\quad k = p_i = 1/n$ \hfill (3.4)

which is just a uniform distribution.

As a result of all this analysis we are able to say that if all we know about a set of probabilities is that they form a complete distribution then the least biased estimate is to set all probabilities equal. Now you probably saw that already and we used it earlier in Table 2.2. What we have done is to set up formally the mechanism that is implicit in Figure 2.2, which shows that maximum entropy in a two state system is when both probabilities have the same value of 0.5.

Although this seems a fairly trivial example it does establish the main points of the procedure. Notice the diagnostic loop. When we found a result that we thought was wrong we were able to augment the problem specification with an appropriate new constraint and get a revised expression for p_i.

Since we shall in this book always require a complete probability distribution ($\Sigma p_i = 1.0$) we may rewrite (3.2) as

$$p_i = k \exp[\alpha g'(p_i) + \beta h'(p_i) + ...] \quad (3.5)$$

where k is the scaling constant chosen to ensure that the probabilities sum to 1.0 and the other constraints, $g(p_i)$, $h(p_i)$ and so on, ensure that the mean or variance or whatever have the required values.

Before moving on recall the main points of methodology encountered in this section:

1. To get the least biased estimate find the probabilities that maximise H subject to any *a priori* constraints, $g(p_i)$, $h(p_i)$, etc., that we wish to impose.

2. Find the optimum expression for p_i from equation (3.5) and Table 3.1.

3. To see how to determine the values of the multipliers α, β, etc., refer to the constraints with which they are associated and choose the values which ensure that they are met.

4. If the result is unsatisfactory decide why this is so and express your finding as another constraint. Go back to step 1 and repeat the process.

3.3 ASSESSMENT WHEN THE MEAN IS KNOWN

Let us now apply what we have found to the problem posed at the beginning of this chapter: that of finding the probability that the coin I choose is a 10p piece. The recommendation from the previous section is that I use a uniform distribution. Since there are seven possible denominations of coin I assign a probability of $1/7 = 0.143$ to each of the coins to give me a probability distribution (Table 3.2 and Figure 3.1) showing how likely it is that a coin chosen at random will be of a particular denomination. In particular, prob(10p coin) = 0.143.

value of coin – x_i	prob(x_i) = 1/7	$x_i$prob(x_i)
1	0.143	0.143
2	0.143	0.286
5	0.143	0.714
10	0.143	1.429
20	0.143	2.857
50	0.143	7.143
100	0.143	14.286
	1.000	26.857

Table 3.2

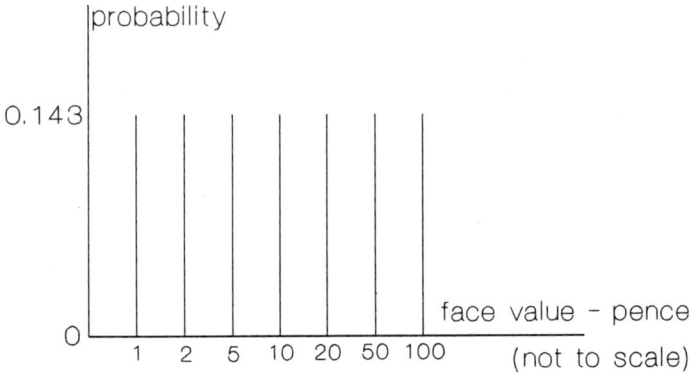

Figure 3.1 Probability distribution nothing known (see Table 3.2)

Does this distribution meet all the requirements given in the problem statement? I have 11 coins worth £2.07 so the mean value per coin is $207/11 = 18.818$p and not the 26.857p of Table 3.2.

Look at step 4 in the procedure described above. We need to add the constraint

$$\sum p_i x_i = \mu \qquad (3.6)$$

From Table 3.1 if

$$g(p_i) = \sum p_i x_i \qquad (3.7)$$

then

$$g'(p_i) = x_i$$

and, from (3.5),

$$p_i = k \exp(\alpha x_i) \qquad (3.8)$$

As before k will be chosen to ensure that the probabilities sum to 1.0. The term $\exp(\alpha x_i)$ permits a series of weights to be applied to the x values that will shift the probability distribution either, if $\alpha < 0$, to increase the probabilities of low values of x and so reduce μ, or else to increase the probabilities of high values and so increase μ. Figure 3.2 gives some idea of what this weighting function looks like.

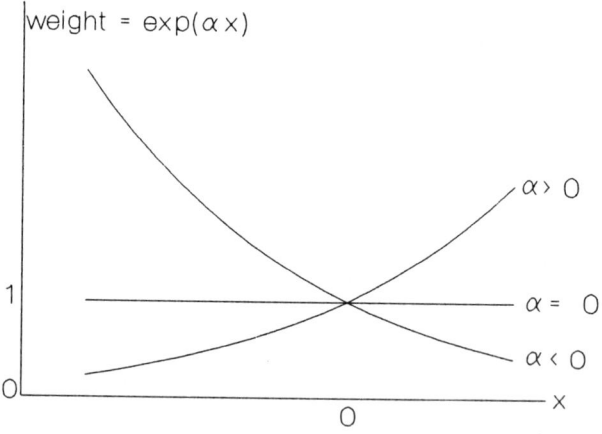

Figure 3.2

Table 3.3 shows the calculation to get a probability distribution for an arbitrarily chosen value of $\alpha = 0.010$. The second column shows the values of the weighted estimate.

Sec. 3.3] Assessment when the mean is known

For instance, for the 5p piece $\exp(\alpha x_i) = \exp(0.01 \times 5) = \exp(0.05) = 1.051$. The next column shows these values scaled so that they sum to 1.0 ($k = 1/9.775 = 0.102$). This is the probability distribution.

value of coin – x_i	$\exp(\alpha x_i)$	$\text{prob}(x_i) = \exp(\alpha x_i)/9.775$	$x_i \text{prob}(x_i)$
1	1.010	0.103	0.103
2	1.020	0.104	0.209
5	1.051	0.108	0.538
10	1.105	0.113	1.131
20	1.221	0.125	2.499
50	1.649	0.169	8.433
100	2.718	0.278	27.808
	9.775	1.000	40.721

Table 3.3 $\alpha = 0.010$

The mean of 40.721p is far too high. It is even higher than for the uniform distribution, $\alpha = 0$, and so a negative value of the parameter is needed. (Figure 3.3 gives you some idea of the relationship between μ and α.)

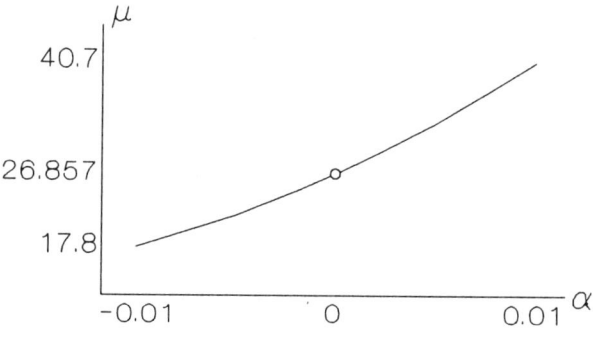

Figure 3.3

It was quite easy to set up the calculation of Table 3.3 on a spreadsheet and, after a little trial and error, find the correct distribution shown in Figure 3.4 and Table 3.4. (If you have done some computing you can probably see how to write a program that will automatically find the correct distribution. Try it.)

Notice the effect of having a negative value of α. The probabilities of higher denominations are now much lower than before and my revised estimate of prob(10p) is 0.159, up from 0.143.

value of coin – x_i	$\exp(\alpha x_i)$	$\text{prob}(x_i) = \exp(\alpha x_i)/5.763$	$x_i \text{prob}(x_i)$
1	0.991	0.172	0.172
2	0.983	0.171	0.341
5	0.958	0.166	0.831
10	0.917	0.159	1.592
20	0.842	0.146	2.921
50	0.650	0.113	5.637
100	0.422	0.073	7.325
	5.763	1.000	18.818

Table 3.4 $\alpha = -0.008625$

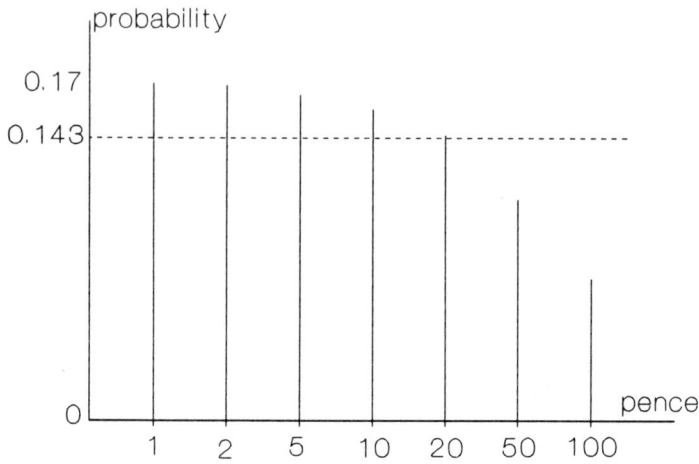

Figure 3.4 Probability distribution mean known (see Table 3.4)

3.4 ASSESSMENT WHEN MEAN AND VARIANCE ARE GIVEN

Suppose that, seeing you to have a statistical bent, I tell you that the variance of the face values of the coins on my desk is 684.876p².

You immediately recognise that you will need to reformulate the maximum entropy model by incorporating a further constraint which allows *a priori* specification of the variance, σ^2.

The formula for variance, (1.17), is

$$\sigma^2 = \Sigma\, p_i x_i^2 - \mu^2 \tag{3.9}$$

Sec. 3.4] Assessment when mean and variance are given

and so the new constraint is

$$\Sigma p_i x_i^2 = \sigma^2 + \mu^2 \tag{3.10}$$

with

$$h(p_i) = \Sigma p_i x_i^2 \tag{3.11}$$

and

$$h'(p_i) = x_i^2$$

Using this in (3.5), together with the constraint (3.7), gives

$$p_i = k \exp(\alpha x_i + \beta x_i^2) \tag{3.12}$$

It will be more convenient to write this as

$$p_i = k \exp[-c(x_i - m)^2] \tag{3.13}$$

where c is related to β and so controls variance while m, being based on α, controls the mean. As we shall see, c and m are not independent and this interaction affects the procedure by which they are calculated.

(If you want to do the algebra yourself add the two terms $\alpha^2/4\beta$ and $-\alpha^2/4\beta$ to those already in the parentheses in (3.12) and then look for the perfect square. The constant term gets absorbed into k leaving $c = -\beta$, and $m = -\alpha/2\beta$.)

The calculation proceeds broadly as before, except that this time we have two parameters, c and m, for which to find appropriate values. Start by setting a value for one of them, say $m = 0$. Find, by trial and error perhaps, a value of c that gives the correct value of σ^2. Now find a value of m that gives the correct value of the mean μ. But this will throw out the value of σ^2 so find c again. Proceed in this way (Table 3.5). The adjustments to c and m will get progressively smaller until both constraints will be simultaneously satisfied.

Table 3.6 shows the final table, with the parameter values, and indicates how to set up a spreadsheet for the calculation. The probability distribution is shown in Figure 3.5. Note that the revised estimate is prob(10p) = 0.162.

m	c	μ	σ^2
0	0.0000947	18.833	684.253
−0.11	0.0000947	18.818	683.513
−0.11	0.00009434	18.840	684.875
−1.180	0.0000925	18.818	684.807

Table 3.5
The first three and the last iterations

value of coin − x_i	A	B	C	D
1	1.000	0.163	0.163	51.89
2	0.999	0.163	0.327	46.21
5	0.996	0.163	0.815	31.11
10	0.989	0.162	1.616	12.57
20	0.959	0.157	3.137	0.22
50	0.785	0.128	6.416	124.78
100	0.388	0.063	6.343	418.04
	6.116	1.000	18.818	684.807

key to columns:
A weight = $\exp[-c(x_i - m)^2]$
B prob(x_i) = A/6.116
C $x_i$prob(x_i)
D prob$(x_i)(x_i - \mu)^2$

Table 3.6 $m = -1.18$, $c = 0.0000925$

The general shape of the weighting function (3.15) is shown in Figure 3.6. The parameter c controls the spread of the curve and m its general position, rather like a mode. In the case we have just looked at notice how the value of m that we finally obtained was outside the range of the variable (coin values) meaning that we were using a portion of the curve to the right of the mode, as shown on the diagram. If you believe that the probability distribution that you seek is likely to be roughly symmetric then setting $m = \mu$ is probably a good starting point.

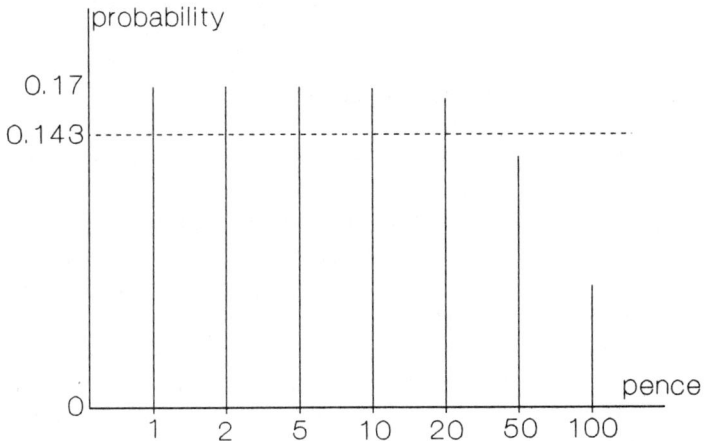

Figure 3.5 Probability distribution mean and variance known (see Table 3.6)

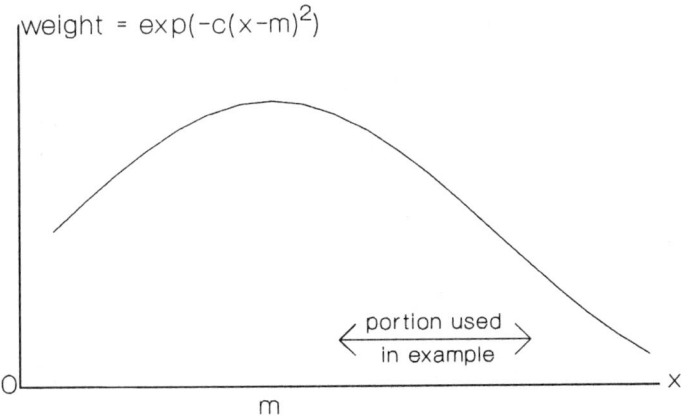

Figure 3.6

3.5 FURTHER READING

Much has been written about the difficulties people face when asked to make probability assessments and how best to help. Much stems from the paper of Tversky and Kahneman (1974). Reviews and developments are given by Hogarth (1975), Kahneman et al. (1982), Nisbett and Ross (1980) and Smith (1988). Miller (1956) and others emphasised the restricted short term memory and information processing capacities of the human brain. There is currently renewed emphasis by psychologists and philosophers in problems of cognition. See, for instance, Goldman (1986).

Following his 1957 paper Jaynes subsequently considered a more general application in decision analysis (Jaynes 1968). The breadth of Jaynes' work can be seen in Jaynes (1983). Some of the many applications are discussed in Levine and Tribus (1979), Smith and Grandy (1985) and Smith and Erickson (1987).

There has been much discussion of Jaynes' principle and from several perspectives. Denbigh and Denbigh (1985, Chapter 5) summarise some of the debate by those whose main concern is with matters thermodynamic. The maximum entropy methodology has obvious interest for philosophers and statisticians interested in knowledge acquisition and inference (Rosenkrantz, 1977, 1981) and in 1985 the journal *Synthese* devoted a whole issue to the topic. From this Shimony (1985) and Skyrms (1985) are among those with reservations. Jaynes (1985) replies.

For other approaches to obtaining subjective probabilities see, for instance, Winkler (1967a, 1980). Lindley et al. (1979) examine how to reconcile probability estimates so as to ensure coherence. Winkler (1967b, 1971) and Savage (1971) discuss the use of scoring rules among other feedback procedures. Bunn (1984, Chapter 8) gives a good brief introduction.

The debate on subjectivity is often found conflated with discussions of the appropriateness of Bayes' Rule for updating probabilistic estimates (a topic to be covered in Chapter 5). Rosenkrantz (1977) calls Jaynes' methods "objective Bayesianism". Although closely connected within a single practical proposal for assessing and modifying probabilistic estimates the two issues are separate.

Recognition of the interaction between the specification of probabilities and the importance of different outcomes has led to the interesting proposal from Guiasu (1977, Chapter 4) for a weighted entropy $\Sigma w_i p_i \ln(p_i)$, in which the weights w_i reflect the importance of different goals.

3.6 EXAMPLES

Example 3.1

So far this season *Bruddersford United* football club have never scored more than four goals in any of their matches. The mean number of goals per match is 2.6. In what proportion of their matches did the team fail to score?

Solution

We use the model (3.8). Since the mean of 2.6 is greater than the unweighted mean score $(0 + 1 + 2 + 3 + 4)/5 = 2$ the value of the parameter α must be positive. Having set up the table below on my spreadsheet I found, after some trial and error, that $\alpha = 0.3124$ describes a probability distribution that has the required mean of 2.6. Here is the calculation and the probabilities:

x	$\exp(0.3124x)$	prob – p	px
0	1.000	0.097	0.000
1	1.367	0.133	0.133
2	1.868	0.182	0.364
3	2.553	0.248	0.745
4	3.489	0.340	1.358
	10.276	1.000	2.600

The best estimate of the proportion of games in which *Bruddersford* failed to score is 0.097.

Example 3.2

Bruddersford, fresh from the previous example, are to play *Rawsley Athletic*. *Rawsley* have scored no more than six goals, with a mean of 2.4 goals per game. What is the probability that *Bruddersford* will win?

Solution

Bruddersford will win if they score more goals than *Rawsley*.

We must make a couple of assumptions in order to proceed. First, that the probability that a team will score a particular number of goals is the same as the proportion of games in which they have scored that many goals. Second, that the number of goals scored is independent of the identity of the opposing team. You may wish to question one or both of these assumptions, but we don't have anything else to go on and to proceed differently would involve importing bias.

In the previous example we found the probability distribution describing the number of goals scored by *Bruddersford* and so the probabilities of the number that they will score in the forthcoming match. Similarly, the probability that *Rawsley* will score a particular number of goals is given by (3.8) with $\alpha = -0.153$. The probabilities are shown at the right hand margin of the table below. The probabilities for *Bruddersford* are shown at the bottom margin of the table.

Given the assumption of independence the probability of, say, *Bruddersford* beating *Rawsley* 2–1 is, by (1.8), just the product of the probabilities that *Bruddersford* will score 2 and that *Rawsley* will score 1:

$0.182 \times 0.185 = 0.034$

The table shows the joint distribution, the probabilities of each of the 35 possible outcomes. The probabilities of those results describing a *Bruddersford* victory are shown in italics. Their sum is 0.477, the required probability.

		\multicolumn{5}{c}{Goals scored by *Bruddersford*}	
		0	1	2	3	4	
Goals scored	0	0.021	*0.029*	*0.039*	*0.054*	*0.073*	0.216
by *Rawsley*	1	0.018	0.025	*0.034*	*0.046*	*0.063*	0.185
	2	0.015	0.021	0.029	*0.039*	*0.054*	0.159
	3	0.013	0.018	0.025	0.034	*0.046*	0.136
	4	0.011	0.016	0.021	0.029	0.040	0.117
	5	0.010	0.013	0.018	0.025	0.034	0.100
	6	0.008	0.011	0.016	0.021	0.029	0.086
		0.097	0.133	0.182	0.248	0.340	

Example 3.3

A postman delivers letters to a particular group of houses. The mean number delivered to each house each day is 1.3 and the coefficient of variation is 0.8. No house has ever received more than ten letters. What is the probability that a house receives no letters?

Solution

The maximum entropy estimate of the probability distribution of the number of letters delivered per day is found using (3.13). We require the mean to be 1.3 and the standard deviation to be $(0.8 \times 1.3) = 1.04$.

Here is the spreadsheet that I used to make the calculation:

x	model	prob – p	px	px^2
0	0.705	0.251	0.000	0.000
1	0.999	0.355	0.355	0.355
2	0.747	0.265	0.531	1.062
3	0.295	0.105	0.314	0.942
4	0.061	0.022	0.087	0.348
5	0.007	0.002	0.012	0.060
6	0.000	0.000	0.001	0.005
7	0.000	0.000	0.000	0.000
8	0.000	0.000	0.000	0.000
9	0.000	0.000	0.000	0.000
10	0.000	0.000	0.000	0.000
	2.815	1.000	1.300	2.772

The second column contains values of $\exp[-c(x-m)^2]$ and so, in this case, the scaling constant is $k = 1/2.815$.

Here is the sequence of successive calculations that I used. The arbitrary starting values were $m = 1$ and $c = 0.2$. Holding m constant the value of c was adjusted to give the correct standard deviation. Then m was adjusted to give the correct mean, and so on. The final distribution is given in the table above from which you can see that the probability that a house gets no letters is 0.251.

m	c	mean	standard deviation
1.000	0.200	1.484	1.231
1.000	0.313	1.277	1.040
1.034	0.313	1.300	1.047
1.034	0.3185	1.294	1.040
1.043	0.3185	1.300	1.042
1.043	0.3195	1.299	1.040
1.045	0.3195	1.300	1.041
1.045	0.3198	1.300	1.040

3.7 EXERCISES

3.1 Four grades of timber joist are produced having different moduli of rupture values of 1650, 1400, 1100 and 850. A parcel of joists contains a random selection of the four types and the mean modulus for the parcel is 1350. What is the composition of the parcel?

(Suggested by Munro, 1979)

3.2 Cars travelling on a motorway contain, on average, 1.9 people. No car carries more than six people. What proportion of cars carry no passengers? What proportion of cars carry more than three people?

3.3 The number of people waiting to see a doctor in an outpatients clinic does not exceed 20. The mean number waiting is 12.3. For what percentage of time is the queue less than 11?

3.4 A clothing mail order company despatches goods to customers. It never receives orders for more than eight items to be despatched in one pack. The mean number of items per pack is 3.1. What proportion of packs contain five or more items?

3.5 What would your answer to the previous exercise have been if it was additionally known that the standard deviation of the number of items per pack was 0.95?

How has the entropy of the distribution of pack sizes changed, and why?

3.6 In a survey of business schools the age distribution of one MBA class is

    minimum age   24
    maximum age   41
    mean   29.4
    standard deviation   3.2

What proportion of this class is younger than 30?

4

More on Unbiased Estimates

In the previous chapter least biased estimates were obtained that were consistent with *a priori* constraints. Quite often, however, some estimate of the whole distribution may also be available. This typically arises when we have some new, but not extensive, data and an older, but more complete, tabulation of the distribution. We will want to *update* the distribution in line with the new estimate of, say, the mean. The first part of this chapter shows how to do this.

We have so far only dealt with variables that take discrete values. The second part of this chapter will describe some maximum entropy density functions that are of use with real valued variables.

4.1 INFORMATION CHANGE

The expression that we have to describe information potential, $H = -\Sigma\, p_i \log(p_i)$, refers only to our information state before receiving some data or message.

Some writers (e.g. Goldman, 1953) prefer to consider information as the difference between our uncertainty before and after receiving the message. The argument here is that information is not an *absolute* quantity, as implied by H, but can only be seen as a *change* in uncertainty.

Let q_i be the assessed probability of the i'th event before receiving the message and, as before, let p_i be the revised probability having received the message. The change in our state of uncertainty is

$$[-\log(q_i)] - [-\log(p_i)] = \log(p_i) - \log(q_i)$$
$$= \log(p_i/q_i)$$

as shown in Figure 4.1. If the message received indicates that the i'th event is certain, $p_i = 1$ and $\log(p_i) = 0$ giving a change in information of $-\log(q_i)$. This is surprisal, which was introduced in Chapter 2.

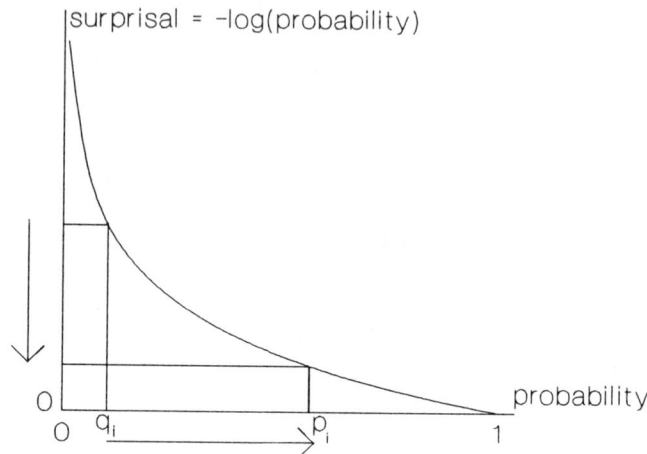

Figure 4.1 Information change

An entropy-like measure is found by taking the expected value of $\log(p_i/q_i)$ to give the expected information gain (relative entropy) as

$$I = \Sigma\, p_i \ln(p_i/q_i) \qquad (4.1)$$

Consider the difference between the estimate with an *a priori* constraint on the mean only (Table 3.4) and the estimate that respected both given mean and variance (Table 3.6). In terms of (4.1) these are q_i and p_i respectively. Table 4.1 shows the calculations.

p_i	q_i	$-p_i\ln(p_i)$	$-q_i\ln(q_i)$	$p_i\ln(p_i/q_i)$
0.163	0.172	0.296	0.303	−0.009
0.163	0.171	0.296	0.302	−0.008
0.163	0.166	0.296	0.298	−0.003
0.162	0.159	0.295	0.292	0.003
0.157	0.146	0.291	0.281	0.011
0.128	0.113	0.263	0.246	0.016
0.063	0.073	0.174	0.191	−0.009
		1.910	1.914	0.002

Table 4.1
Measuring entropy change

The difference between the Shannon measures is

$H(p) - H(q) = 1.914 - 1.910 = 0.004$

while the difference measured by (4.1) is $I = 0.002$. This is because I treats q_i as an initial estimate of p_i while the difference between the H measures treats the two distributions as independent of each other.

Note that when the probability distribution $(q_1, q_2, q_3, ...)$ is uniform then $I = H$.

The measure I was originally introduced by Kullback and Leibler (1951). Hobson and Cheng (1973) discuss reasons, mathematical and otherwise, for preferring I to H and vice versa.

4.2 USING AN EXISTING DISTRIBUTION

Think again about the problem you had in the previous chapter concerning the change in my pocket. You could have decided that the best guess you could make about the coins I had was related in some way to the distribution of all coins in circulation. That being so you could obtain from the Royal Mint a useful little booklet giving just that information which is shown here as Table 4.2.

face value pence	no. in circulation millions	proportion in circulation
1	5600	0.354
2	3500	0.221
5	2320	0.147
10	1510	0.095
20	1295	0.082
50	676	0.043
100	911	0.058
	15812	1.000

(Source: The Royal Mint, Performance Statistics)

Table 4.2
Coins circulating in the UK – 31 March 1990
Mean value is 12.02p/coin

This relative frequency distribution forms an initial guess, the q_i's above. As before, you know the required mean and that it is different from that of the initial guess. Using the maximum entropy principle will give the least biased estimate subject to the constraint $\mu = 18.818$p. Now we wish to make use of the existing distribution. What principle will enable us to do this?

The maximum entropy principle says that our estimates ought to stay as near uniform (flat) as possible. Given the *a priori* estimates, q_i, it is natural to wish our revised estimates to resemble the *a prioiri* estimates as closely as possible. This is achieved by minimising the expected information gain, I.

We can now return to the problem of incorporating the data from The Royal Mint into the estimate of the coins in my pocket.

In Appendix A1 it is shown that the optimal estimate for the general problem of minimising I subject to the same sort of requirements as we used in the previous chapter is

$$p_i = kq_i \exp[\alpha g'(p_i) + \beta h'(p_i) + ...] \qquad (4.2)$$

When a given mean is required this gives

$$p_i = kq_i \exp(\alpha x_i) \qquad (4.3)$$

Comparing this with (3.8) shows that the initial estimate (the $q_i(s)$ provides additional weights in the model. If we had no initial estimate from the Royal Mint, or anywhere else, then our *a priori* estimate would be a uniform distribution. Each q_i would be just a constant value (= 1/7) and (4.3) would be identical to (3.8).

The calculations are set out in Table 4.3. The second column shows the initial assessment of the distribution (from Table 4.2), the next column shows the raw estimate of the adjusted distribution which is then scaled to give, in the penultimate column, the updated assessment (Figure 4.2). As in Chapter 3 the value of the parameter α which gave the required mean value was found by trial and error.

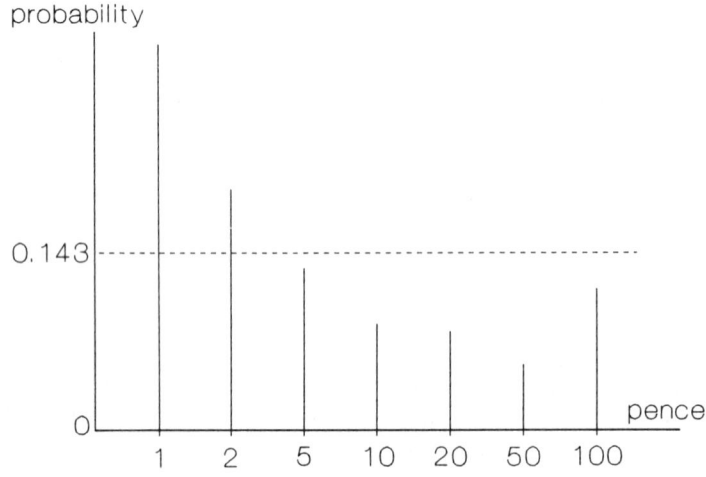

Figure 4.2 Probability distribution using existing distribution (see Table 4.2)

Sec. 4.3]   **Using information about parts of the distribution**   79

Thinking again about the question posed in the previous chapter you now have prob(10p) = 0.091.

face value x_i pence	q_i	$q_i \exp(\alpha x_i)$	p_i	$p_i x_i$
1	0.354	0.357	0.314	0.314
2	0.221	0.225	0.198	0.396
5	0.147	0.153	0.135	0.673
10	0.095	0.104	0.091	0.914
20	0.082	0.097	0.085	1.708
50	0.043	0.066	0.058	2.883
100	0.058	0.136	0.119	11.932
	1.00	1.138	1.000	18.819

Table 4.3
Updated distribution, $\alpha = 0.008575$

4.3   USING INFORMATION ABOUT PARTS OF THE DISTRIBUTION

I now tell you that eight of the eleven coins in my pocket have face values no higher than 10p so that prob($X \le$ 10p) = 8/11 = 0.73. How can this be used?

So that we can focus on just this problem we shall in this section ignore all other *a priori* requirements. Incorporating them is straightforward.

We begin, as before, with the national data of Table 4.2 and equation (4.2). The constraint we now have is

$$g(p_i) = \sum_{i=1}^{4} p_i = 0.73 \qquad (4.4)$$

and so, from Table 3.1,

$$g'(p_i) = 1 \; ; \quad 1 \le i \le 4$$
$$ = 0 \; ; \quad 4 < i$$

and from (4.2)

$$p_i = kq_i \exp(\alpha)$$
$$ = kcq_i \qquad (4.5)$$

where $c = \exp(\alpha)$ is just a scaling constant acting on the first four probabilities to ensure that (4.4) is respected. The constant k ensures that all probabilities sum to 1.0.

In this case, however, it may be more convenient to ensure that the first four probabilities sum to 0.73 and then that the remainder sum to 0.27. The calculations are set out in Table 4.4.

q_i	ensure $\sum_{i=1}^{4} p_i = 0.73$		ensure $\sum_{i=5}^{7} p_i = 0.27$		p_i
(1)	(2)	(3)	(4)	(5)	(6)
0.354	0.354	0.316			0.316
0.221	0.221	0.198			0.198
0.147	0.147	0.131			0.131
0.095	0.095	0.085			0.085
0.082			0.082	0.121	0.121
0.043			0.043	0.063	0.063
0.058			0.058	0.085	0.085
1.000	0.818	0.730	0.182	0.270	1.000

Table 4.4

Column 1 shows the national estimate. Column 2 shows the first four terms of that estimate. These are scaled in column 3 to ensure that they sum to 0.73 as required. Columns 4 and 5 repeat this for the remaining probabilities. Finally, in column 6, the scaled estimates from columns 3 and 5 are brought together to give the required estimate of the coins in my pocket. In particular, the probability of choosing a 10p coin is 0.085.

Suppose that I now tell you, in addition, that all but one of the coins have a face value greater than 2p so that

$$\text{prob}(X \geq 5p) = 10/11 = 0.91$$

requiring

$$\sum_{i=3}^{7} p_i = 0.91 \qquad (4.6)$$

as well as (4.4).

Applying the same arguments as before we have, for the general case with any number of such constraints on parts of the distribution,

$$p_i = k q_i c_1 c_2 c_3 \ldots \qquad (4.7)$$

where each of the c's is a scaling constant which ensures a particular constraint is met. Table 4.5 shows the calculations: first probabilities are adjusted so that (4.4) is

Sec. 4.3] Using information about parts of the distribution

met, then another adjustment ensures that (4.6) is met and, finally, we ensure that the sum of all probabilities is 1.0. Because the ranges of probabilities covered by the constraints overlap each adjustment undermines previous adjustments and so we must repeat the process (Table 4.6) with the revised p_i's from one iteration becoming the starting points of the next. The calculations continue in this way until all constraints are simultaneously respected. I stopped when they were all met to within 1% as shown in Table 4.7.

q_i	ensure $\sum_{i=1}^{4} p_i = 0.73$		ensure $\sum_{i=3}^{7} p_i = 0.91$		ensure $\sum_{i=1}^{7} p_i = 1.0$	
0.354	0.354	0.316			0.316	0.222
0.221	0.221	0.198			0.198	0.139
0.147	0.147	0.131	0.131	0.299	0.299	0.210
0.095	0.095	0.085	0.085	0.195	0.195	0.137
0.082			0.082	0.187	0.187	0.131
0.043			0.043	0.098	0.098	0.069
0.058			0.058	0.132	0.132	0.092
1.000	0.818	0.730	0.399	0.910	1.424	1.000

Table 4.5

start	ensure $\sum_{i=1}^{4} p_i = 0.73$		ensure $\sum_{i=3}^{7} p_i = 0.91$		ensure $\sum_{i=1}^{7} p_i = 1.0$	
0.222	0.222	0.229			0.229	0.179
0.139	0.139	0.143			0.143	0.112
0.210	0.210	0.217	0.217	0.303	0.303	0.237
0.137	0.137	0.141	0.141	0.197	0.197	0.154
0.131			0.131	0.184	0.184	0.143
0.069			0.069	0.096	0.096	0.075
0.092			0.092	0.129	0.129	0.101
1.000	0.708	0.730	0.650	0.910	1.282	1.000

Table 4.6

start	ensure $\sum_{i=1}^{4} p_i = 0.73$		ensure $\sum_{i=3}^{7} p_i = 0.91$		ensure $\sum_{i=1}^{7} p_i = 1.0$	
0.061	0.061	0.061			0.061	0.061
0.038	0.038	0.038			0.038	0.038
0.381	0.381	0.382	0.382	0.385	0.385	0.382
0.248	0.248	0.249	0.249	0.251	0.251	0.249
0.122			0.122	0.123	0.123	0.122
0.064			0.064	0.064	0.064	0.064
0.086			0.086	0.087	0.087	0.086
1.000	0.729	0.730	0.902	0.910	1.009	1.000

Table 4.7

4.4 THE ANSWER!

Time to put you out of your misery. The 11 coins that I emptied onto my desk are shown in Table 4.8.

face value – pence	number of coins in my pocket
1	0
2	1
5	1
10	6
20	2
50	0
100	1

Table 4.8
The coins in my pocket

The probability of me choosing a 10p coin was $6/11 = 0.545$ which is much higher than any of the values obtained above. This simply emphasises the difference between an assessment correctly arrived at, as were all those that we have used, and one that proves, in retrospect, to have been numerically correct. In any case, you shouldn't be too disillusioned. I enjoy a lunchtime game of pool and the table on which I play only accepts 10p pieces so I tend to hold on to those coins as I get them.

This would have been useful information for you to have had, but how would constraint equation as we did for the mean and variance to reflect my lunchtime habits. About the best that you could do, I think, is to have arbitrarily changed the you have incorporated it into your assessment? Clearly, it isn't possible to write a data in Table 4.2 by doubling, or whatever, the number of 10p coins. This is unsatisfactory and illustrates that not all data that you may have about a problem are capable of incorporation into a formal assessment procedure because they are not sufficiently numerically specific. In such cases you will probably have to ignore them. You may be tempted to adopt some rough and ready means of using them. If you are, be very careful of overestimating how much you think you know, for this is where we came in with Jaynes' argument about bias.

Confirm for yourself that if the updated assessment had to respect a variance constraint as well as a mean you would use (c. f. (3.13))

$$p_i = kq_i \exp[-c(x_i - m)^2] \qquad (4.8)$$

in place of (4.3).

4.5 MAXIMUM ENTROPY DENSITY FUNCTIONS

All the discussion so far has been about discrete valued variables. The equations that have been found to describe these probabilities may also be used, by analogy here, to define density functions for real valued variables. For instance, equation (3.4) describes the uniform probability distribution that is our maximum entropy estimate when we are not prepared to specify any constraints. Figure 4.3 shows the corresponding result for a real valued variable. This is the *rectangular* density function:

$$f(x) = 1/r \qquad (4.9)$$

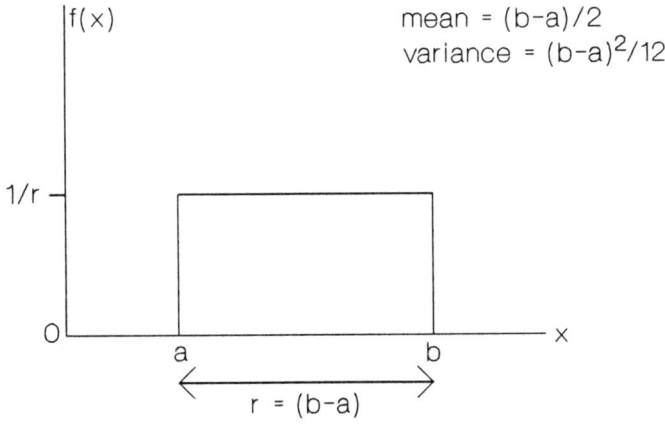

Figure 4.3 Rectangular pdf

The entropy for the discrete case (equation (3.4)) is, from Chapter 2, $H = \log(n)$ so that entropy increases as the number of possible values increases. In the real valued case (4.9) entropy is the logarithm of the range, $H = \log(r)$. This means that, for example, when the range is 1.0 then $H = 0$. Further, when the range is less than 1.0 then the entropy H is negative. This illustrates a fundamental difference: in the discrete case H may have the *absolute* interpretations of Chapter 2 while in the real case it may only be considered as a *relative* measure. These different interpretations do not affect its use in deriving probability distributions.

When the mean is specified the density function for $a \leq x \leq b$ is, from (3.8),

$$f(x) = k \exp(\alpha x) \qquad (4.10)$$

as shown in Figure 4.4.

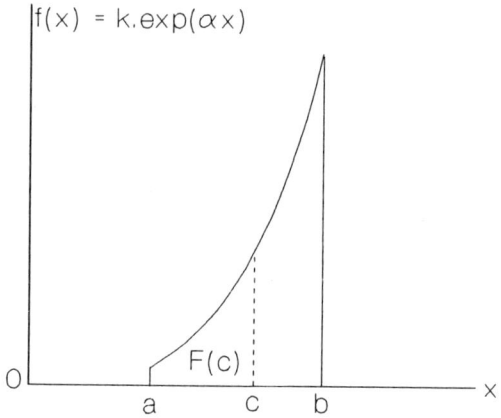

Figure 4.4 Maximum entropy pdf mean given

The area under a curve $y = \exp(\alpha x)$ between the limits $x = a$ and $x = b$ is

$$[\exp(\alpha b) - \exp(\alpha a)]/\alpha \qquad (4.11)$$

For (4.10) to be a density function, therefore,

$$k = \alpha/[\exp(\alpha b) - \exp(\alpha a)] \qquad (4.12)$$

and so

$$f(x) = \frac{\alpha \exp(\alpha x)}{\exp(\alpha b) - \exp(\alpha a)} \qquad (4.13)$$

For some intermediate value, $x = c$, the distribution function can be found using (4.11) to be

$$F(c) = \text{prob}(X \le c) = \frac{\exp(\alpha c) - \exp(\alpha a)}{\exp(\alpha b) - \exp(\alpha a)} \quad (4.14)$$

As before, the value of the parameter α is chosen so that the distribution has the required mean. For this density function the mean is

$$\mu = \frac{[(\alpha b - 1)\exp(\alpha b) - (\alpha a - 1)\exp(\alpha a)]}{\alpha[\exp(\alpha b) - \exp(\alpha a)]} \quad (4.15)$$

Since it is not obvious how to find α given μ using (4.15) we shall have to proceed by trial and error. Before doing so note that when the mean is equal to the mid-range ($\mu = (a + b)/2$) then we have the rectangular density (4.9) and $\alpha = 0$. In this case the formulae (4.11) to (4.15) are unusable since they would involve division by zero.

If the required mean is greater than the mid-range then we need a positive value of α, while if $\mu < (a + b)/2$ then $\alpha < 0$, as shown in Figure 4.5.

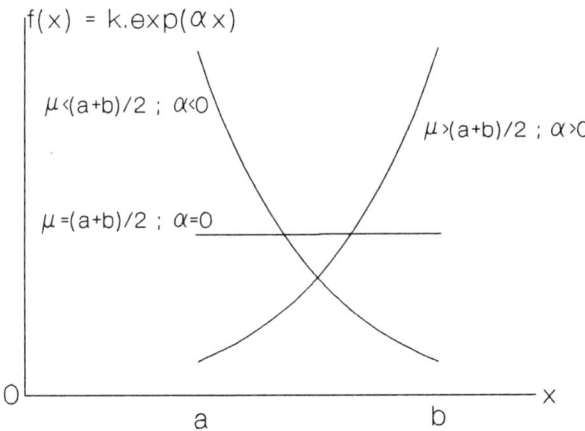

Figure 4.5 Maximum entropy densities for given means

Suppose that we have to estimate the growth in demand for a certain product during the next year and that after some consideration believe that this will be between 2% and 6% and that our best guess, a mean, is 5%. Clearly a density function of the form (4.13) will describe the estimate quite nicely so all that remains is to find the appropriate value of α from (4.15) given $a = 2$, $b = 6$ and $\mu = 5$.

For instance, we could use a spreadsheet to do this by tabulating values of α and μ with an increasingly fine increment until finally we get an acceptable value for μ and so for α. Alternatively, we could just blunder around with different values for α, and this may well work quite satisfactorily.

A more orderly procedure, which is easily automated, is to employ the logarithmic search which was introduced in Chapter 2. Begin by choosing two values of α to give values of μ which bracket the required value. We know that since the mean is to be greater than the mid-range then α must be positive, so choose $\alpha = 0$ (in fact a very small positive value) and, arbitrarily, $\alpha = 3$ as satisfactory starting points. Now choose a value halfway between, i.e. $\alpha = 1.5$. This gives $\mu = 5.34$ and so now replaces $\alpha = 3$ as the new upper estimate (Table 4.9). Proceed in this way until a sufficiently accurate value for μ is obtained. This eventually gives $\alpha = 0.89868$ and the density and distribution functions shown in Figure 4.6. As you would expect from (4.10) and (4.14) these functions have the same shape.

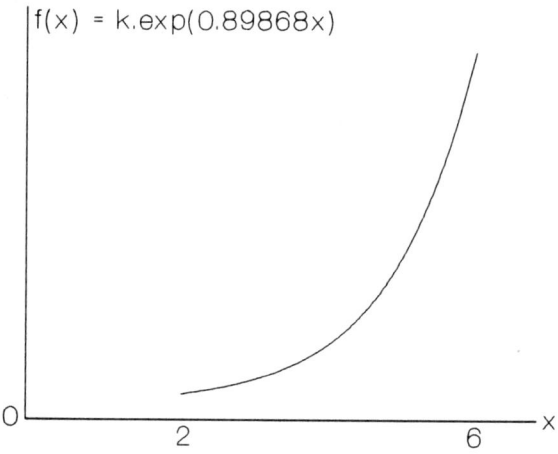

Figure 4.6(a) Density function; mean = 5

Sec. 4.5]   Maximum entropy density functions   87

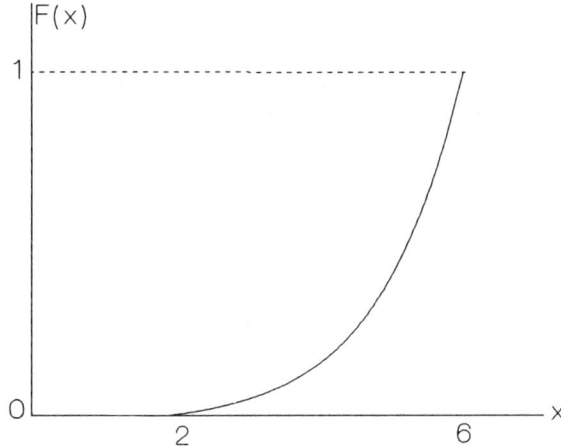

Figure 4.6(b)   Distribution function; mean = 5

	α	μ	% error
iteration 1	0.000	4.00	−20.0
	1.500	5.34	6.9
	3.000	5.67	13.3
iteration 2	0.000	4.00	−20.0
	0.750	4.88	−2.5
	1.500	5.34	6.9
iteration 3	0.750	4.88	−2.5
	1.125	5.16	3.1
	1.500	5.34	6.9

Table 4.9
Logarithmic search for α

If we now want to know the probability that the growth will not exceed 3.5% we have, from (4.14),

$$F(3.5) = \frac{\exp(0.89868 \times 3.5) - \exp(0.89868 \times 2)}{\exp(0.89868 \times 6) - \exp(0.89868 \times 2)}$$

$$= \frac{\exp(3.145) - \exp(1.797)}{\exp(5.392) - \exp(1.797)}$$

$$= \frac{23.229 - 6.034}{219.660 - 6.034}$$

$$= 0.080$$

Check for yourself that $F(4.5) = 0.239$ and so the probability that growth will be between 3.5% and 4.5% is

$$F(4.5) - F(3.5) = 0.239 - 0.080 = 0.159$$

4.6 THE NEGATIVE EXPONENTIAL DISTRIBUTION

A real variable having unbounded non-negative values has a lower limit, a, of zero and an upper limit, b, of infinity. Since the mean must by definition be less than the mid-range (infinity) then $\alpha < 0$. This gives, from (4.13), the density function

$$f(x) = \alpha \exp(-\alpha x) \qquad \alpha > 0 \qquad (4.16)$$

which is called the *negative exponential* density function.

The corresponding distribution function is, from (4.14),

$$F(x) = 1 - \exp(-\alpha x) \qquad (4.17)$$

and (4.15) gives the mean as

$$\left. \begin{array}{l} \mu = 1/\alpha \\ \text{Additionally,} \\ \sigma = \mu \end{array} \right\} \qquad (4.18)$$

The negative exponential distribution is shown in Figure 4.7.

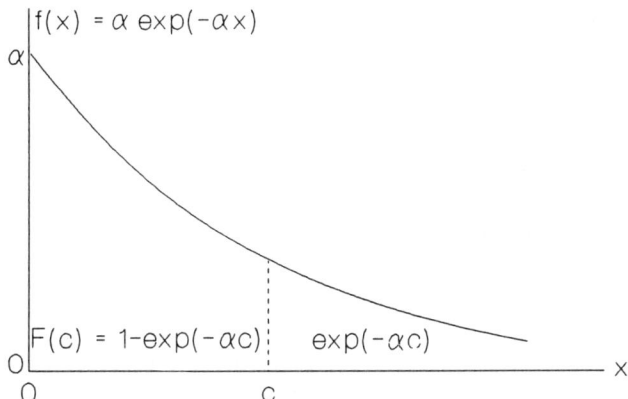

Figure 4.7(a) Negative exponential density function

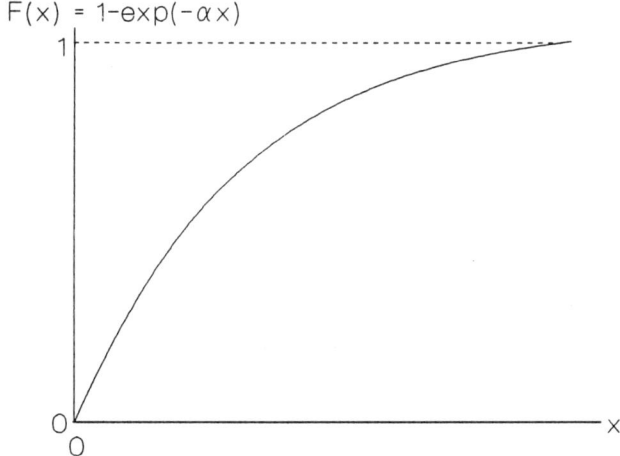

Figure 4.7(b) Negative exponential distribution function

You are waiting for a bus. They never arrive early but are pretty good at time keeping, given the amount of congestion, with an average delay of 2 minutes. Some, however, suffer extreme delays. What is the probability that you will have to wait longer than 4 minutes?

The negative exponential distribution gives a suitable estimate with, from (4.18),

$$\alpha = 1/2 = 0.5$$

so that

$$\begin{aligned}\text{prob(delay} > 4) &= 1 - F(4) \\ &= \exp(-4\alpha) = \exp(-2) \\ &= 0.14\end{aligned}$$

4.7 THE NORMAL DISTRIBUTION

We shall now consider assessments for real variables when both the mean and variance are given. By analogy with (3.13) the density function is of the form

$$f(x) = k \exp[-c(x - m)^2] \qquad (4.19)$$

A particularly ubiquitous density is given when the variable is unbounded. Figure 4.8 shows its symmetrical bell-like shape. Notice the effect that the parameter c has on the spread of the distribution. Because of its ubiquity this is called the *Normal* distribution. When X is unbounded the parameters of (4.19) are

$$\left.\begin{aligned} m &= \mu \\ c &= 1/2\sigma^2 \\ k &= 1/[\sigma(2\pi)^{0.5}] \end{aligned}\right\} \qquad (4.20)$$

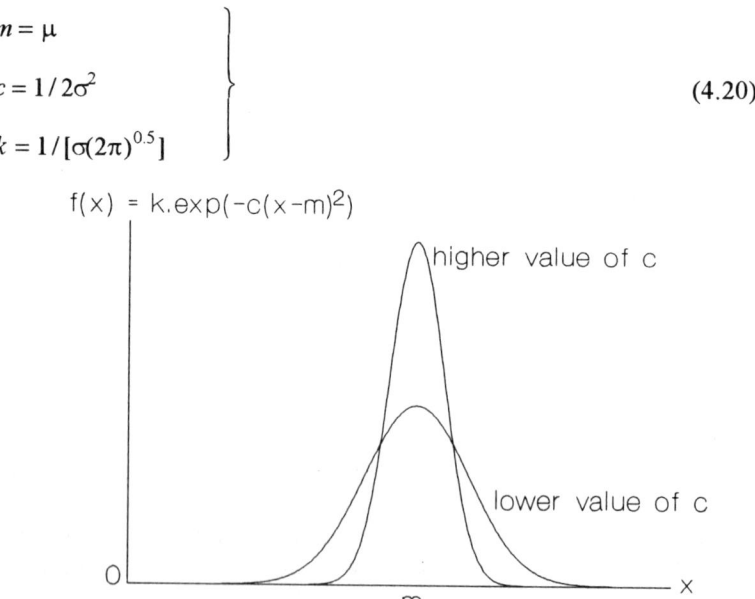

Figure 4.8(a) Normal density function

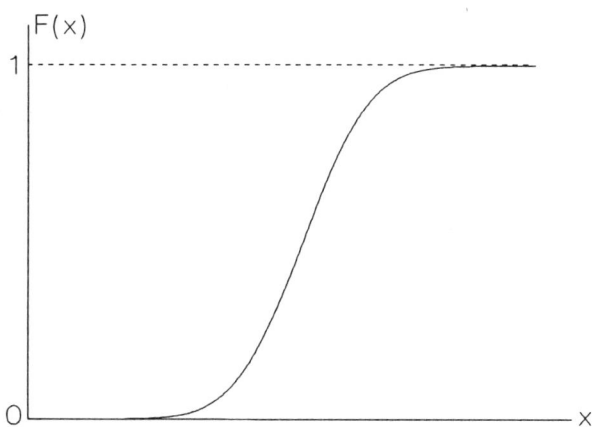

Figure 4.8(b)   Normal distribution function

Although the parameters may be found directly from the mean and variance the distribution function cannot be written down as a simple formula and so we must use tables which have been found by other computational methods.

There are an infinite number of Normal distributions corresponding to different (μ, σ) pairs. Figure 4.9 shows one of them, called the *standardised* Normal distribution. It measures values not in their natural units but as the number of standard deviations from the mean so that

$$z = (x - \mu)/\sigma \qquad (4.21)$$

where x is a value of any Normally distributed variable with arbitrary mean and standard deviation and z is the value of the standardised Normal variate. The standardised distribution has a mean of 0 and a standard deviation of 1.0.

Although it is not general for any distribution function it is a property of the Normal distribution that

$$F(x) = F(z) \qquad (4.22)$$

Given (4.21) and (4.22) we can now make probability statements about any Normally distributed variable that we wish using the tabulations of $F(z)$ given in Appendix A5.3.

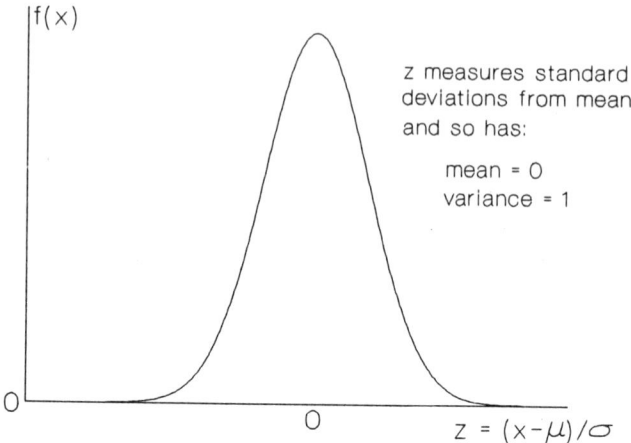

Figure 4.9 Standardised Normal density

Suppose that the amounts spent in a supermarket by shoppers is Normally distributed with a mean of £21.62 and a standard deviation of £7.14 and that we want to know the probability that a customer will spend no more than £25.00. First, use (4.21) to find

$$z = \frac{25.00 - 21.62}{7.14} = 0.47$$

So £25 is 0.47 standard deviations greater than the mean expenditure. From Appendix A5.3,

$$F(0.47) = 0.681$$

4.8 BOUNDED REAL VARIABLE: MEAN AND VARIANCE GIVEN

Often a variable will have some natural or imposed bounds so that, given values of μ and σ, we will have the density function (4.19) but without the convenient set of relationships (4.20). This means that we must find values of the parameters c and m by trial and error, as we did earlier (see Table 3.5). Remember that c controls the variance and m the mean (noting that m is the mode might help here).

An upper bound to the value of variance that you can specify is given by that of the rectangular distribution (Figure 4.3), which is $(b - a)^2/12$, where a and b are the lower and upper limits of the variable.

An easy way of actually calculating the mean and variance of the real variable is to divide the range of values into a number of classes and then take the mid–point of each class as a point value for the purpose of making calculations such as those shown in Table 3.6. This approximation can easily be carried out on a spreadsheet so using a satisfactorily large number of classes, say a hundred or more, presents no problems. Appendix A3 gives another, more accurate, method.

As an example suppose that we are trying to estimate the proportion, π, of a population in a town that proposes to vote for a particular party at the next election. Your estimate has, you assess, a mean of 0.3 and a standard deviation of 0.2. Satisfy yourself that the resulting distribution, shown in Figure 4.10, has parameters $c = 7$, $m = 0.2$ and that prob$(\pi > 0.5) = 0.17$.

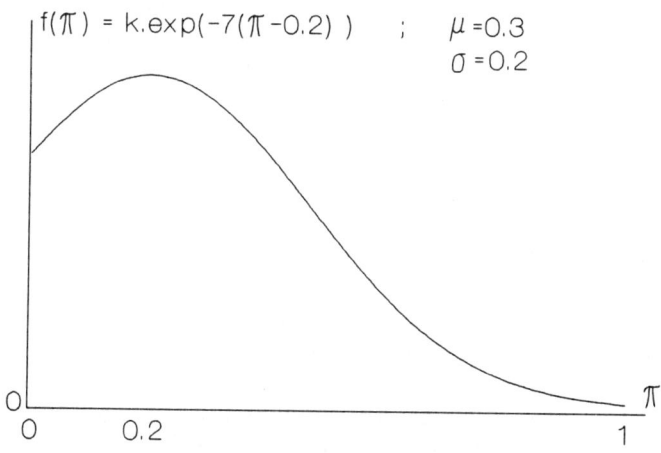

Figure 4.10 pdf for proportion – π

4.9 SOME PRACTICAL CONSIDERATIONS

What sorts of prior requirements should we incorporate into the entropy maximising model? Jaynes (1968) describes a requirement as *testable* if it can be determined unambiguously whether or not the probability distribution agrees with the requirement. This gives a pretty large number of possibilities.

On a practical level, however, we need to specify prior conditions that are differentiable (for any function $g()$ we can obtain $g'()$) and so in the last example, for instance, we specified the mean and standard deviation. But we may not normally think in such terms and so feel a bit uncomfortable with making specifications in this way, even though we agree with the entropy maximising criterion. Let us review where we are in defining average and spread.

In many respects spread is the more problematic simply because the form that is mathematically convenient, variance and standard deviation, is certainly not the way that most readily springs to mind if we are asked to talk about spread. We would most naturally specify ranges and the like. Fortunately we have seen how to deal with this within the entropy maximising framework. Equation (4.7) shows how to deal with the situation when we have histogram-like constraints on parts of a distribution. With no *a priori* distribution and a specification of probabilities in mutually exclusive and exhaustive classes we have

$$p_i = c_1 c_2 c_3 \ldots \qquad (4.23)$$

where each of the c_j's is chosen to ensure that the probability in the i'th class is as specified: in other words a uniform distribution within each class.

If, in addition, we wished to specify the mean we would have

$$p_i = \exp(\alpha x_i) c_1 c_2 c_3 \ldots \qquad (4.24)$$

Example 4.6 below shows an application of this model.

It is clear that if we are able to give the variance then we have made an extremely powerful specification. This is most likely to be possible after access to data of some sort, rather than as a result of subjective introspection.

Of the three different measures of average we have already seen how to incorporate a given value for the mean. The median is just a special case of (4.24): two equiprobable classes.

Dealing with the mode within this framework may be tricky. If no variance is specified then the distributions which result, (3.8) and (4.10), have modes only at extreme values. If a variance is specified then specifying a mode as well directly sets the parameter m in (3.13) or (4.19) and can give a distribution with a hump such as that shown in Figure 4.10. Note, however, that the maximum variance that may be specified is reduced. Remember that specifying the mean rather than the mode does not necessarily result in a humped distribution (Figure 3.6) and permits a higher variance. Giving a modal constraint is quite powerful; be careful.

We generally associate one parameter uniquely with one constraint or requirement. However, when modelling opinion it is as well to check for consistency by generating probabilities from the derived distribution to see that they are agreeable and, if they are not, to incorporate what you have learned in further constraints and so get a modified model.

If you are stuck in making a specification it is almost certainly better to extemporise than give up, but don't forget the consistency checks.

Sec. 4.10] Examples

4.10 EXAMPLES

Example 4.1

The table below shows 178 measurements of the carbon content of a mixed powder fed to a chemical plant.

If the mean carbon content at another plant is 4.5% what would be the distribution for that plant?

carbon – x%	frequency
4.10–4.19	1
4.20–4.29	2
4.30–4.39	7
4.40–4.49	20
4.50–4.59	24
4.60–4.69	31
4.70–4.79	38
4.80–4.89	24
4.90–4.99	21
5.00–5.09	7
5.10–5.19	3

(Source: Davies, 1961, p6)

Solution

The data are tabulated as given in the original. It is reasonable to interpret the class definitions as $4.1 \leq x \leq 4.2$, $4.2 < x \leq 4.3$ and so on. The mid-points are then 4.15, 4.25, ... Here are the data re-presented:

x_i	frequency	probability p_i	$p_i x_i$
4.15	1	0.006	0.023
4.25	2	0.011	0.048
4.35	7	0.039	0.171
4.45	20	0.112	0.500
4.55	24	0.135	0.613
4.65	31	0.174	0.810
4.75	38	0.213	1.014
4.85	24	0.135	0.654
4.95	21	0.118	0.584
5.05	7	0.039	0.199
5.15	3	0.017	0.087
	178	1.000	4.703

The mean carbon content is 4.703. To obtain a distribution with a mean of 4.5 use equation (4.3) with the probabilities given above as the q_i values. Since the required mean is less than the existing mean α will be negative. Try $\alpha = -2$.

x_i	q_i	$q_i \exp(\alpha x_i)$ $\alpha = -2$	p_i	$p_i x_i$
4.15	0.006	1.396	0.016	0.065
4.25	0.011	2.286	0.026	0.109
4.35	0.039	6.551	0.074	0.321
4.45	0.112	15.325	01172	0.767
4.55	0.135	15.056	0.169	0.771
4.65	0.174	15.922	0.179	0.833
4.75	0.213	15.980	0.180	0.854
4.85	0.135	8.263	0.093	0.451
4.95	0.118	5.919	0.067	0.330
5.05	0.039	1.615	0.018	0.092
5.15	0.017	0.567	0.006	0.033
	1.000	88.881	1.000	4.625

(note: $q_i \exp(\alpha x_i)$ multiplied by 10^6)

This is still too high. After some trial and error find the correct value of $\alpha = -5.370$.

x_i	q_i	$q_i \exp(\alpha x_i)$ $\alpha = -5.370$	p_i	$p_i x_i$
4.15	0.006	1.178	0.063	0.261
4.25	0.011	1.377	0.074	0.312
4.35	0.039	2.817	0.150	0.654
4.45	0.112	4.704	0.251	1.118
4.55	0.135	3.300	0.176	0.802
4.65	0.174	2.491	0.133	0.618
4.75	0.213	1.785	0.095	0.453
4.85	0.135	0.659	0.035	0.171
4.95	0.118	0.337	0.018	0.089
5.05	0.039	0.066	0.004	0.018
5.15	0.017	0.016	0.001	0.005
	1.000	18.730	1.000	4.500

(note: $q_i \exp(\alpha x_i)$ multiplied by 10^{12})

Example 4.2

The distribution of times between successive arrivals at an automatic cash dispenser follows a negative exponential distribution with mean 0.37 minutes. What is the probability of a gap being between 0.2 and 0.5 minutes?

Solution

Use (4.17) and (4.18) to give $F(x) = 1 - \exp(-x/0.37)$ and so

$$\begin{aligned}\text{prob}(0.2 \leq X \leq 0.5) &= F(0.5) - F(0.2) \\ &= [1 - \exp(-0.5/0.37)] - [1 - \exp(-0.2/0.37)] \\ &= \exp(-0.2/0.37) - \exp(-0.5/0.37) \\ &= \exp(-0.541) - \exp(-1.351) \\ &= 0.582 - 0.259 = 0.323\end{aligned}$$

Example 4.3

The time taken for workers to complete a certain assembly task is Normally distributed with mean 5.3 minutes and standard deviation 0.3 minutes. What is the probability that any one such assembly will

(a) take no longer than 5.5 minutes?
(b) take between 5.0 minutes and 5.5 minutes?

Solution

(a) The number of standard deviations by which 5.5 exceeds the mean is

$$z = \frac{5.5 - 5.3}{0.3} = 0.67$$

and so from the table in Appendix A5.3

$$\text{prob}(\text{time} \leq 5.5 \text{ minutes}) = F(Z = 0.67) = 0.749$$

(b) A similar calculation gives

$$\begin{aligned}\text{prob}(\text{time} \leq 5.0 \text{ minutes}) &= F(Z = (5.0 - 5.3)/0.3) \\ &= F(Z = -1.00)\end{aligned}$$

and, by symmetry,

$$F(Z = -1.0) = 1.0 - F(Z = 1.0) = 1 - 0.841 = 0.159$$

Finally,

$$\text{prob}(5.0 \le \text{time} \le 5.5) = F(\text{time} = 5.5) - F(\text{time} = 5.0)$$
$$= 0.749 - 0.159 = 0.590$$

Example 4.4

As part of the feasibility study for the setting up of a new manufacturing plant abroad we need to estimate the proportion of defectives produced by each work group per shift. We know that at existing plants the mean rate is 0.05 (5%). To take account of the different working practices that we expect to encounter our best guess is that the defective rate will have a mean of 0.08. What is the probability that a group will produce more than 10% defectives?

Solution

Setting equation (4.15) into a spreadsheet with $a = 0$ and $b = 1$ find, by trial and error, that $\alpha = -12.5$ gives $\mu = 0.08$.

From (4.14)

$$\text{prob}(X > 0.1) = 1 - F(0.1)$$

$$= 1 - \frac{\exp(-12.5 \times 0.1) - \exp(-12.5 \times 0)}{\exp(-12.5 \times 1) - \exp(-12.5 \times 0)}$$

$$= 1 - \frac{[\exp(-1.25) - 1]}{[\exp(-12.5) - 1]}$$

$$= 0.287$$

Example 4.5

Consider again the previous example. Suppose that in the existing plants the distribution of the proportion of defectives per shift has a coefficient of variation of 20% and that we think it reasonable that our estimate has this same value. What now is the probability that a group will produce more than 10% defectives?

Solution

The diagram below shows my spreadsheet set to calculate equation (4.19) and, as shown, with parameter values that give $\mu = 0.08$.

Examples

lower: 0.000			model is $f(x) = k.\exp(-c(x-m)^2)$			
upper: 1.000			$c = 1.000$			
			$m = -6.090$			

mean = 0.080
var = 0.006
s.d. = 0.079
H = 0.489

x	unscaled	p	cum	p.x	p.x^2	−p.ln(p)
0.001	0.000	0.024	0.024	0.000	0.000	0.091
0.003	0.000	0.024	0.048	0.000	0.000	0.089
0.005	0.000	0.023	0.071	0.000	0.000	0.087
0.007	0.000	0.024	0.094	0.000	0.000	0.086

The column headed "unscaled" contains values of $\exp[-c(x-m)^2]$. The next column shows the probabilities (the entry in the previous column divided by the sum of that column) and the rest are self-explanatory. The range $0 \le X \le 1$ was divided into 500 classes.

We now require that, as well as the mean being 0.08, the standard deviation should be $(0.08 \times 0.2) = 0.016$. After some trial and error we get $c = 1950$ and $m = 0.08$. The two distributions are shown in Figure 4.11. Notice the dramatic effect of specifying the standard deviation.

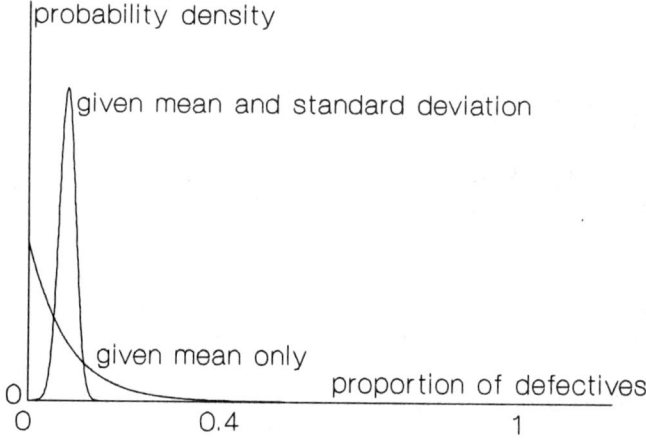

Figure 4.11

The results of applying this model and from the previous example are:

constraints	entropy – H	prob($X \geq 0.1$)
model 1: $\mu = 0.08$	4.69	0.287
model 2: $\mu = 0.08$ $\sigma = 0.016$	3.49	0.095

Note how the extra constraint has reduced entropy.

Example 4.6

Suppose that we had been unhappy in modifying Example 4.4 by specifying variance or the coefficient of variation and instead decided that the only constraint about spread that we wished to impose was that the probability of the defect rate falling below 5% was 0.2 (rather than the 0.47 which results from Example 4.4). How would the estimate have changed?

Solution

The model required is that shown in (4.10) but modified according to (4.24).
First, a general result. If the range of a real variable, X, is divided into a number of non-overlapping classes then the mean is

$$\sum_i p_i \mu_i$$

where p_i is the probability that X lies within the i'th class

and μ_i is the conditional mean of the i'th class, i.e. the mean value of X for those values of X in the i'th class

In this example we have specified the mean and two classes. The unconditional class means are given by (4.15) with a and b as the lower and upper class limits, 0 and 0.05, and 0.05 and 1. The density function is

$$f(x) = k_1 \exp(\alpha x), \qquad 0 \leq x \leq 0.05$$
$$= k_2 \exp(\alpha x), \qquad 0.05 \leq x \leq 1$$

with, by the same argument as (4.12) but scaling to give the p_i's rather than 1.0,

$k_1 = 0.2\alpha/[\exp(0\alpha) - \exp(0.05\alpha)]$

$k_2 = 0.8\alpha/[\exp(0.05\alpha) - \exp(1\alpha)]$

The parameter α is, as before, chosen so that the mean value of X over the whole range of its values is 0.08.

The following table shows the calculation with $\alpha = -22.0$ found by trial and error, as usual.

class limits		p_i	k_i	μ_i	$p_i \mu_i$
lower	upper				
0.00	0.05	0.2	6.595	0.021	0.004
0.05	1.00	0.8	52.873	0.095	0.076
					0.080

Figure 4.12 shows this density function together with that from Example 4.4 which is shown as the broken line.

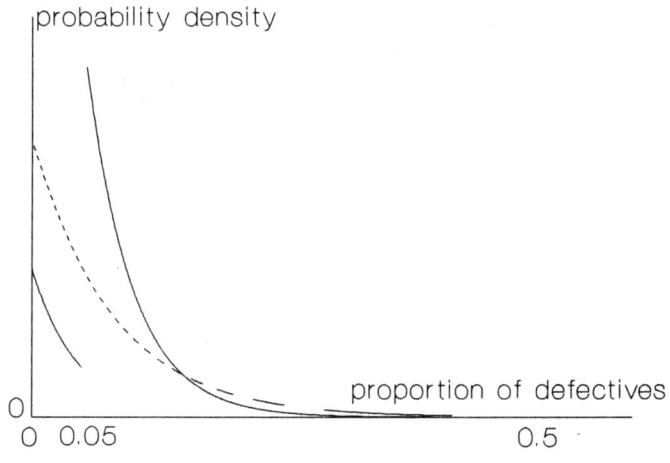

Figure 4.12

Now we need $\text{prob}(X \geq 0.1) = 1 - F(0.1)$. The value in which we are interested, 0.1, lies in the second part of the function and so we have, from (4.14) and remembering to multiply by $p_2 = 0.8$,

$$\text{prob}(0.05 \leq X < 0.1) = 0.8 \frac{[\exp(-22 \times 0.1) - \exp(-22 \times 0.05)]}{[\exp(-22 \times 1.0) - \exp(-22 \times 0.05)]}$$

$$= 0.534$$

Now,

$$F(0.1) = \text{prob}(X < 0.1)$$
$$= \text{prob}(0.0 \le X \le 0.05) + \text{prob}(0.05 \le X < 0.1)$$
$$= 0.2 + 0.534 = 0.734$$

and, finally,

$$\text{prob}(X \ge 0.1) = 1.0 - F(0.1)$$
$$= 1.0 - 0.734 = 0.266$$

4.11 EXERCISES

4.1 In 1971 the distribution of household size in this country was:

number of persons in household	percentage of households
1	17
2	31
3	19
4	18
5	8
6 or more	6

(Source: General Household Survey 1987)

The mean household size was 2.91. By 1987 this had fallen to 2.55. Estimate the new distribution of household size.

4.2 In a survey of how managers use their non-work time some were found to do household duties. The table shows the distribution (in per cent) of how much time was spent in these duties.

In a group of managers it is known that the mean amount of time spent on household duties is 3.5 hours per week. What proportion of managers spend no more than 4 hours per week on this work?

average number of hours per week	per cent of managers
0	24.0
1	9.7
2	21.9
3	8.4
4	10.7
5	11.4
6	5.5
7	2.6
8	4.5
9	
10	0.5
11	
12	0.8

(Source: Fletcher et al., 1993)

4.3 How would your answer to the previous question have changed if it was also known that 20% spent 7 hours per week or more on household duties?

4.4 Plastic balls are made to fit into the tops of bottles of deodorant. The specification is for balls of 2.5 cm diameter. The manufacturing process is such that the diameters of the balls are Normally distributed with a mean of 2.5 cm and a standard deviation of 0.2 mm. Balls are rejected if they are either 0.4 mm too small or 0.45 mm too large. What percentage are rejected?

4.5 A company fills packets of soap powder. The nominal weight of each packet is 1 kg. A characteristic of the machinery used for this task is that the actual weight of powder put into packets is Normally distributed with a coefficient of variation of 0.5%. Legislation permits a tolerance of 2% of the nominal weight and the company interprets this to mean that the probability of falling below this tolerance should be no more than 1%. What will be the mean weight of powder per packet if this requirement is just met?

4.6 The cost overruns on building refurbishment contracts follow a negative exponential distribution with a mean of 4.6%. What is the probability that a job will overrun by

(a) more than 5%
(b) less than 4%?

(c) What is the median overrun?

4.7 How would your answers to the previous exercise change if all overruns were known to lie between 1.5% and 9% but with the same mean?

4.8 A piece of new machinery is to be installed at a manufacturing plant. Like any machine it may break down. Discussions with other companies already operating this machine may be summarised as:

"The average number of breakdowns per year is three. There have never been more than ten. There is a 50 per cent chance of between two to five breakdowns in a year."

What is the probability of seven or more breakdowns in a year?

4.9 An airline is considering beginning operations on a new route. A planning meeting is called to consider the likely share of traffic on the route that the airline will be able to attract. After some discussion it is agreed that the best estimate (to be taken as the mean) is 30%. Further, it is thought that the probability of market share being below 20% is 0.3 and that the probability that the share may exceed 50% is 0.2. What is the probability that the share may be greater than 60%?

4.10 How would your answer to the previous exercise alter if the meeting had agreed an estimate characterised by a mean of 30% and a standard deviation of 20%?

5

Bayes' Rule

We have seen how the principle of entropy maximisation can be used to obtain unbiased probability assessments and also to update earlier estimates in the light of new information about the values of mean and variance.

There will be occasions when the data available for updating take a different form. For instance, we have an estimate of the proportion of defectives produced by some process and then take a sample of which a number of items are found to be defective. How should we revise our estimate in the light of this finding? Bayes' Rule provides a framework for doing this which takes account of the mechanism that generated the data rather than just the data themselves.

5.1 BACKGROUND AND BASIC STATEMENT

You are setting out on a journey and have to decide whether or not to take precautions against rain by carrying umbrellas and raincoats. You would rather not have to add these items to your baggage, provided the chance of rain is small. The weather forecaster on your local radio station says that it will be dry, but you know that the predictions are not perfect. More precisely you recall hearing a news item that said dry days had been correctly forecast 90% of the time while wet days had been correctly predicted only 80% of the time. What probability should you give to it being a rainy day?

Table 5.1 shows the information that you have about the accuracy of the forecasts.

state	forecaster's prediction		sum
	wet	dry	
wet	0.8	0.2	1.0
dry	0.1	0.9	1.0

Table 5.1
Likelihood distributions

The information is written as two conditional probability distributions showing how likely a particular prediction is *given* a particular (subsequent) weather state: prob(prediction | state). These distributions are, naturally enough, called *likelihood distributions*. They describe the accuracy of the data source that we wish to use. Alternatively, in the language of Figure 2.5, they indicate the amount of noise in the system. It is fundamental that before we use any source of data, speculation or whatever, we have some idea of the likelihood distributions, for how else are we to know the importance to attach to what we measure or hear?

The forecaster has predicted "dry" and so we focus on the second column of Table 5.1. It seems intuitively reasonable that the probability you give to each weather state is proportional to the likelihood that that state generated, as it were, the message or forecast that has been received. The chance of a wet day rather than a dry day is in the ratio 0.2 to 0.9. These likelihood values can be scaled to give a probability distribution:

$$\text{prob(state = wet | forecast = dry)} = 0.2/(0.2 + 0.9) = 0.18$$

$$\text{prob(state = wet | forecast = wet)} = 0.9/(0.2 + 0.9) = 0.82$$

What we have done seems sensible enough; it has a theoretical justification too.

Equation (1.7) gives the basic rule for prob(A and B) as

$$\text{prob}(A \text{ and } B) = \text{prob}(A|B)\text{prob}(B) = \text{prob}(B|A)\text{prob}(A)$$

from the last two terms of which we have

$$\text{prob}(B|A) = \frac{\text{prob}(A|B)\text{prob}(B)}{\text{prob}(A)} \tag{5.1}$$

The significance of this equation becomes clear if we let A stand for the data, forecast or other evidence and B for the system state in which we are interested. The main probability distributions in (5.1) can now be described using the names commonly given to them:

prior   prob(B) is the probability of the state before any data are collected

likelihood   prob($A|B$) is the probability of the data given a particular system state

posterior   prob($B|A$) is the revised probability of the state after the data have been taken into account

where "probability of the state" means the assessment we make that the system state has occurred or will occur.

Background and basic statement

Equation (5.1) now becomes

POSTERIOR is proportional to PRIOR × LIKELIHOOD

or   POSTERIOR = k(PRIOR × LIKELIHOOD)   (5.2)

where k is, as usual, a scaling factor to ensure that the probabilities sum to 1.0.

Let us review the weather forecast we have just made in the light of (5.2). We already know from Table 5.1 what the likelihoods are but what about the prior probabilities? If you have no reason or wish to make an independent assessment of the weather then the least biased estimate is, from (3.4), a uniform distribution:

prob(state = wet) = prob(state = dry) = 0.5

The calculation for (5.2) is shown in Table 5.2.

state	prior	likelihood	prior × likelihood	posterior
wet	0.5	0.2	0.10	0.18
dry	0.5	0.9	0.45	0.82
			0.55	1.00

Table 5.2

We get the same result as before. Notice, however, that we have allowed for the possibility of you having an opinion about the weather independently of that expressed by the forecaster. Suppose, for example, that before listening to the radio you had inspected your seaweed and telephoned Aunt Agatha, as a result of which you came to the opinion that there was a 70% chance that it would rain. The calculation, with this new prior distribution, is shown in Table 5.3.

state	prior	likelihood	prior × likelihood	posterior
wet	0.7	0.2	0.14	0.34
dry	0.3	0.9	0.27	0.66
			0.41	1.00

Table 5.3

Because you were predisposed to believe that it would rain your posterior probability that it will be wet has increased from 0.18 to 0.34. It is this ability to combine an initial, perhaps subjective, assessment with data to give a revised assessment that makes this procedure so useful.

In Tables 5.2 and 5.3 we have considered only those probabilities from the likelihood distribution that have been needed for calculating revised probabilities, and this will be our normal practice. Table 5.4 shows the complete set of probabilities for the last problem above. The figures in the body of the table are the products (prior × likelihood). The row marginal probabilities (0.70, 0.30) are the prior distribution. The column marginals (0.59, 0.41) show the probabilities that the forecaster, when asked, will say "wet" or "dry". This is called the *predictive distribution* and is the denominator in (5.1). The predictive probabilities describe the estimated likelihood of occurrence of each piece of data or evidence *before* it is received.

state	forecaster's prediction		
	wet	dry	
wet	0.56	0.14	0.70
dry	0.03	0.27	0.30
	0.59	0.41	

Table 5.4

Before proceeding here are two points about which you may need to be careful.

First, if you assign a prior probability of zero to some value then, because the updating is multiplicative, the posterior probabilities will always be zero too, whatever data are collected. If in doubt put some small value rather than exactly zero in the prior distribution.

Second, it is implicit that the source of the prior opinion and that of any subsequent data are statistically independent. If they are not you will need to modify the procedures described in this chapter. Failure to do so will lead to spurious precision in your posterior estimates. To take an extreme case you could just keep processing and reprocessing the same data to give an infinitely precise (zero variance) estimate on the basis of very little evidence.

The principle given in (5.2) is called *Bayes' Rule* and is shown schematically in Figure 5.1.

Sec. 5.2]   **Bayes' Rule with a Binomial likelihood**   109

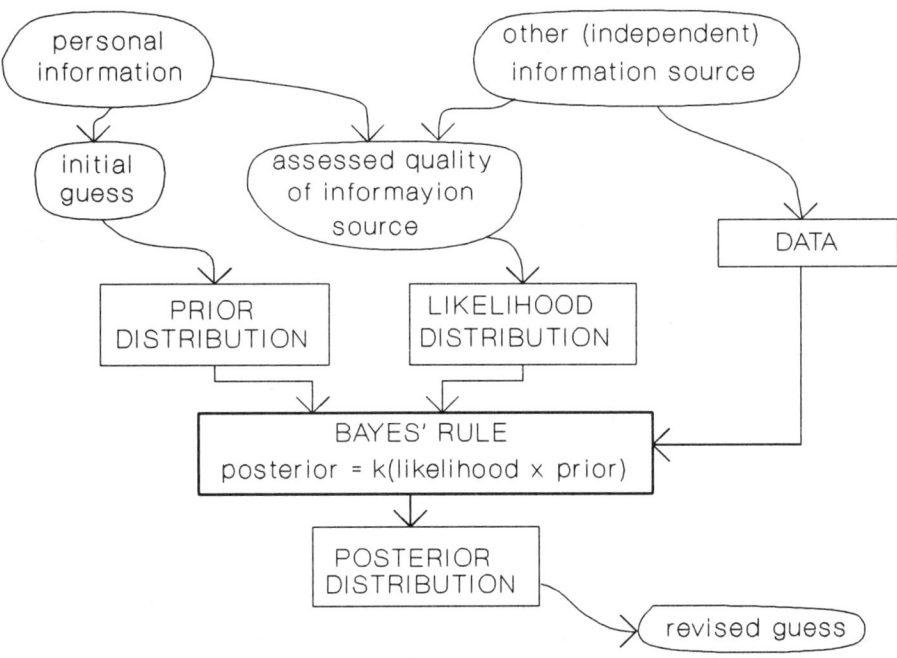

Figure 5.1   Revision of opinion by Bayes' Rule

5.2   BAYES' RULE WITH A BINOMIAL LIKELIHOOD

At the end of the previous chapter we estimated the proportion, π, of voters likely to vote for a particular party and obtained the maximum entropy distribution shown in Figure 4.10. Suppose now that we interview five people and that three of them say that they will vote for the party in question. How do we revise our opinion?

The prior distribution has already been found (Figure 4.10) so that now we need the likelihood that describes the process of selection and interview that we have followed to obtain our data.

The following assumptions describe selection procedures like the one that we used:

1.   Objects (interviewees in this example) are selected at random.

2.   Each object has the same probability, p, of possessing the attribute of interest (an intention to vote for a particular party).

3.   The possession of the attribute by any object is independent of its possession by any other object (how I vote is unaffected by how you vote).

If these assumptions hold then the probability that out of a sample of n interviewees exactly x will vote for the party is

$$\text{prob}(X = x) = \frac{n!}{x!(n-x)!} p^x (1-p)^{n-x}, \quad x = 0, 1, 2, \ldots, n \quad (5.3)$$

This is called the *Binomial distribution*. Its derivation is given in Appendix A4. The notation $n!$ is read as "n factorial" and is just the product of the integers from 1 to n so that

$$4! = 1 \times 2 \times 3 \times 4 = 24$$

A special case is $0! = 1$.
For the Binomial distribution

mean = np
variance = $np(1-p)$

We start by setting the probability p equal to the proportion, π, of people intending to vote for the party. We know that three out of the five interviewed said that they would vote for the party, so $n = 5$ and $x = 3$. The probability of these data if, say, $p = \pi = 0.4$ is

$$\text{prob}(X = 3) = \frac{5!}{3!2!} 0.4^3 0.6^2 = \frac{120}{6 \times 2} \times 0.064 \times 0.36$$

$$= 0.230$$

Alternatively use the table of the Binomial distribution function in Appendix A5.1 to give

$$\text{prob}(X = 3) = F(3) - F(2) = 0.91296 - 0.68256$$
$$= 0.230$$

This is the likelihood of the data given $\pi = 0.4$.
Using the Binomial likelihood distribution and the maximum entropy prior the posterior distribution is, from Bayes' Rule,

$$f(\pi) = k \exp[-7(\pi - 0.2)^2] \pi^3 (1 - \pi)^2 \quad (5.4)$$

The constant part of the Binomial distribution has been absorbed into the scaling factor k.

Sec. 5.2] Bayes' Rule with a Binomial likelihood

This posterior distribution is shown in Figure 5.2. As expected the distribution has shifted to the right because in the sample 60% were going to vote for the party whereas the mean of the prior was only 0.3. Notice too that the variance of the posterior is less than that of the prior: as we get more data the precision of our estimate increases.

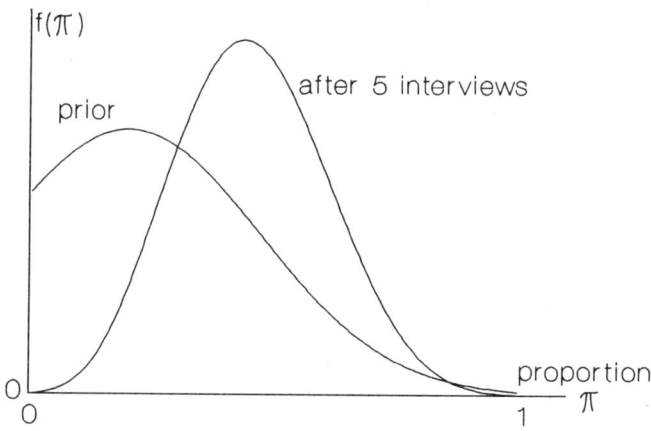

Figure 5.2 Original and revised estimates

To emphasise this suppose that another 15 people were interviewed of whom 11 said that they intended to vote for the party. Using (5.4) as the prior and a Binomial likelihood with $n = 15$ and $x = 11$ gives a posterior distribution of

$$f(\pi) = k\{\exp[-7(\pi - 0.2)^2]\pi^3 (1-\pi)^2\}\{\pi^{11}(1-\pi)^4\}$$

$$= k \exp[-7(\pi - 0.2)^2]\pi^{14}(1-\pi)^6$$

(5.5)

Figure 5.3 shows the new estimate. The mean and the precision have both increased again.

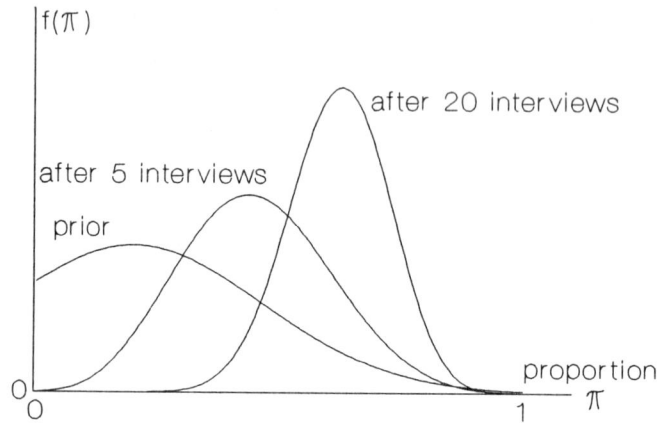

Figure 5.3 Estimate after 20 interviews

The powers of π and $(1 - \pi)$ in the last two terms of (5.5), $\pi^{14}(1-\pi)^6$, are the result of two surveys and were found by adding the results of both. The same posterior distribution would have been obtained had the larger survey been considered first or if we had just had one survey of 20 interviewees. This demonstrates an important characteristic of Bayesian revision when more than one update is performed: the resulting posterior distribution is independent of the order in which new data are considered.

Remember that we have assumed that the data are statistically independent.

5.3 CONVENIENCE PRIORS: THE BETA DISTRIBUTION

Revised distributions of the form (5.5) present no real problems now that most people have fairly ready access to some form of computer. Numerical methods such as those described in Appendix A3 enable probabilities to be found quite easily. For some likelihood distributions, however, alternative prior distributions are often used which make life easier still because both prior and likelihood share the same algebraic form making Bayesian revision a simple adjustment of parameter values. Such distributions are called *convenience* or *conjugate* priors.

These distributions are not usually justified via the entropy maximising methodology (though see section 5.7 below) and so are not, in that sense, unbiased. On the other hand, it is usually the case that with volumes of data commonly found in, for example, market research surveys the posterior distributions are pretty much the same whether a convenience or a maximum entropy prior is used. The ease of use of

Sec. 5.3] Convenience priors: the Beta distribution

the former is then generally thought to outweigh the correctness of the latter. Remember, though, that if the sample size is small, and especially if you are concerned with probabilities at the extremes of the range, the results obtained may significantly depend on which of the alternative distributions you choose. In these cases you should use the maximum entropy priors.

For the Binomial likelihood the convenience prior is the *Beta* density function

$$f(\pi) = k\pi^a(1-\pi)^b, \quad 0 \leq \pi \leq 1 \tag{5.6}$$

where, as before, the constant k ensures that the area under the pdf is 1.0.

Figure 5.4 shows some Beta densities. The relative values of a and b control the skew of the distribution; when they are equal the distribution is symmetrical. The absolute values control the spread, so that as a and b increase the variance decreases. When $a = b = 0$ the density is flat and suitable for a non-informative prior.

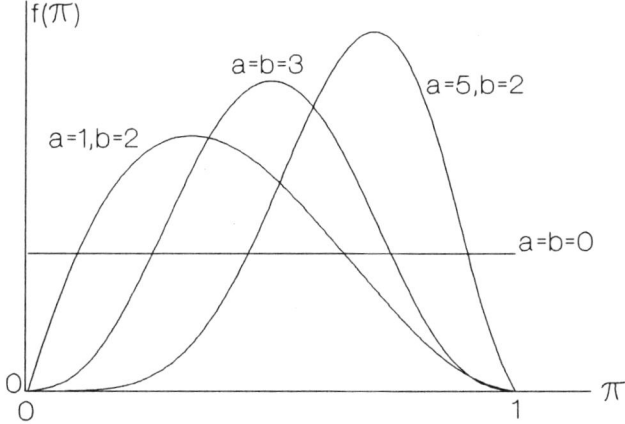

Figure 5.4 Some Beta density functions $f(\pi) = k\pi^a(1-\pi)^b$

Summary measures for the Beta distribution are

$$\text{mode} = \frac{a}{a+b} \tag{5.7}$$

$$\text{mean} = \mu = \frac{a+1}{a+b+2} \tag{5.8}$$

$$\text{variance} = \sigma^2 = \frac{(a+1)(b+1)}{(a+b+3)(a+b+2)^2} \tag{5.9}$$

To specify a Beta prior with the same constraints as the maximum entropy prior of Figure 4.10 we need to express the parameters a and b in terms of the mean and variance. From (5.8) and (5.9) we can get

$$a = \mu\left[\frac{\mu(1-\mu)}{\sigma^2} - 1\right] - 1 \qquad (5.10)$$

$$b = \frac{a+1}{\mu} - a - 2 \qquad (5.11)$$

Substituting $\mu = 0.3$ and $\sigma^2 = 0.04$ gives $a = 0.275$ and $b = 1.975$. This Beta prior and the corresponding maximum entropy prior are shown in Figure 5.5(a).

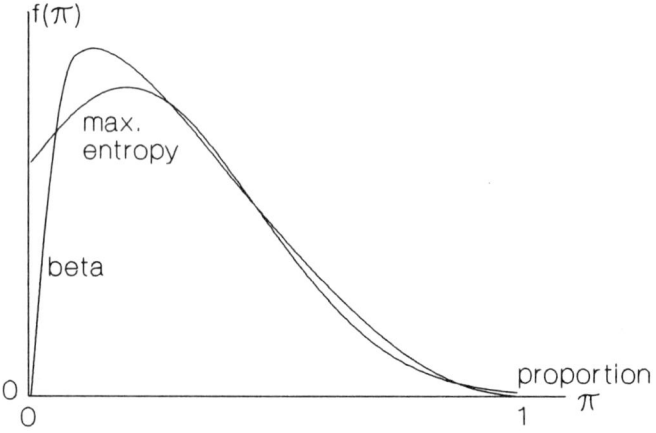

Figure 5.5(a) Using maximum entropy and convenience priors
the prior distributions

Applying Bayes' Rule the posterior distribution is

posterior $= k$[Beta prior][Binomial likelihood]

$$= k[\pi^a(1-\pi)^b][\pi^x(1-\pi)^{n-x}]$$

$$= k\pi^{a+x}(1-\pi)^{b+(n-x)} \qquad (5.12)$$

which is just another Beta distribution with parameters $[a + x]$ and $[b + (n - x)]$. The Bayesian revision of opinion is achieved by just adding the parameters of the prior and of the likelihood. After the five interviews where three people said they would

Sec. 5.3]   **Convenience priors: the Beta distribution**   115

vote for the party ($n = 5$ and $x = 3$) the revised estimate of π is given by the Beta distribution

$$f(\pi) = k\pi^{3.275}(1 - \pi)^{3.975}$$

Figure 5.5(b) shows the posterior distributions after five interviews using both priors. You can see how similar the distributions are becoming. This effect is even more marked in Figure 5.5(c) after 20 interviews ($n = 20$, $x = 14$ as before).

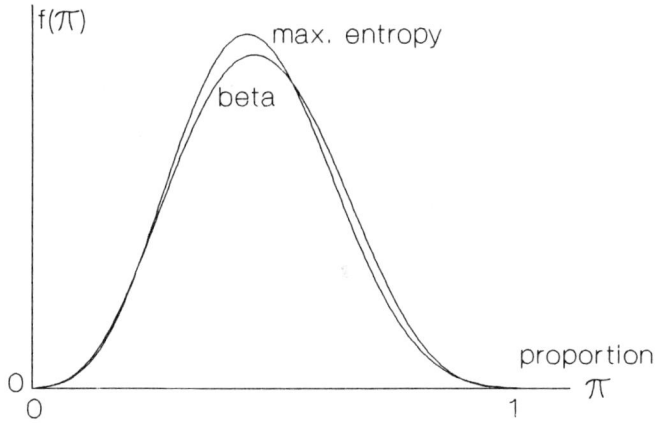

Figure 5.5(b)   Using maximum entropy and convenience priors posteriors after five interviews

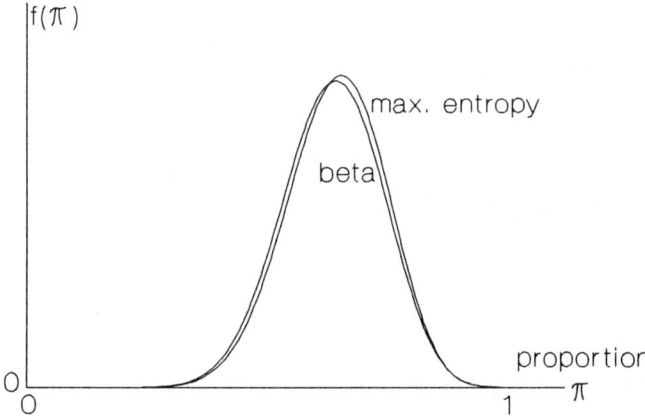

Figure 5.5(c)   Using maximum entropy and convenience priors posteriors after 20 interviews

In practice we would use much larger samples thus making the differences between using the maximum entropy or convenience prior negligible.

Suppose now that you and I had both started with different prior opinions concerning the proportion voting for the party, yours described as above by the Beta distribution with $a = 0.275$ and $b = 1.975$ and mine by a Beta distribution with $a = 5$ and $b = 2$. Our opinions are significantly different since your prior has a mean of 0.3 and mine has a mean of 0.67. We are now both faced with the results of interviews with 50 people of whom 30 said that they would vote for the party. The prior and posterior distributions are given in Figure 5.6. See how our opinions are converging in the face of the evidence.

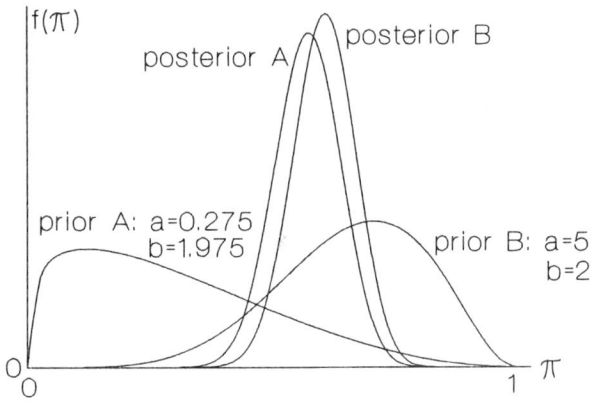

Figure 5.6 Convergence after sample of 50

Common sense tells you that this is what should happen. The mechanism can be seen in (5.12). As the weight of evidence increases so do the values of x and $(n - x)$ until eventually they completely outweigh the prior opinions reflected by a and b, whatever those opinions are.

You could think of the information in the prior as being equivalent to another sample so that, for example, my prior with $a = 5$ and $b = 2$ is numerically equivalent to a sample of seven people of whom five said that they would vote for the party. This is a fairly small sample compared to the 50 that were in fact interviewed.

Appendix A5.4 describes how to find probabilities using a Beta density function. Confirm for yourself that accepting my prior the probability that the proportion voting for the party is no more than 0.4 is $\text{prob}(\pi \leq 0.4) = 0.05$.

5.4 POISSON LIKELIHOOD AND GAMMA PRIOR

To help in the physical redesign of a bank we need to estimate the rate at which customers arrive during the busiest time of the day. The time period over which we measure the rate is arbitrary and we choose to consider arrivals per minute.

Having some knowledge of the part of town in which this branch is located our estimate of the average number of arrivals per minute is described by a distribution with a mean of 2 and a variance of 1.

When passing the branch we stop for a minute and observe five customers entering. What is our revised estimate of the mean arrival rate?

The likelihood distribution for problems such as this must describe a relationship between the mean number of arrivals in the chosen period, m, and the number arriving in any particular period, x. The *Poisson* distribution does just this:

$$\text{prob}(X = x) = \frac{m^x \exp(-m)}{x!}, \qquad x = 0, 1, 2, \ldots \tag{5.13}$$

The Poisson distribution is a special approximation to the Binomial distribution. A person will either arrive or not arrive at the bank while we are counting. The number of people who *may* arrive is huge and includes all the people who have accounts there, people with accounts at other branches wishing to cash cheques, and so on. The value of n in (5.3) is therefore very large and the value of p correspondingly small. Given these conditions Appendix A4 shows that (5.3) becomes (5.13). Note that we no longer need to know the values of n and p, only the average arrival rate m ($= np$).

For the Poisson distribution

mean = variance = m

If the mean arrival rate is $m = 3$ arrivals/min the probability of $x = 5$ arrivals in any one minute is

$$\text{prob}(X = 5) = \frac{3^5 \exp(-3)}{5!}$$

$$= \frac{243 \times 0.0498}{120} = 0.101$$

Tables of the Poisson distribution are given in Appendix A5.2.

An appropriate convenience prior for the mean m is the *Gamma* density function

$$f(m) = km^a \exp(-bm) \tag{5.14}$$

Figure 5.7 shows some Gamma distributions. The parameter b is just a scaling factor depending on the (usually arbitrary) units of measurement that we have adopted so that the degree of skew depends only on *a*. Notice the special case when $a = 0$ giving the negative exponential distribution (4.16).

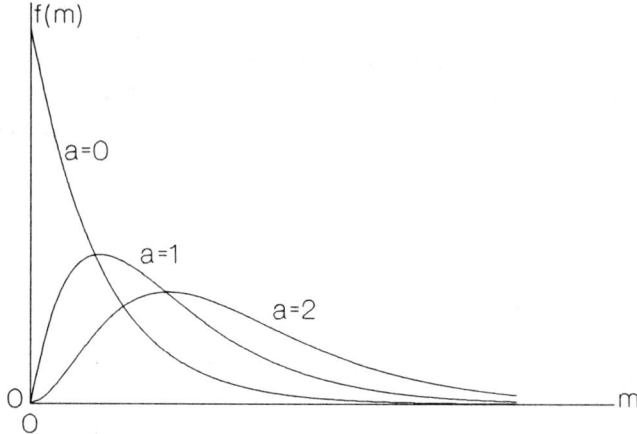

Figure 5.7 Some Gamma density functions
$f(m) = km^a \exp(-bm)$

Summary measures for the Gamma distribution are

$$\text{mode} = \frac{a}{b} \qquad (5.15)$$

$$\text{mean} = \mu = \frac{a+1}{b} \qquad (5.16)$$

$$\text{variance} = \sigma^2 = \frac{a+1}{b^2} = \frac{\mu}{b} \qquad (5.17)$$

To find the values of the parameters *a* and *b* that give a Gamma distribution with required mean and variance we have, from (5.16) and (5.17),

$$b = \frac{\mu}{\sigma^2} \qquad (5.18)$$

and $\quad a = \mu b - 1 = (\mu/\sigma)^2 - 1 \qquad (5.19)$

Sec. 5.4]   **Poisson likelihood and Gamma prior**   119

In our case we have $\mu = 2$ and $\sigma^2 = 1$ giving $b = 2$ and $a = 3$.

A non-informative Gamma prior, one with no impact on the posterior estimate, would have $a = b = 0$.

Applying Bayes' Rule, and amalgamating the constant factors,

$$\text{posterior} = k[\text{Gamma prior}][\text{Poisson likelihood}]$$

$$= k[m^a \exp(-bm)][m^x \exp(-m)]$$

$$= km^{a+x} \exp[-(b+1)m] \qquad (5.20)$$

This is a Gamma distribution with new parameter values found by adding the number of arrivals to the original value of a and incrementing b by 1 (= one time period) to give a posterior Gamma distribution with

$a = 3 + 5 = 8$
$b = 2 + 1 = 3$

Suppose now that we waited outside the bank for another minute and counted y arrivals. With (5.20) as the prior we would have

$$\text{posterior} = k\{m^{a+x} \exp[-(b+1)m]\}\{m^y \exp(-m)\}$$

$$= m^{a+x+y} \exp[-(b+2)m]$$

You can see from this that, in general, after a number of observations

$a = $ (prior value of a) + (total number of arrivals)

$b = $ (prior value of b) + (number of time periods surveyed)

Notice again that the order in which we consider the data is irrelevant.

Suppose that we counted for another 12 minutes and saw a further 45 arrivals; then the Gamma posterior would have parameters

$a = 3 + (5 + 45) = 53$
$b = 2 + (1 + 12) = 15$

The results are shown in Figure 5.8.

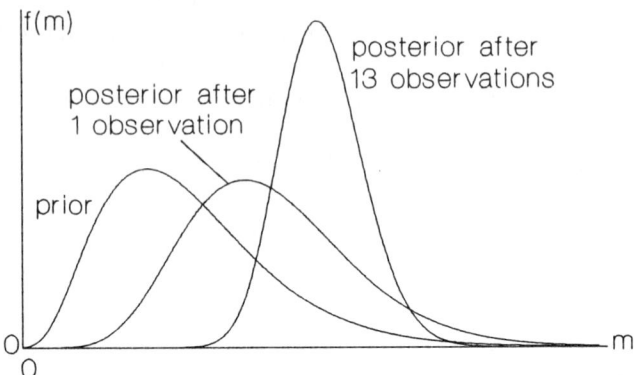

Figure 5.8 Bayes estimate of Poisson mean – m

Appendix A5.5 tells you how to find probabilities from a Gamma density. Check for yourself that using the Gamma prior with $a = 3$ and $b = 2$ the probability of m not exceeding 2.5 is 0.735.

There are relationships between the Poisson distribution and others (see Haight, 1967, for a full discussion of the Poisson). The probability of no events or arrivals per unit of time is given by (5.13) with $x = 0$ to be $\exp(-m)$. The probability of a gap of t units between events is just the probability of t successive gaps of unit time with no arrivals in any of them and so is the product of t probabilities each of $\exp(-m)$, which is $\exp(-mt)$. This means that the probability of an event occurring after time t has elapsed is $\exp(-mt)$. Now look at Figure 4.7(a) to see that the probability distribution of the duration of gaps between successive Poisson events follows the negative exponential distribution.

More generally, the Gamma distribution $f(x) = kx^{n-1}\exp(-mx)$ shows the probability that the n'th Poisson event will occur at time x, given that the mean of the Poisson distribution is m.

5.5 NORMAL LIKELIHOOD AND PRIOR

When using a Normal likelihood distribution the appropriate convenience prior is also a Normal distribution. This is one occasion when the convenience and maximum entropy priors are the same.

We observe a value, x. The likelihood of occurrence of this value is determined by a Normal likelihood with mean m and standard deviation s, which is, from (4.19) and (4.20),

$$\text{likelihood} = f(x) = k \exp[-(x - m)^2/2s^2]$$

Normal likelihood and prior

The likelihood depends on the value of m. Suppose that our prior estimate of m is also Normally distributed with mean μ and standard deviation σ so that

$$\text{prior} = g(m) = k \exp[-(m - \mu)^2/2\sigma^2]$$

The revised estimate is found using Bayes' Rule

$$\begin{aligned}\text{posterior} &= kf(x)g(m) \\ &= k \exp[-(x - m)^2/2s^2]\exp[-(m - \mu)^2/2\sigma^2] \\ &= k \exp[-(x - m)^2/2s^2 - (m - \mu)^2/2\sigma^2]\end{aligned}$$

which, after some manipulation, becomes

$$\text{posterior} = k \exp[-(m - \mu_R)^2/2\sigma_R^2]$$

This is another Normal distribution with

$$\text{revised mean} = \mu_R = (\mu\sigma^{-2} + xs^{-2})/(\sigma^{-2} + s^{-2}) \quad (5.21)$$

$$\text{revised variance} = \sigma_R^2 = 1/(\sigma^{-2} + s^{-2}) \quad (5.22)$$

The posterior mean is just the weighted average of the means of the prior and likelihood distributions, the weights being the reciprocal of the variances so that the smaller the variance the greater the weight. A non-informative prior with infinite variance would have no effect on the posterior estimate. The reciprocal of the variance is called the *precision*. In this Bayes' revision precisions are additive.

Imagine that I have to assess the temperature of a liquid. I have two thermometers of different construction, both of which have been calibrated by comparing the temperatures that they register with the known temperatures of certain substances. In both cases measurement error was Normally distributed with zero mean and a standard deviation of 0.5°C for thermometer A and 0.3°C for thermometer B. (Normally distributed error with $\mu = 0$ is sometimes called *white noise*.)

Measuring the temperature of my liquid with thermometer A gives a reading of 38.6°C while thermometer B shows 39.4°C.

I have no *a priori* estimate of the temperature so use a non-informative prior with the data from thermometer A. This gives a posterior estimate with the same characteristics as the data:

mean = 38.6°C
standard deviation = 0.5°C

Considering now the additional data from thermometer B, the posterior estimate is Normally distributed with

$$\frac{1}{\text{variance}} = \frac{1}{0.5^2} + \frac{1}{0.3^2} = 15.111$$

so   variance $= 1/15.111 = 0.0662\,°C^2$

and   standard deviation $= 0.257\,°C$

also $\text{mean} = \dfrac{\dfrac{38.6}{0.5^2} + \dfrac{39.4}{0.3^2}}{\dfrac{1}{0.5^2} + \dfrac{1}{0.3^2}} = 39.188°C$

As a result of combining the measurements from both thermometers we have an estimate which is a Normal distribution with a mean value between the readings and with a standard deviation lower than the calibration error on either instrument so that, as you might expect, combining estimates reduces error.

5.6   NORMAL APPROXIMATIONS

As well as being of use in its own right the Normal distribution may also be used as an approximation for other distributions. The posterior distributions in Figure 5.3 look like Normal distributions. Figures 5.4 and 5.6 indicate that Beta distributions also look like Normals when the mean is close to 0.5 (*a* and *b* roughly equal) or when the variance is low (*a* + *b* high). When the mean of a Gamma distribution starts to be large it too can be approximated with a Normal (Figure 5.8).

The Normal distribution may also be used as an approximation to discrete distributions such as the Binomial and Poisson when the expected value (mean) is high. Each discrete value, x, is approximated in the Normal by the range $x \pm 0.5$. Example 5.3 below illustrates the method.

5.7   FITTING DISTRIBUTIONS

Earlier in this chapter it has been suggested that an appropriate way of setting the parameters of the Beta, Gamma and Normal distributions was first to specify the mean and standard deviation, say, and then calculate the corresponding parameter values. This is somewhat in contrast to the procedures in Chapters 3 and 4 where the parameter values were chosen to maximise entropy subject to *a priori* requirements. Now, as we saw in (4.20), with the Normal distribution there is no problem because the mean and variance are exactly the *a priori* requirements that lead directly to that distribution.

Sec. 5.7] Fitting distributions

Suppose that we wish to specify both the mean of X and also the mean of $\ln(X)$. We wish to maximise H subject to

$$\left.\begin{array}{l} \Sigma\, p_i = 1 \\ \Sigma\, p_i \ln(x_i) = \mu \\ \Sigma\, p_i \ln(x_i) = \gamma \end{array}\right\} \qquad (5.23)$$

This gives, in the usual way,

$$\begin{aligned} p_i &= k \exp[\alpha x_i + \beta \ln(x_i)] \\ &= k x_i^a \exp(b x_i) \end{aligned} \qquad (5.24)$$

(with $a = \beta$ and $b = \alpha$) and when X is continuous

$$f(x) = k x^a \exp(bx) \qquad (5.25)$$

which is identical to (5.14): the Gamma distribution is obtained as the maximum entropy distribution given the constraints (5.23). Now, it is not at all clear that we would ever wish to specify the mean of $\ln(X)$ when expressing an opinion about the distribution of X, but it does provide an alternative way of fitting the distribution to data, as Example 5.5 illustrates.

In similar fashion the Beta distribution is found as the result of maximising H subject to

$$\left.\begin{array}{l} \Sigma\, p_i = 1 \\ \Sigma\, p_i \ln(x_i) = \gamma \\ \Sigma\, p_i \ln(1 - x_i) = \eta \end{array}\right\} \qquad (5.26)$$

to give

$$\begin{aligned} p_i &= k \exp[\alpha \ln(x) + \beta \ln(1 - x_i)] \\ &= k x_i^a (1 - x_i)^b \end{aligned} \qquad (5.27)$$

which, when $0 \leq X \leq 1$ is continuous, gives (5.6).

5.8 BAYESIAN LEARNING

The revision of opinion via Bayes' Rule represents an acquisition of knowledge via induction. Some authors argue, plausibly, that Bayes' Rule does not describe the way in which humans do actually update their beliefs and, less plausibly, that it is therefore not a secure base for reasoning about what we know of the world around us.

It is easy to agree that Bayes' Rule is not behaviourally descriptive: it was never meant to be. What it does, especially when used with the entropy maximising formalism for obtaining unbiased priors, is to provide a consistent and coherent methodology for belief revision. We may think of this as a process of learning. The extent to which ignorance is reduced by this process is measured by changes in the entropy, H.

Suppose that we are monitoring components manufactured by a complex sequence of operations and that we are monitoring only the final product, not at intermediate points in the chain of manufacture. It looks as if something is going wrong since the proportion of defective products appears unusually large. It is known that there are three modes of failure of the manufacturing system. To stop production and remedy the fault is expensive so it is important to have a clear idea of what is wrong. The three failure modes produce 10%, 20% and 30% defectives so diagnosing the fault is equivalent to correctly estimating the proportion of defectives.

We have no *a priori* idea of what is wrong and so believe all three possible proportions equally likely. A sample of ten components is collected of which one is found to be defective. Using Binomial likelihoods our revised estimate is that

prob(proportion defective = 10%) = 0.50

prob(proportion defective = 20%) = 0.35

prob(proportion defective = 30%) = 0.16

as shown in the second line of Table 5.5. We have learnt something about the system and as a result our ignorance has reduced from $H = 1.10$ to $H = 1.00$. The table shows continued learning (H reducing) as more samples are collected. Notice that sample number 8 causes a small hiatus because of the unusually high number of defectives found but that H continues to decrease thereafter.

sample	sample size	no. of defects	probability estimate for possible proportion			H
			0.10	0.20	0.30	
prior			0.33	0.33	0.33	1.10
1	10	1	0.50	0.35	0.16	1.00
2	10	0	0.81	0.17	0.02	0.56
3	10	2	0.73	0.24	0.02	0.66
4	10	1	0.81	0.19	0.01	0.52
5	10	1	0.86	0.14	0.00	0.42
6	10	0	0.95	0.05	0.00	0.19
7	10	2	0.93	0.07	0.00	0.26
8	10	3	0.79	0.21	0.00	0.52
9	10	0	0.92	0.08	0.00	0.27
10	10	1	0.95	0.05	0.00	0.21

Table 5.5
Revision and learning: uniform prior

We may have been able to spot something about the operation of the manufacturing system that enabled us to be pretty sure of what was wrong. Table 5.6 shows just such a case. The initial level of ignorance is, of course, now reduced and the subsequent learning follows a similar pattern. Both situations are shown in Figure 5.9.

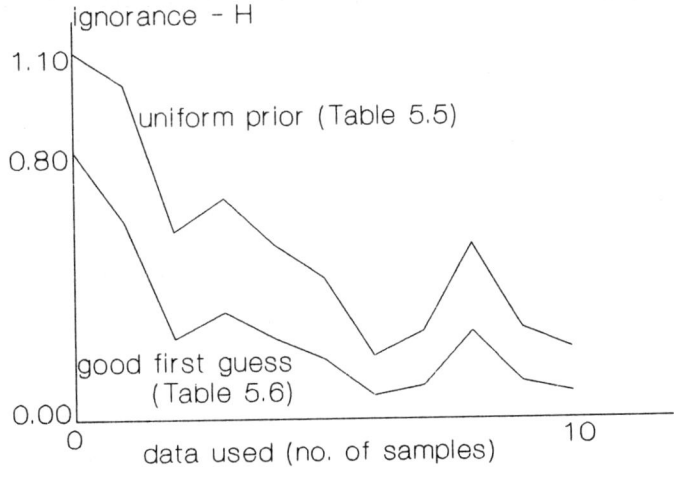

Figure 5.9 Learning as ignorance reduction

sample	sample size	no. of defects	probability estimate for possible proportion			H
			0.10	0.20	0.30	
prior			0.70	0.20	0.10	0.80
1	10	1	0.80	0.16	0.04	0.59
2	10	0	0.94	0.06	0.00	0.24
3	10	2	0.91	0.09	0.00	0.32
4	10	1	0.94	0.06	0.00	0.24
5	10	1	0.96	0.04	0.00	0.18
6	10	0	0.99	0.01	0.00	0.07
7	10	2	0.98	0.02	0.00	0.10
8	10	3	0.93	0.07	0.00	0.26
9	10	0	0.98	0.02	0.00	0.11
10	10	1	0.98	0.02	0.00	0.08

Table 5.6
Revision and learning: good first guess

The danger with a non-uniform intuitive first guess is that it may be wrong. We already know that using Bayes' Rule convergence to a correct estimate is assured given enough data. Table 5.7 and Figure 5.10 show just such a case. Notice how the initially skewed distribution first of all becomes flatter, thereby increasing H, until it is skewed in the other (correct) direction and thereafter learning occurs as before. Following Watanabe (1969) we may think of this as a two stage process: the incorrect initial belief has first to be unlearned before learning proper can take place.

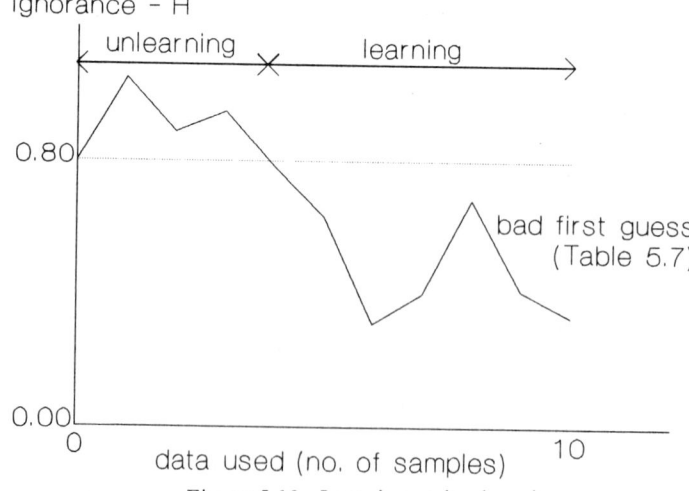

Figure 5.10 Learning and unlearning

sample	sample size	no. of defects	probability estimate for possible proportion			H
			0.10	0.20	0.30	
prior			0.10	0.20	0.70	0.80
1	10	1	0.22	0.30	0.48	1.05
2	10	0	0.62	0.27	0.11	0.89
3	10	2	0.53	0.35	0.11	0.95
4	10	1	0.65	0.30	0.04	0.78
5	10	1	0.75	0.24	0.02	0.63
6	10	0	0.91	0.09	0.00	0.31
7	10	2	0.87	0.13	0.00	0.40
8	10	3	0.65	0.35	0.01	0.68
9	10	0	0.86	0.14	0.00	0.41
10	10	1	0.90	0.10	0.00	0.33

Table 5.7
Revision and learning: bad first guess

5.9 FURTHER READING

Bayes' Rule is named after the Rev. Thomas Bayes and his now famous paper (Bayes, 1763). (Interestingly, in an abridged collection of the papers of the Royal Society (Hutton *et al.*, 1809) Bayes' paper is omitted as being "much too long and intricate to be at all materially and practically useful". Such is fame.)

The revival of interest in Bayesian statistics began with Schlaiffer (1959, 1961) and has grown quickly, finding application in a number of fields. Press (1989) gives a good brief survey and also lists some software which helps in making the sometimes complex calculations required by Bayesian analysis.

Introductory texts are provided by Philips (1973), Schmitt (1969), Iversen (1984) and Lee (1989) while those of Lindley (1965) and Press (1989) are mathematically more explicit. Discussions of the Bayesian methodology are often to be found in books primarily concerned with decision theory such as those by Berger (1985), French (1986), Lindley (1971), Raiffa (1968), Smith (1988), and Winkler (1972).

Winkler (1981) provides further discussion on non-independent data sources.

Surveys of Bayesian statistics with issues and applications may be found in Aykac and Brumat (1977), Bernardo *et al.* (1980, 1985, 1988), *Journal of the Institute of Statisticians* (1983) and Lindley (1972).

Bayesian methods in economics are to be found in Zellner (1971, 1980). Hey (1983) provides an introductory text in this area. Morris (1974, 1977) gives a Bayesian approach to the combination of expert opinions.

Applications of Bayes' Rule in which priors are an explicit function of past observations, rather than opinion based, and so are open to a relative frequency interpretation are called empirical Bayes methods (see, for instance, Maritz, 1970).

Proponents of Bayesian induction include various philosophers such as Aune (1991) and Rosenkrantz (1977, 1981). Earman (1992) adopts a more cautious stance. See also Horwich (1981).

Behaviourist objections are exemplified by, for instance, Goldman (1986).

5.10 EXAMPLES

Example 5.1

A manufacturer of outdoor clothing has three factories each of which produces the same type of waxed cotton jacket. The defect rates at the factories are 3%, 5% and 4%. What is the probability that a faulty jacket, returned by a customer, was made in the third factory?

How would your estimate change if it was known that the factories produce 30%, 20% and 50% respectively of all jackets sold?

Solution

This requires a straightforward application of Bayes' Rule. The likelihoods are just the defect rates and in the first case, since there are no data about the relative size of the factories, we assume a uniform prior. Here is the calculation:

factory	likelihood	prior	(likelihood × prior)	posterior
A	0.03	1/3	0.03/3	0.250
B	0.05	1/3	0.05/3	0.417
C	0.04	1/3	0.04/3	0.333
		1.0	0.04	1.000

So, the probability that the defective came from factory C is 0.33.

Knowing the relative outputs of the factories it is sensible to use these proportions as priors. This means that *a priori* we would think that the probability that a defective jacket came from a certain factory was just the proportion of jackets made in that factory. Here is the revised calculation:

factory	likelihood	prior	(likelihood × prior)	posterior
A	0.02	0.3	0.006	0.167
B	0.05	0.2	0.010	0.278
C	0.04	0.5	0.020	0.556
		1.0	0.036	1.001

Sec. 5.10]   **Examples**   129

Our estimate has increased to 0.556 because C produces the most jackets.

Example 5.2

Table 5.8 shows the number of words contained in 200 sentences from each of three Sunday newspapers published on 12 September 1993. Sentences made up in whole or substantial part of quotation were omitted.

words per sentence	*News of the World*	*The Mail on Sunday*	*The Independent on Sunday*
4	2	1	2
5	2	1	2
6	4	1	3
7	3	1	4
8	9	2	3
9	7	3	8
10	9	7	3
11	6	7	8
12	16	2	3
13	9	9	3
14	6	4	7
15	15	6	12
16	16	8	10
17	12	8	6
18	12	13	9
19	12	16	12
20	3	10	7
21	12	7	9
22	7	9	4
23	9	14	8
24	10	14	7
25	9	5	6
26	1	5	11
27		15	6
28	3	4	6
29	1	5	4
30		3	3
31	2	4	6
32		5	1
33		3	3
34	1	1	1
35		1	3
36		1	4
37	1		1
38		1	4
39			3
40			2
41		1	1
42			1
43		1	2
44	1		1
45		1	
46			1
47			
48		1	

Table 5.8

The following passage is taken from a different edition of one of the papers.

"Foreign programmes are among the most popular on British television. *The Simpsons* cartoon consistently draws the highest audiences on satellite, with more than a million viewers. Channel 4's most popular series include three US imports – *Roseanne, The Golden Palace* and *The Cosby Show. Quantum Leap,* from the US, has finished a run which sometimes saw it ranked first among BBC 2 programmes and new episodes of the Australian soap *Neighbours* (13 million viewers) remain a BBC 1 staple.

Despite the volume of imported material, Grant McKee, controller of programmes at Yorkshire-Tyne Tees, said that quality, home-produced programmes would always attract audiences and that terrestrial channels' commitment to them was as strong as ever."

Make a suitably probabilistic estimate of the source of this extract.

Solution

In this example text and source newspaper are connected via the survey of sentence length. To make the required inference use Bayes' Rule.
    Consider first the prior probabilities. I have no reason to believe that the extract is more likely to come from any one paper rather than another and so set all prior probabilities to 1/3. It may be that, having read the text, you think that you recognise the style and so want to incorporate this into your prior. Be careful here. Remember that data must be independent and that we shall be considering sentence length. This means that only stylistic data independent of sentence length are permissible when forming your prior, and this may be difficult to ensure.
    The evidence provided by the extract is the length of the sentences used. I have taken names such as BBC 1 to be one word and so obtained lengths

    10 16 17 33 35

Finally, form the likelihood distributions. The frequencies found in the survey may be converted directly to probabilities so that, for instance,

    prob(sentence of 23 words | *News of the World*) = 9/200

Since the products of which these likelihoods form a part will be scaled anyway the frequencies may be used without conversion to probabilities.
    No sentences of length 33 or 35 words were found in the sample from the *News of the World* so how can likelihoods be found? This draws attention to the more general point that we have samples of *only* 200 and so may have reservations about

the precision with which the underlying frequencies are estimated. This situation may be ameliorated by grouping the data from Table 5.8 as shown in Table 5.9. Classes were defined arbitrarily but, I hope, sensibly. Were you to group the data differently you might obtain a slightly different result for this exercise.

words per sentence	News of the World	The Mail on Sunday	The Independent on Sunday
1– 5	4	2	4
6–10	32	14	21
11–15	52	28	33
16–20	55	55	44
21–25	47	49	34
26–30	5	32	30
31–35	3	14	14
36–40	1	2	14
41 or more	1	4	6

Table 5.9

Finally, form the revised probability estimates as the product of the prior probabilities and the likelihoods for each of the five sentence lengths. The likelihoods are taken as the grouped frequencies in Table 5.9.

Here are the calculations:

paper	prior	likelihoods					prior × likelihoods	revised
NoW	1/3	32	55	55	3	3	290400	0.05
MoS	1/3	14	55	55	14	14	2766867	0.48
IoS	1/3	21	44	44	14	14	2656192	0.47
	1.00						5 713 459	1.00

Table 5.10

We conclude that the text is unlikely to have come from the *News of the World* but is equally likely to have come from either of the other two papers.

This is an extremely simple example of the way in which cases of disputed authorship and the like are settled. If this interests you try Holmes (1985), Mosteller and Wallace (1984) and O'Brien and Darnell (1982).

Example 5.3

A university lecturer is to recommend a textbook to a class of 120 students and tells the University Bookshop so that they may ensure that copies of the book are available for students to buy. Based on experience the manager of the shop believes that about 30% of students buy recommended texts.

(a) What is the probability that 40 books will be sold?

(b) What is the probability that at least 40 books will be sold?

(c) The manager wishes to order a number of books that will ensure that the probability of having stock left is no more than 15%. How many should the manager order? The book is unlikely to be bought by anyone except the students to whom it has been recommended.

Solution

A suitable model of the number of books sold is provided by the Binomial distribution with $n = 120$ and $p = 0.3$. Because n is high and p not too extreme we may use a Normal approximation with

$$\text{mean} = np = 120 \times 0.3 = 36$$

and

$$\begin{aligned}\text{standard deviation} &= [np(1-p)]^{0.5} \\ &= [120 \times 0.3 \times 0.7]^{0.5} \\ &= 5.02\end{aligned}$$

(a) If X is the Normal variate with these characteristics and remembering that we are modelling a discrete distribution

$$\begin{aligned}\text{prob}(40 \text{ books sold}) &= \text{prob}(39.5 \leq X \leq 40.5) \\ &= F(40.5) - F(39.5)\end{aligned}$$

To look up the tables in Appendix A5.3 we need the values of Z, the standard Normal variate, which correspond to the X values.
When $X = 40.5$,

$$Z = (40.5 - 36)/5.02 = 0.90$$

and, from the table,

$$F(X = 40.5) = F(Z = 0.90) = 0.81594$$

Similarly,

$$F(X = 39.5) = F(Z = (39.5 - 36)/5.02)$$
$$= F(Z = 0.70) = 0.75804$$

So,

$$\text{prob}(40 \text{ books sold}) = F(40.5) - F(39.5)$$
$$= 0.81594 - 0.75804 = 0.0579$$

(It may be of interest to note that calculating the Binomial probability gives 0.0565, so the approximation is pretty good.)

(b) $\quad \text{prob}(\text{at least } 40 \text{ books sold}) = 1 - F(X = 39.5)$
$$= 1 - 0.75804$$
$$= 0.24196$$

(c) We need to find the number of books demanded, x, such that $F(x) \leq 0.15$. By looking in the body of the table of the standard Normal distribution function in Appendix A5.3 we need to find that value of Z, z, that gives $F(z) = 0.15$. But values of $F(z)$ are only tabulated for $z \leq 0$ and so, by symmetry, we look for $F(z) = 1 - 0.15 = 0.85$ and find $z = 1.04$. The value of X that we require is 1.04 standard deviations below the mean:

$$x = 36 - (1.04 \times 5.02) = 30.78$$

The manager ought to order 31 copies of the book.

Example 5.4

Here are data showing the number of passengers carried on a sample of 291 flights between London and a Far East destination in the early 1980's. The data show only those passengers with a certain class of ticket. In the table x is the number of passengers on the flight and $f(x)$ is the observed frequency.

We wish to fit a model to these data.

x	f(x)	x	f(x)	x	f(x)
0	0	16	14	32	1
1	0	17	14	33	5
2	1	18	15	34	4
3	0	19	14	35	1
4	1	20	16	36	3
5	2	21	15	37	4
6	0	22	10	38	2
7	5	23	11	39	3
8	2	24	16	40	2
9	5	25	12	41	0
10	8	26	13	42	1
11	7	27	10	43	2
12	12	28	3	44	1
13	13	29	8	45	0
14	13	30	6	46	0
15	12	31	3	47	1

Solution

The distribution is skewed and the variable is non-negative so a model of the form (5.24) is appropriate:

$$p_i = kx_i^a \exp(bx_i)$$

This model is a result of the constraints (5.23) and so is fitted by iteratively choosing values of a and b that ensure, respectively, that values of $E[\ln(X)]$ (the mean of the logarithm of the number of passengers) and $E[X]$ (the mean number of passengers) are the same for the model as for the data. It was noted earlier that models are often fitted to respect mean and variance constraints even though these had not been used to derive the model. As illustration this strategy for finding values of a and b was also used. Here are the results:

	data	fitted mean, $E[X]$ and $E[\ln(X)]$	fitted mean, $E[X]$ and variance
mean	20.780	20.780	20.780
variance	66.412	69.097	66.412
H		3.491	3.472
I		0.068	0.068
parameters: a		4.72000	5.00831
b		−0.271630	−0.285911

The dissimilarity between the two modelled distributions and the data, I, is the same. (More is said about testing for dissimilarity in Chapter 7.)

As would be expected the model fitted to the correct maximum entropy constraints has the higher entropy, H.

The two models are shown in Figure 5.11. The vertical bars depict the data. For visual convenience only the models are shown as continuous lines though they are, of course, discrete. The model fitted to the mean and variance has the higher peak at the mode. Draw your own conclusions about these results. It doesn't seem as if much harm will be done if parameters for Beta and Gamma priors are found from mean and variance estimates, especially given the subsequent effect of data via Bayes' Rule.

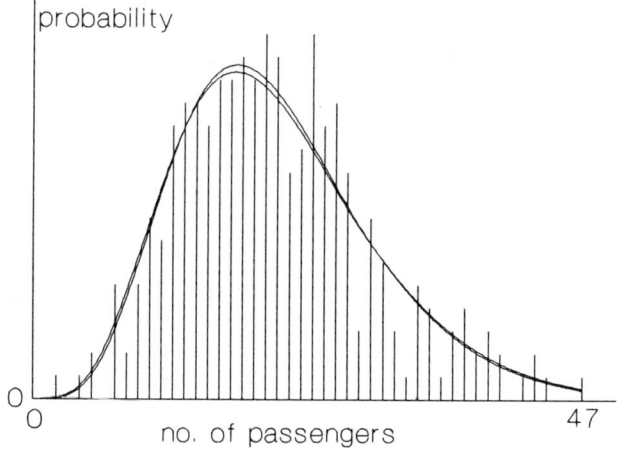

Figure 5.11

You may wish to fit model (3.13), the one that correctly comes from a specification of mean and variance.

Example 5.5

Your company makes turbine generators. As part of their activities your sales staff note that a rival company, new to the industry, seems to be making inroads into your market share and so you wish to estimate the number of generators that it has sold. Discussion in the office leads to the unanimous opinion that it could not conceivably have sold more than 20. It is known that its machines have been given serial numbers starting at 1 for the first machine built, 2 for the second, and so on. One of your sales staff recalls seeing one of these machines with the serial number 7 and another with the serial number 3. What is your estimate of the number of machines sold by your new competitor?

(This example is based on one given in Tribus, 1969, p249.)

Solution

Without knowing the serial numbers our *a priori* view was that the number of machines sold, N, was not more than 20 and so the maximum entropy prior distribution is uniform with

$$\text{prob}(N = n) = 1/21 = 0.0476, \qquad n = 0, 1, 2, \ldots, 20$$

Now, think about the likelihood of observing a particular serial number, x, given a particular value of n. If the number of machines sold is less than x then the probability of observing x is zero: machines that have not yet been made and sold cannot be seen. If, on the other hand, the number of machines sold is greater than or equal to x then we must assume that the probability of any particular machine being seen is the same for all machines. The likelihood distribution is therefore

$$\text{prob}(x \mid n) = 0 \qquad \text{for } n < x$$
$$\phantom{\text{prob}(x \mid n)} = 1/n \qquad \text{for } n \geq x$$

The table below shows the prior and likelihood distributions and the subsequent calculations. The revised distribution is proportional to the product of the prior and likelihoods for $x = 3$ and $x = 7$. This product is shown in the penultimate column and is then scaled to give, in the last column, the revised estimate, which is depicted in Figure 5.12.

Sec. 5.10]   Examples   137

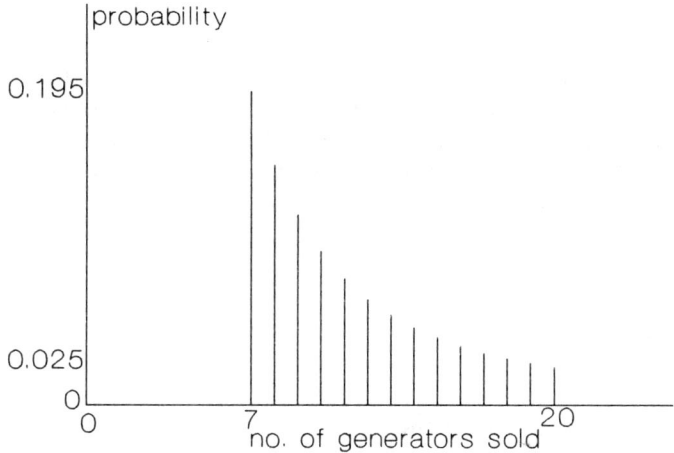

Figure 5.12

n	prior	likelihood = prob($x\|n$)		prior × likelihoods	revised
		$x = 3$	$x = 7$		
0	0.0476	0.0000	0.0000	0.0000	0.0000
1	0.0476	0.0000	0.0000	0.0000	0.0000
2	0.0476	0.0000	0.0000	0.0000	0.0000
3	0.0476	0.3333	0.0000	0.0000	0.0000
4	0.0476	0.2500	0.0000	0.0000	0.0000
5	0.0476	0.2000	0.0000	0.0000	0.0000
6	0.0476	0.1667	0.0000	0.0000	0.0000
7	0.0476	0.1429	0.1429	0.0010	0.1948
8	0.0476	0.1250	0.1250	0.0007	0.1491
9	0.0476	0.1111	0.1111	0.0006	0.1178
10	0.0476	0.1000	0.1000	0.0005	0.0954
11	0.0476	0.0909	0.0909	0.0004	0.0789
12	0.0476	0.0833	0.0833	0.0003	0.0663
13	0.0476	0.0769	0.0769	0.0003	0.0565
14	0.0476	0.0714	0.0714	0.0002	0.0487
15	0.0476	0.0667	0.0667	0.0002	0.0424
16	0.0476	0.0625	0.0625	0.0002	0.0373
17	0. 0476	0.0588	0.0588	0.0002	0.0330
18	0.0476	0.0556	0.0556	0.0001	0.0295
19	0.0476	0.0526	0.0526	0.0001	0.0264
20	0.0476	0.0500	0.0500	0.0001	0.0239
				0.0050	1.0000

Example 5.6

A general practitioner reads that the proportion of patients that are overweight in GP practices in the region is distributed with mean 0.2 and standard deviation 0.1.
 Of the six patients that have most recently been seen two were overweight. What is the best estimate of the proportion of patients that are overweight? What is the probability that more than 40% are overweight?

Solution

Since we are considering a proportion the convenience prior is a Beta distribution with, using (5.10) and (5.11),

$$a = 0.2[(0.2 \times 0.8)/0.01 - 1] - 1 = 2$$

and $b = 3/0.2 - 2 - 2 = 11$

From (5.12) the revised distribution is also Beta with

$$a = 2 + 2 = 4$$

and $b = 11 + 4 = 15$

From Appendix A5.4 the probability that more than 40% of patients are overweight is

$$1 - F_{BETA}(0.4 \,|\, 2,11) = F_{BINOMIAL}(4 \,|\, 20, 0.4) = 0.051$$

5.11 EXERCISES

5.1 Sitting in the pub pondering how best to spend your redundancy money it occurs to you that there is no builder or plumber in your village. Being fairly handy yourself you decide that you could set up as a small business in this line of work. Thinking deeply you consider that about 30% of the households in the village would make use of your services. Not being entirely confident of this figure you decide that the probability that 30% of households are potential clients is 0.5 while the probability that it is 20% is 0.3 and that it is 40% is 0.2, and you decide to restrict yourself, for the moment, to these three possible values.

Having heard of market research you ask the 20 people in the bar whether they would employ you as a builder/plumber and seven say that they would.

What now are the probabilities that you assign to the three proportions?

Sec. 5.11] **Exercises** 139

5.2 Recalculate your answer to Exercise 3.2 but this time assuming that the number of passengers carried follows a Binomial distribution. Is this a sensible assumption? By how much has this recalculation changed the entropy of the distribution of people per car? How do you account for this change?

5.3 Police in a particular town use a breathalyser to detect drivers whose alcohol level exceeds the permitted limit. The instruments are 95% accurate, by which is meant that 95% of those truly over the limit are correctly identified and, similarly, 95% of those under the limit are correctly identified. It is known that in the town 5% of drivers are over the limit. What is the probability that a driver chosen at random and shown by the breathalyser to be over the limit really is guilty? (Make a quick guess at the answer before calculating the probability and then see how close you were.)

Suppose that each person shown by the breathalyser to be over the limit is subjected to a second test by another similar instrument. What is the probability that they will then be shown to be over the limit?

(This exercise is based on one given in Hey, 1983, p42.)

5.4 Your estimate of the mean number of arrivals at a supermarket carpark, m arrivals per minute, is

  prob($m = 1.0$) = 0.4
  prob($m = 1.5$) = 0.4
  prob($m = 2.0$) = 0.2

In three 1 minute periods you observe 1, 2 and 2 arrivals. What is your revised estimate of m?

5.5 Suppose that in the previous exercise you had not restricted yourself to just three possible values of m but had characterised your prior estimate by a Gamma distribution with mean 2 and variance 1. After the three observations what do you think is the probability that m is greater than 2?

5.6 A few years ago a company made a limited run of 120 special calibration instruments. All were sold.

The sales manager is interested to know how many of these instruments are still being used and guesses about 40.

Most maintenance of the instruments is easily done by third party engineering companies but some tasks may only be carried out back at the factory at which they were made. Last year the factory undertook seven such maintenance jobs.

If the probability that any instrument is returned to the factory in a given year is 0.05, how ought the sales manager's estimate be revised and, in particular, (a) what is the probability that at least 100 machines are still in use and (b) what is the median estimate of the number of machines still in use?

(This exercise is based on an example given in Tribus, 1969, p252.)

5.7   A pharmaceutical company packs pills in boxes of 200. It is known that 0.5% of pills are defective. What proportion of boxes has no defectives?

5.8   A surveyor has available two instruments for measuring distance. The measurements given by the instruments are unbiased and have Normally distributed errors measured by coefficients of variation of 1% and 3%. Measurements of the distance between two buildings obtained by using the two instruments are 110.0 m and 112.3 m respectively. What is the probability that the true distance is greater than 111 m?

5.9   A company's switchboard has six incoming lines. The probability that a line is busy is 0.1. What is the probability that three lines are busy? What is the most likely number of busy lines and how probable is that state?

5.10 I tell you that a prize is under one of three upturned cups and ask you to guess which. When you have guessed I lift one of the other two cups and reveal that the prize is not under it. Clearly, I would never reveal the prize. I now ask you if you wish to alter your guess. What is your reply, and why?

(Thanks to Vince Kwasnica for introducing this problem to me.)

6

Decision Theory

Having seen how we can combine data and opinion and express the result as a probability distribution we turn briefly in this chapter to look at a framework for using these estimates in decision problems.

6.1 DECISIONS

We are now able to make probabilistic estimates of quantities that are of interest to us, but what will we do with such estimates, how will we use them? It seems safe to assume that making the estimate was part of making a decision: if there is a high probability that it will rain I shall carry an umbrella, if it is likely that this project will not make a sufficiently large profit then I shall not invest in it, and so on. Let us see how we can approach making such decisions.

A small general haulage company is a bit short of work. The man who owns the company has the opportunity to acquire some old, but still serviceable, coaches which could be used for local trips and for other journeys too, though he wouldn't like to use them for touring holidays. He has plenty of space in which to garage them and reckons that to keep them taxed, insured and maintained ready to use would cost about £10 per day, including an allowance for their purchase. He and some of his staff and friends would be readily available as drivers so that on the days that the coaches are used the (variable) operating cost for what he assesses to be an average mileage would be £110. For such work he thinks that he can charge £150 per day and so obtain a daily profit of £40, excluding the fixed costs.

After finding out a little about the local market for coach hire the owner of the company does not believe that demand for his coaches would ever exceed three per day and that the average daily demand will be for one coach. How many coaches should he buy? His decision will be affected by the level of demand and also by the costs and revenues of operation so we need a framework that will take account of both.

The probability that D coaches will be needed on any given day is determined by the maximum entropy distribution (3.8). You can verify that to give $\mu = 1$

$$\text{prob}(D = d) = k \exp(-0.42d), \qquad d = 0,1,2,3 \qquad (6.1)$$

and so

$$\text{prob}(D = 0) = 0.42$$
$$\text{prob}(D = 1) = 0.28$$
$$\text{prob}(D = 2) = 0.18$$
$$\text{prob}(D = 3) = 0.12$$

From the financial assessments, if C coaches are bought the daily profits are

$$\left. \begin{array}{ll} \text{profit} = 40D - 10C, & D \leq C \\ = 40C - 10C = 30C, & D > C \end{array} \right\} \quad (6.2)$$

These results are shown in Table 6.1 and constitute a complete description of the decision problem as it has been described.

		\multicolumn{4}{c}{daily demand – D coaches}			
		0	1	2	3	E[profit]
coaches	0	0	0	0	0	0.0
bought	1	–10	30	30	30	13.1
C	2	–20	20	60	60	15.2
	3	–30	10	50	90	10.0
prob(D)	=	0.42	0.28	0.18	0.12	

Table 6.1
Entries in table show daily profit – £

If one coach is bought the expected profit is

$$(-10 \times 0.42) + (30 \times 0.28) + (30 \times 0.18) + (30 \times 0.12) = 13.1$$

This and the other expected profits are shown at the right hand margin of the table and would seem to indicate that buying two coaches is the best decision.

Notice that each of the actions is optimal under some circumstances (some demand level). For example, if it was known that demand would be for two coaches then the optimal action is to buy two coaches.

Any action which is never optimal is said to be *dominated* by some other action or set of actions and so would be eliminated from the analysis. There are no such actions in this problem.

Now this is all fine, but the owner feels some reservation about this result. He had expected either to buy a single coach or not to do anything and yet the recommendation, based on his own data, is to buy two coaches.

The company isn't doing too well. In better times he would have no reservations about pressing ahead with the purchase of two vehicles but, given his rather thin reserves, some of the losses in Table 6.1 are quite large. He feels in need of some further analysis which explicitly recognises that these losses, if realised, might well result in severe difficulties for his firm.

We can conceptualise these worries in the following way. The value or worth of an extra £50 of profit will depend on whether, for example, it increases profit from £300 to £350 or decreases loss from £250 to £200: the latter has more worth to the owner than the former. This worth, which is a non-linear function of profit, is called *utility*.

6.2 UTILITY

You may be used to the idea of utility being synonymous with monetary value or some general idea of usefulness. We need here a richer concept, a utility measure constructed to ensure the correctness or *coherence* of our decision making by reflecting our attitude to *risk*. The approach was proposed by von Neumann and Morgenstern (1947) and is best explained in terms of a hypothetical lottery, A, and a corresponding certain return, B.

I offer you the following choice:

A: I will toss a coin. If it shows heads I will give you £90 and if it shows tails you will give me £30.

B: I give you £6.

Which will you choose? If you urgently need £6 for an important purchase and have no other way of getting it, or if you do not have £30 to give me, you are likely to choose B, otherwise you may well choose A and have a flutter. Which option did you choose?

Whichever you preferred it seems that in principle I can find an amount to offer you in B such that you are indifferent between A and B. I am fairly confident about this since if I were to offer you £100 then you would certainly prefer B because you can get no more with A, whatever happens. On the other hand if I were to demand £50 from you under B then you would certainly prefer A, since you can lose no more than £30 by so doing. I assume that at some point in between you are indifferent between A and B. Let us suppose that you are indifferent when the sure offer, B, is £10.

In this uncertain, risky, situation you are clearly not responding just to money values since the expected return on the lottery A is £30 [= $(0.5 \times -30) + (0.5 \times 90)$]. You are willing to pay £20 (the difference between the expected return and your indifference value) to avoid the uncertainty of the lottery; more precisely to avoid the risk of losing £30. This difference is called a *risk premium*, and those willing to pay such a premium are said to be *risk averse*.

We can represent what is going on here by using a quantity related to money values, but not equal to them. Let us call this *utility* and write the utility of amount £x as $u(x)$. We now define utility to be a function such that at your indifference point of £10 we can write

$$u(10) = 0.5u(-30) + 0.5u(90)$$

The units that we choose for utility are arbitrary so let us set $u(-30) = 0$ and $u(90) = 1$ which gives $u(10) = 0.5$. I could now proceed by finding the sure value at which you were indifferent between receiving either £10 or £90 on the toss of a coin. The utility of that sure amount is then $0.5u(10) + 0.5u(90) = 0.75$. Continuing in this way I might end up with the graph in Figure 6.1.

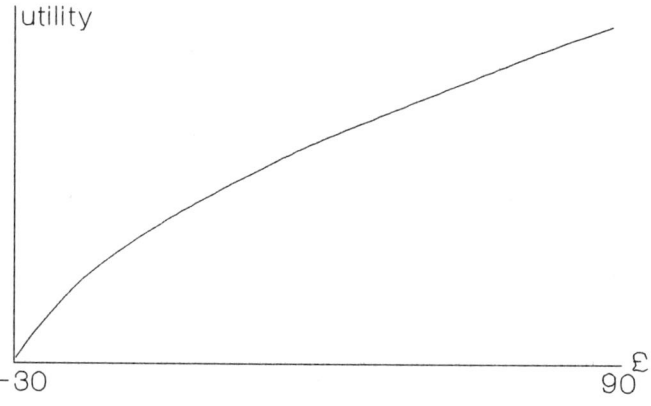

Figure 6.1 Utility curve

In general, then, the two options are:

A: receive H with probability p or receive L with probability $(1 - p)$

B: receive x for sure

Initial values of L and H are chosen so that all payoffs for a particular problem lie between them (i.e. $L \leq x \leq H$). It is usually convenient to set L equal to the minimum outcome and H equal to the maximum, as was done here.

When we are indifferent between A and B

$$u(x) = pu(H) + (1 - p)u(L) \qquad (6.3)$$

In the example above we chose to set $p = 0.5$ and then seek an appropriate value of x. This has the considerable merit that most people have a pretty good intuitive feel for equally likely outcomes. You can see, however, that we could fix upon a value of x and then vary the probability p. This is often suggested but runs the danger that the perception of the whole range of probability values by most (untrained) assessors is likely to be poor.

We are only going to use utility values to help us choose between alternative actions and so the units that we use are arbitrary. Here is an analogous situation. If I know from a map the height of two hills then I can say which is the higher. This does not depend on whether their height is measured in feet or metres nor on the datum chosen by the map makers from which to measure such heights. In the UK heights are given in metres and the datum is the mean sea level at Newlyn in Cornwall. If I wish to change the units of measurement from, say, metres to feet I multiply by 3.28. If, in addition, I wish to measure from a datum 200 feet higher than Newlyn then I subtract 200. To convert from a standard height in metres, M, to an adjusted height in feet, F, I use the formula

$$F = 3.28M - 200$$

Decisions such as picking the highest of a group of hills are not affected by this linear conversion.

Utility is like height in this sense; comparative judgements are indifferent to a linear transformation. What this means is that I can arbitrarily choose two values on the utility scale. I shall choose, as I did above, $u(H) = 1$ and $u(L) = 0$ and so from (6.3) we see that the utility $u(x)$ is just the probability of obtaining the maximum payoff, H.

Rather than constructing the utility curve by repeated interrogations it will often be easier to fit an appropriate algebraic form, once the general shape is agreed.

A simple shape is given by noting that the ratio $(x - L)/(H - L)$ lies between 0 and 1 so that raising this to some power, a, induces the sort of behaviour in which we are interested (Figure 6.2). We therefore use the utility function

$$u(x) = [(x - L)/(H - L)]^a \qquad (6.4)$$

with $a < 1$ describing risk averse behaviour
      $= 1$ describing risk neutral behaviour
      > 1 describing risk prone behaviour

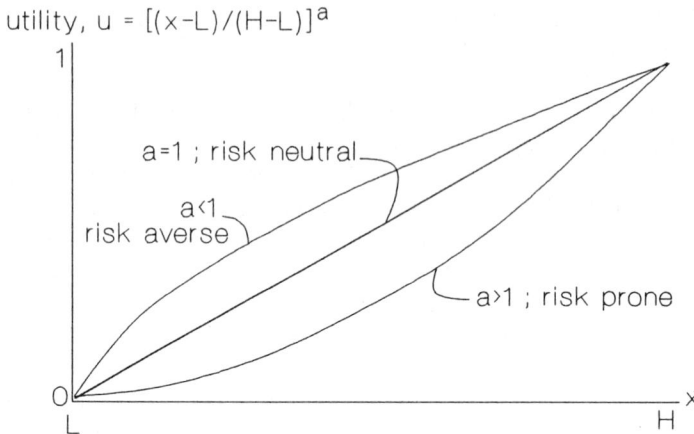

Figure 6.2 Simple utility function

(This is by no means the only possible formula. See the Further Reading section for sources of alternatives.)

This convenient and simple taxonomy of attitudes to risk may miss some of the subtleness or perversity of human behaviour. For instance, empirical investigations by Swalm (1966) seem to indicate a function of the form shown in Figure 6.3.

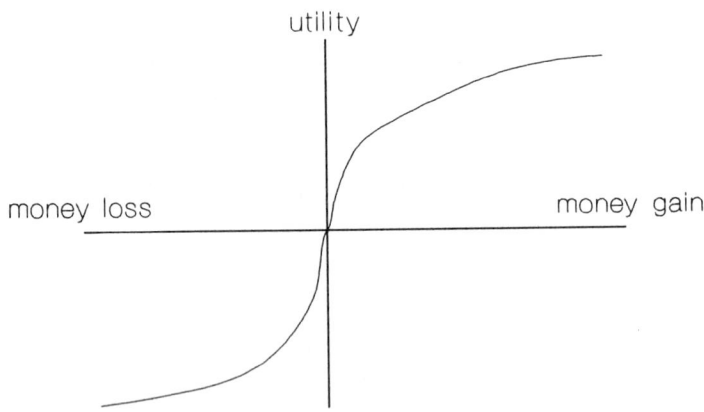

Figure 6.3 A utility curve sometimes found in practice
(Source: Swalm, 1966)

To find a we know that the utility of the indifference value $x = 10$ is 0.5 and so

$$0.5 = u(10) = [(25 + 30)/(90 + 30)]^a = (0.33)^a$$

Therefore

$$a = \ln(0.5)/\ln(0.33) = 0.63$$

and

$$u(x) = [(x - L)/(H - L)]^{0.63} \qquad (6.5)$$

Sample values are shown in Table 6.2. Substituting utilities for daily profits gives the description of the problem shown as Table 6.3.

profit – £x	$(x - L)/(H - L)$	$u(x) = [(x - L)/(H - L)]^{0.63}$
−30	0.00	0.00
−20	0.08	0.21
−10	0.17	0.32
0	0.25	0.42
10	0.33	0.50
20	0.42	0.58
30	0.50	0.65
40	0.58	0.71
50	0.67	0.77
60	0.75	0.83
70	0.83	0.89
80	0.92	0.95
90	1.00	1.00

Table 6.2
Values of the utility function

		daily demand – D coaches				E[utility]
		0	1	2	3	
coaches	0	0.42	0.42	0.42	0.42	0.42
bought	1	0.32	0.65	0.65	0.65	0.51
C	2	0.21	0.58	0.83	0.83	0.50
	3	0.00	0.50	0.77	1.00	0.40
prob(D)	=	0.42	0.28	0.18	0.12	

Table 6.3
Entries in table show utility

6.3 A DECISION RULE

Using utilities instead of just profit means that attitudes to risk become part of the problem formulation. To see the correct decision rule remember that a utility is just the probability of gaining the maximum profit, H.

Now think of the situation when we have bought two coaches and the demand on a particular day is for one coach. Using subscripts to indicate the number of coaches call the outcome (C_2, D_1). The utility of this outcome is $u_{21} = 0.58$. This means that the outcome (C_2, D_1) is equivalent to a lottery with probability 0.58 of obtaining H.

$$u_{21} = \text{prob}(H \mid C_2, D_1) = 0.58$$

But the probability of D_1 is 0.28 and so the probability of H and D_1 if C_2 is chosen is

$$\begin{aligned}\text{prob}(H.D_1 \mid C_2) &= \text{prob}(H \mid C_2, D_1).\text{prob}(D_1) \\ &= u_{21}.\text{prob}(D_1) \\ &= 0.58 \times 0.28 = 0.16\end{aligned}$$

Taking account of all possible demand levels we can find the probability of obtaining H if action C_2 is taken as

$$\text{prob}(H \mid C_2) = \text{prob}(H.D_0 \mid C_2) \text{ OR } \text{prob}(H.D_1 \mid C_2) \text{ OR } \text{prob}(H.D_2 \mid C_2) \text{ OR } \ldots$$

$$= \Sigma \, \text{prob}(H.D_i \mid C_2)$$

But this is just the expected utility of C_2, $EU[C_2]$:

$$\begin{aligned}EU[C_2] &= \text{prob}(H \mid C_2) = \Sigma \, \text{prob}(H.D_i \mid C_2) \\ &= \Sigma \, u_{2i} \text{prob}(D_i) \\ &= (0.21 \times 0.48) + (0.58 \times 0.28) + (0.83 \times 0.18) \\ &\quad + (0.83 \times 0.12) \\ &= 0.50\end{aligned} \qquad (6.6)$$

All the expected utilities are shown at the right-hand margin of Table 6.3.

Remembering that the expected utility of an action is just the probability of obtaining H if that action is chosen the decision rule is obvious. Since the only rational plan is to maximise the probability of obtaining H, choose the action with the highest expected utility.

This leads to the choice (just) of action C_1. Notice that because of the risk aversion of the owner of the company the optimal number of coaches has been reduced from two (when maximising expected profit) to one.

The expected utility of this optimal action is 0.51. To find the money equivalent use (6.5):

$$0.51 = [(x + 30)/120]^{0.63}$$

and so

$$x = 120(0.51)^{1/0.63} - 30 = 11.21$$

The expected profit from the optimal action is £11.21/day.

You will see now the close connection between assessed probabilities and utility and why some writers recommend that both should be developed together rather than independently. The further reading recommended in Chapter 3 contains some relevant references.

6.4 LOSS

Rather than using utilities to consider the direct effects of outcomes there is another way of thinking about the results of our actions.

Imagine how the owner of the coaches would feel if demand on a given day is for one coach. If he had bought just one coach he would be happy using all his resources profitably. If, on the other hand, he had made any other decision, bought any other number of coaches, he would, on this occasion, have acted suboptimally. For example, if two coaches had been bought the utility obtained would be 0.58 rather than 0.65. The difference of 0.07 is called the *loss* associated with the outcome.

A table of losses is easily obtained from the utilities in Table 6.3 by taking one column at a time and subtracting each utility from the maximum utility in that column. The results are shown in Table 6.4. Notice that the loss for an optimal action is zero.

		\multicolumn{4}{c}{daily demand – D coaches}			
		0	1	2	3	E[loss]
coaches	0	**0.00**	0.23	0.41	0.58	0.21
bought	1	0.10	**0.00**	0.18	0.35	0.12
C	2	0.21	0.07	**0.00**	0.17	0.13
	3	0.42	0.15	0.06	**0.00**	0.023
prob(D)	=	0.42	0.28	0.18	0.12	

Table 6.4
Losses

It is probably intuitively obvious that the correct decision to make is to choose the action with the smallest expected loss. This is correct, and here is the reason.

Let the maximum utility obtainable under any given system state, j, be M_j, which is just the largest utility in column j. The loss for action i and system state j is then $M_j - u_{ij}$ and so

$$\text{Expected loss for action } i = \sum_j p_j(M_j - u_{ij})$$

$$= \sum_j p_j M_j - \sum_j p_j u_{ij} \qquad (6.7)$$

$$= K - \sum_j p_j u_{ij}$$

since $\sum_j p_j M_j$ is just a constant (= 0.63 in this case)

So, in general,

$$\text{expected loss} = \text{constant} - \text{expected utility} \qquad (6.8)$$

and minimising expected loss is equivalent to maximising expected utility.

This idea of loss, a sort of opportunity loss rather than an absolute amount, is a convenient way of thinking about the consequences of one action compared to others. It is often this relative worth that we think about when making decisions.

6.5 FURTHER READING

The above is an elementary description of the most basic kind of decision analysis. The extensions and applications of these ideas have led to a wide literature. Here are some starting points.

Books on decision theory have already been given in the Further Reading section of Chapter 5. Other useful texts are those by Clemen (1986), Goodwin and Wright (1991), Jones (1977), LaValle (1978), Moore (1972), Moore and Thomas (1976) and Watson and Buede (1987). French (1986) gives a full mathematical treatment and is generally a good text to read. Some important readings are collected by Edwards and Tversky (1967) and Kaufman and Thomas (1977).

Practical cases showing the application of decision theory are given in French (1989) and Moore et al. (1976). An interesting application is given by Rescher (1985) who casts Pascal's decision about whether or not to believe in God as a problem in decision theory.

As with Bayes' Rule the criterion of maximising expected utility is prescriptive, not descriptive. It is a recommendation for rational action, not necessarily a description of observed behaviour. Some, notably economists, have taken the latter view. Schoemaker (1980) discusses the resultant difficulties and some alternatives. Fishburn (1981) provides a review. A full mathematical treatment is given by Fishburn (1970, 1988). Hull et al. (1973) discuss problems of measurement.

Fairly catholic collections on various aspects of decision making are given by Arkes and Hammond (1986), Humphreys et al. (1983), Scholz (1983) and Wendt and Vlek (1975).

Complex problem frameworks involving multiple outcomes, conflicting objectives and other complications are discussed by Bell *et al.* (1977), Easton (1973), Keeney (1972), Zeleny (1982) and Zionts (1978). Moore (1983) discusses risk in a business context.

Risk analysis (Hertz and Thomas, 1983, 1984) uses some of the same basic ideas. The adjacent area of risk assessment concentrates on the likelihood, severity and perception of physical risks often associated with public health hazards, plant failures and the like. Some of the many texts are those of Covello *et al.* (1983), Fischhoff *et al.* (1981), Schrader-Frechette (1985) and Waller and Covello (1984).

Decision theory has been used to examine the impact of uncertainty in finance and economics. You can get a flavour of the scope in Crum *et al.* (1981), Dreze (1974) and Strong and Walker (1987).

6.6 EXAMPLES

Example 6.1

Work on an urban construction site involves some excavation. The ground conditions, and in particular the degree of obstruction offered by earlier construction, underground pipes and the like, are unknown.

Four different types of excavator are available for hire, each performing differently with different ground conditions. These differences are reflected in the following table which shows hire plus operating costs ($£10^3$) for each combination of excavator and ground.

		Ground Condition		
		clear	slightly obstructed	heavily obstructed
	A	55	120	190
Excavator	B	90	100	180
	C	100	140	150
	D	150	160	200

The site engineer estimates the probability that the ground is clear of obstruction to be 0.4. Which excavator should be used?

(This example was suggested by a problem in Coyle, 1972, p9.)

Solution

In looking at the table of costs notice that whatever the ground condition excavator D is never the cheapest alternative and so may be ignored. This option is *dominated* by the others and so will not be considered further.

We have no estimates of the probability that the ground is obstructed and so, by the principle of maximum entropy, assume that the two obstructed conditions are equally likely: prob(slightly obstructed) = prob(heavily obstructed) = (1 − 0.4)/2 = 0.3.

The expected cost of using excavator A is

$$(55 \times 0.4) + (120 \times 0.3) + (190 \times 0.3) = 115$$

(i.e. £115000). Similarly, the expected costs of B and C are 120 and 127 respectively so use A because this minimises the expected cost of excavation.

Example 6.2

In the previous example suppose that it would cost £20000 to determine the state of the ground. Would it be a price worth paying?

Solution

Why buy information? To avoid the consequences of uncertainty.

We have seen that minimising expected loss is an equivalent criterion to maximising expected utility and also that loss measures the consequences of taking actions that, in retrospect, were not optimal. Here is the previous example expressed in terms of expected loss.

		Ground Condition			
		clear	slightly obstructed	heavily obstructed	expected loss
	A	0	20	40	18
Excavator	B	35	0	30	23
	C	45	40	0	30
probability:		0.4	0.3	0.3	

Although A is the optimum action we still expect to spend £18000 more than would have been necessary if the ground conditions had been known at the time the excavator was selected. Since the knowledge that would reduce expected loss to zero would cost £2000 more than that optimal expected loss it is not worth buying.

The optimal expected loss, £18000, is sometimes called the *expected value of perfect information* and is a function both of the consequences of an action and of the prevailing probability assessment.

Note that the optimal expected loss could alternatively have been calculated using (6.7) as

$$(55 \times 0.4) + (100 \times 0.3) + (150 \times 0.3) - 115 = -118$$

Example 6.3

A piece of machinery has to be replaced. It may be replaced now, at a cost of £100000, or else it may be replaced next year. Deferring the purchase for 1 year effectively reduces the current value of that expenditure to £90000. If, however, the existing machine suffers a serious failure this year the cost of the subsequent disruption and remedial action is £100000 in addition to the cost of a replacement. The probability of such a failure is thought to be 5%. The warranty period of a new machine means that no costs would be incurred were the machine to fail during the period under consideration.

In thinking about the decision the production manager believes him- or herself to be indifferent between (a) incurring a certain expenditure of £175000 and (b) incurring expenditures of either £90000 or £200000 with equal probability. Should the machine be replaced now?

Solution

In considering risk aversion it may be more convenient to think of costs as negative profit. I find this clearer but you may prefer to work in costs directly.

The problem may be expressed in thousands of pounds and represented in the usual way as:

	no failure	failure	expected profit
replace now	−100	−100	−100
wait	−90	−200	−95.5
probability:	0.95	0.05	

The optimal decision is to wait.

Risk aversion may be modelled by the utility function (6.4) with $x = -175$, $u(x) = 0.5$, $L = -200$ and $H = -90$. So,

$$0.5 = [(-175 + 200)/(-90 + 200)]^a$$
$$= (25/110)^a$$

and $\quad a = \ln(0.5)/\ln(25/110) = 0.468$

By definition $u(-200) = 0$ and $u(-90) = 1$. In addition

$$u(-100) = [(-100 + 200)/110]^{0.468} = 0.956$$

Recasting the problem in terms of utilities gives

	no failure	failure	expected utility
replace now	0.956	0.956	0.956
wait	1.000	0.000	0.950
probability:	0.95	0.05	

Incorporating the aversion to risk has reversed the decision: it is better to replace the machine now.

6.7 EXERCISES

6.1 A factory is to be built to make bread. The factory may be either small or large. The amount made and sold, and so the profit, will be determined by the level of demand for the bread according to the following table:

		Demand low	medium	high
Factory size	small	0.50	0.70	0.70
	large	0.00	0.80	1.02

Operating Profit – £m

It is estimated that the probability that demand will be low is 0.2 and that it will be high is 0.3. Which factory should be built, and why?

6.2 How would your answer to the previous exercise have changed if it had been determined that the managing director, who will take the decision, had a utility for money proportional to the square root of profit?

6.3 Concrete for a particular job is specified to have a compressive strength of 3500 psi. A sample is tested and gives an indicated strength of 3590 psi. The various sources of variation in the production and testing procedures mean that the estimated true strength of the concrete is Normally distributed with the test value as the mean and a standard deviation of 90 psi.

If the concrete is rejected at this stage but further tests reveal that it did meet the specification then unnecessary costs of £1200 will have been incurred. If, on the other hand, the concrete is accepted but subsequently proves not to have been strong enough remedial works costing £82000 will have to be undertaken. There are no costs associated with either rejecting a truly weak product or accepting concrete that meets the specification.

Should this concrete be accepted?

Below what indicated test strength ought concrete to be rejected?

6.4   Recalculate the solution to the problem of Table 6.1, but this time with no limit on demand and modelling demand by a Poisson distribution. Does the recommendation change?

6.5   A specialist car restorer requires a large number of components to be on hand in store if delays are not to be incurred. Components are ordered every 3 months, the rate at which they are subsequently used being determined by the nature of the work undertaken. The demand for one such component, a particular type of shock absorber, is known to follow a Poisson distribution with a mean of 5.8 per 3 month period.

If components are not used during the period the extra storage and other costs that are incurred may be modelled by £$10x$, where x is the number left over at the end of the period.

If the restorer runs out of this component then the work is delayed, the workshop space is occupied for longer than necessary, and the customer is inconvenienced. The sum of these costs is modelled by £$10y^4$, where y is the number of shock absorbers required in excess of those in store at the start of the 3 month period.

How many shock absorbers ought to be in store at the start of each period?

6.6   Refer back to Exercise 6.1. How much would you pay for a perfect forecast of the level of demand?

6.7   Commenting on the arrangements for betting on horse racing and football results de Finetti (1974, p194) writes:

> "...those who write down ridiculous forecasts, that by chance turn out to be correct, receive fantastic prizes; whereas those who write down forecasts which could reasonably be thought probable receive, if they win, only very small amounts, since the prize, in this case, will presumably have to be shared with many others."

Given £50 to bet on a horse race how would you choose to place the bet, and why? Would your strategy change if you had £5000?

When you have reached a conclusion you might care to look at Henery (1985).

6.8 A small company makes, among other things, 230000 kitchen knife sharpeners a year and sells them for £1.05 each. It is considering increasing revenue by increasing its price to £1.10 or £1.15.

A new recruit has recently joined the marketing department, brandishing her newly won MBA. She suggests that a price cut to, say, 95p may be more beneficial. In a supporting paper she explains that economists commonly use a model linking the price charged, P, to the amount sold, Q, that is of the form

$$Q = aP^b$$

where b, which is negative, is called the *elasticity* of demand. If the price is changed from P_0 to P_1 then the volume sold will change from Q_0 to

$$Q_1 = Q_0(P_1/P_0)^b$$

Since the elasticity value is not known with any precision she suggests the following estimate:

prob($b = -0.8$) = 0.2
prob($b = -1.0$) = 0.5
prob($b = -1.2$) = 0.3

She also draws attention to the possibility that, in view of its financial position, the company may prefer to avoid taking too much of a risk in making this decision. Her feelings are confirmed in a discussion with Peter, the finance director, about this price change, during which he expresses himself indifferent between, on the one hand, a certain revenue of £250000 and, on the other hand, the probability of obtaining £220000 with probability 0.25 or £270000 with probability 0.75.

What price should be set and what is the money value to the finance director of this decision?

6.9 This is an exercise in which you should construct your own utility curve. You have to travel to your local airport to catch a flight that leaves at 10.05 a.m. You do not enjoy spending time waiting in airports. At what time will you leave home and why?

6.10 Finally, here is a puzzle.

On the table in front of you are two boxes, one of glass and the other of wood. You can see that there is £10 in the glass box and you know that in the wooden box there is either £100 or nothing. You must choose between either taking the wooden box only or taking both boxes.

The contents of the wooden box have been determined by your friend Charles whom you believe to be uncannily good at predicting your behaviour in circumstances such as this. He tells you this much about what he has done: if he predicted that you would take the wooden box only then he put the £100 in it while if he predicted that you would take both boxes he left the wooden box empty.

Determine which action to take using the principle of:

(a) dominance
(b) maximisation of expected gain.

This problem is known as *Newcomb's paradox* and is discussed by Eells (1982), Nozick (1969) and Sorensen (1985).

7

Using Probability Estimates

The previous chapter showed how probabilistic estimates could be used to make decisions. We continue by considering some particular situations where the decision to be made is more that of deciding how to report a finding for the benefit of third parties, possibly unknown to you. The task here is to formulate a report that captures the essential uncertainties of your investigation while being intelligible to people whose knowledge of statistics may be slight.

7.1 AVERAGES

In Chapter 1 the three basic interpretations of the idea of the average were introduced: the mode, mean and median. We can now see that they are the results of particular decision problems.

Suppose that we have a probability distribution describing our estimate of the value of some real variable, X, and that we must characterise this estimate by a single value, A. If the value that we choose is wrong then we incur some loss either because of our annoyance at being incorrect or because of the consequences of actions taken based on our incorrect estimate. The resulting loss function may be linear and symmetrical such as that shown in Figure 7.1. Choosing A to minimise the expected loss, the optimum value of A is the median of the probability distribution.

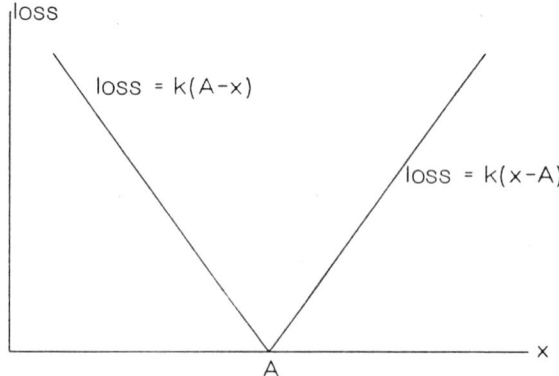

Figure 7.1   Linear loss giving an average A = median

Sec. 7.2] **Intervals** 159

Similarly, if the loss function is quadratic (Figure 7.2) then we should choose the mean of X for the value of A, while if the loss function is as shown in Figure 7.3 the optimum value of A is the mode of the distribution. The calculations are shown in Appendix A6.

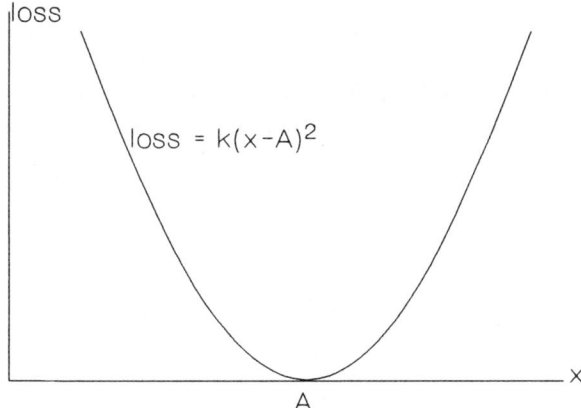

Figure 7.2   Quadratic loss giving an average A = mean

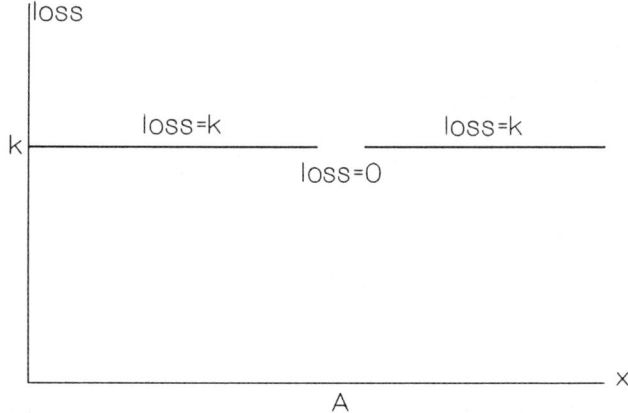

Figure 7.3   Constant loss with a gap giving an average A = mode

7.2 INTERVALS

Rather than summarising a probabilistic estimate with just a single value, as we have just done, we would generally prefer to have a summary that in some sense conveys the uncertainty inherent in our estimate. Quoting the standard deviation or other measure of spread would achieve this to some extent but it seems more natural, and more intelligible to third parties, to quote a range. We could say, for instance, that we believe that it will take between 50 and 70 minutes to make a particular journey. Such a simple statement conveys a number of things: that the journey takes about an hour, that it would be sensible to allow 70 minutes for travelling if the time of arrival is important, and so on.

How may we arrive at such a range given a probability distribution? We could say that the journey either will be instantaneous or will last for ever. While the probability that this statement is correct is 1.0 it is of no practical use. We could, on the other hand, say that the journey will take exactly 65 minutes. The apparent precision of this statement makes it seem very useful, but the probability that the journey will take *exactly* any particular time is zero because the area under the density function is zero. Between these two extremes we can find a range such that the probability that the statement "x lies between L and H" is true is c. The situation is shown in Figure 7.4.

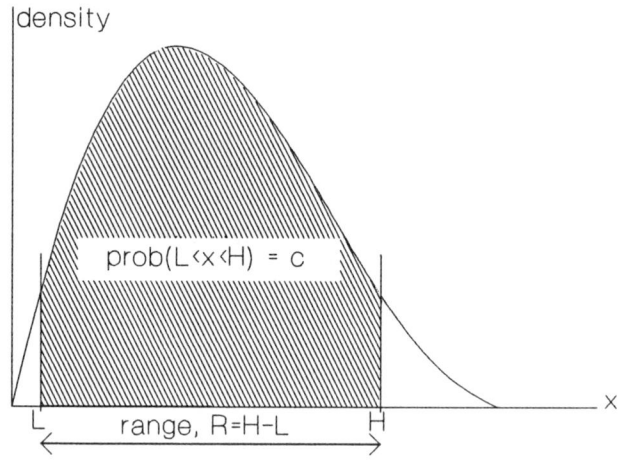

Figure 7.4 A reporting interval

The procedure now is straightforward. We first choose the acceptable probability c and then choose the appropriate range. Let us think about these two steps.

We have to choose c, the probability that the statement is true, to be some value acceptable to us for the case in hand. This implies some consideration of the consequences of actions which are informed by our statement. We have seen in the previous chapter how to use expected utility for reaching decisions when the full decision problem is known, but when choosing reporting intervals we can usually go no further than to arrive at some appropriate solution by introspection. This is clearly less satisfactory but is inevitable if the decisions which the report might inform are not known to us. It may often be more useful to focus on $1 - c$, the probability that what we say is wrong. Saying that there is a 90% chance of sun may convey to you that it is pretty certain to be sunny. Saying that there is a 1 in 10 chance that it will be dull or wet may give a quite different impression, yet they are equivalent statements.

For any value of c there are many ranges that could be chosen (many areas under the graph of the density function that have an area of c) so we must have some criterion to pick just one. It is reasonable to assume that we wish to be as precise as possible, given c, and so we will choose that interval having the smallest range $H - L$ (Figure 7.5a). Such intervals are called *credible intervals*. For a credible interval the value of the density function, $f(x)$, at the two end points is equal: $f(L) = f(H)$. If this is not intuitively clear Appendix A7 gives the proof.

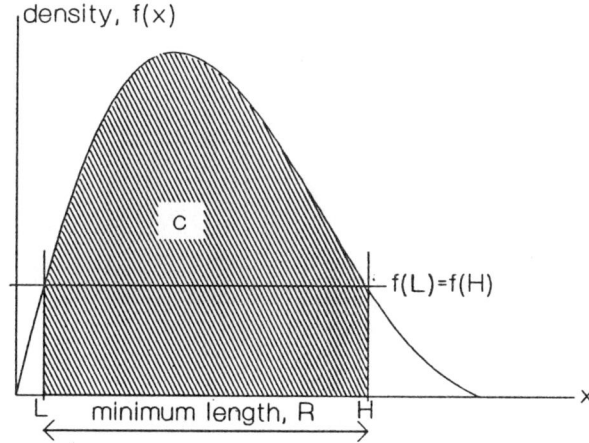

Figure 7.5(a)   Credible interval: minimum interval length

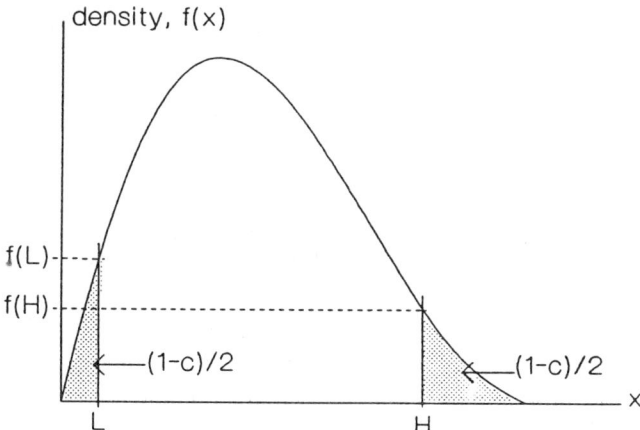

Figure 7.5(b)   Confidence interval: equal tail areas

Some people prefer to choose a range such that if the true value falls outside the range (i.e. we are wrong) it is as likely to be below the low end of the range as above the high end. Such intervals are called *confidence intervals* (Figure 7.5b). If the density function is symmetric, the Normal distribution for example, then the two intervals are coincident.

Suppose we have an estimate of a proportion that is described by a Beta distribution with parameters $a = 4$ and $b = 2$ (Figure 7.6) and that we wish to find a 70% credible interval (i.e. $c = 0.7$). One way to proceed is to pick a value for L and find the corresponding value for H. Repeating this will enable us to locate a minimum value for $H - L$ and so the credible interval.

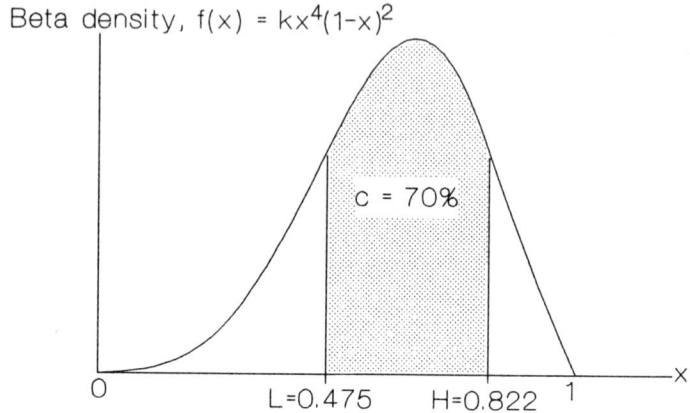

Figure 7.6 70% credible interval for a Beta distribution

First use Appendix A5.4 to find the value of the distribution function for $L = 0.4$.

$$F_{BETA}(0.4 \mid a = 4, b = 2) = 1 - F_{BINOMIAL}(2 \mid n = 7, p = 0.4)$$

$$= 1 - 0.90374 = 0.09626$$

For a 70% interval we now require a higher limit, H, such that $F(H) = 0.79626$.

$$F_{BETA}(H \mid a = 4, b = 2) = 1 - [1 - F_{BINOMIAL}(2 \mid n = 7, p = 1 - H)]$$

$$= F_{BINOMIAL}(2 \mid n = 7, p = 1 - H)$$

Scanning the appropriate line in the tables of the Binomial distribution we see that the required value for $1 - H$ lies between 0.2 ($F(0.2) = 0.85197$) and 0.25 ($F(0.25) = 0.75641$). Interpolating gives

$$1 - H = 0.25 - 0.05 \frac{(0.79626 - 0.75641)}{(0.85197 - 0.75641)} = 0.22915$$

and so $H = 0.77085$. The interval is written as $x = (0.40, 0.77)$. This and some other results are shown in Table 7.1.

L	H	$H - L$
0.35	0.75	0.40
0.40	0.77	0.37
0.45	0.80	0.35
0.50	0.85	0.35
0.54	0.96	0.42

Table 7.1
Some 70% intervals

The minimum value of $H - L$ falls between $L = 0.45$ and $L = 0.50$ so by interpolation we find the required 70% credible interval to be approximately $x = (0.475, 0.875)$. Alternatively, pursuing the calculations further and using a finer interval for the search gives $x = (0.475, 0.822)$.

Check for yourself that the 70% confidence interval is $x = (0.45, 0.80)$. This is, of course, easier to find than the credible interval and you may therefore wish to use it as a good enough approximation unless the distribution is badly skewed. In any case the confidence interval would be a good start when finding a credible interval.

The only difficulty that you are likely to face in constructing a credible interval is in determining the credibility level, c. It has been argued that when one is simply reporting the results of an experiment for general dissemination it is more important that everyone sticks to the same value of c, thereby ensuring a consistency in communication, than worrying too much what particular value of c is chosen. In these circumstances $c = 95\%$ is usually used. Nonetheless the choice of levels of credibility is not free of dispute either in principle or when it comes to a particular level such as the ubiquitous 95% (see Morrison and Henkel, 1970, especially Chapters 17 and 19). It is important that you choose a credibility level with which you feel happy, although you will see that this is easier said than done.

7.3 ESTIMATING THE POPULATION MEAN

Opinion pollsters try to estimate voting intentions of the United Kingdom electorate based upon interviews with a thousand or so individuals. Market researchers attempt to find how we will all react to a new brand of breakfast cereal by asking people in the street a few questions about it. Production engineers monitor the quality of chocolates by examining some now and again.

These are all examples of a fairly common decision problem. There is some group of individuals or objects in which we are interested and this interest takes the form of wanting to know the value of some characteristic of the group. We do not have access to the whole group but only to a proportion of its members. How do we decide what to say about the value in which we are interested? What we say is the *inference* that we make about the population based upon the sample.

The group in which we are interested is called a *population* and those members to which we have access constitute a *sample*. In what follows it is assumed that the sample is only a very small proportion of the population. We shall also assume that the members of the sample have been selected by simply choosing them at random.

Let us take as an example the problem of estimating the mean of some population. Market researchers may wish to estimate the mean consumption per capita of breakfast cereal, the production engineer may want to know the mean weight of the boxes of chocolates produced in a shift, and so on.

The basic situation is shown in Figure 7.7. Each member of the population has a measurable attribute. From all the possible samples of a given size, *n* objects, one such is chosen. Had a different *n* objects been chosen then the sample characteristics would also have been different and so would the inference based upon them. This variation due to choosing one of a number of possible samples is analogous to the noise in a communication system. The analytical task is to decode the message contained in the sample and communicate it, via an appropriate inference, to a final consumer.

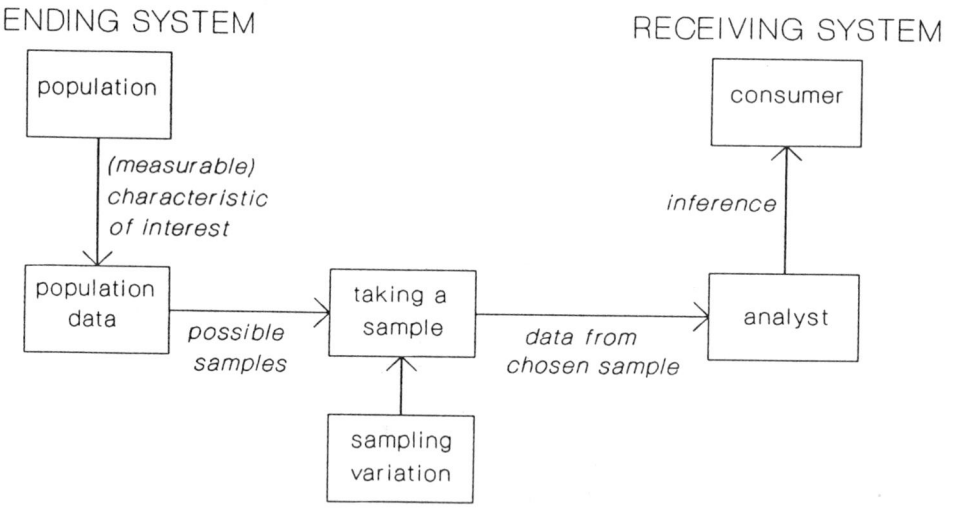

Figure 7.7 Sampling and inference as a communication system (compare with Figure 2.5)

Figure 7.8 shows a simpler representation and the commonly used notation. Note that Greek letters are used for population characteristics to distinguish them from sample characteristics. We know neither μ nor σ: that is why we are taking a sample.

Sec. 7.3] Estimating the population mean

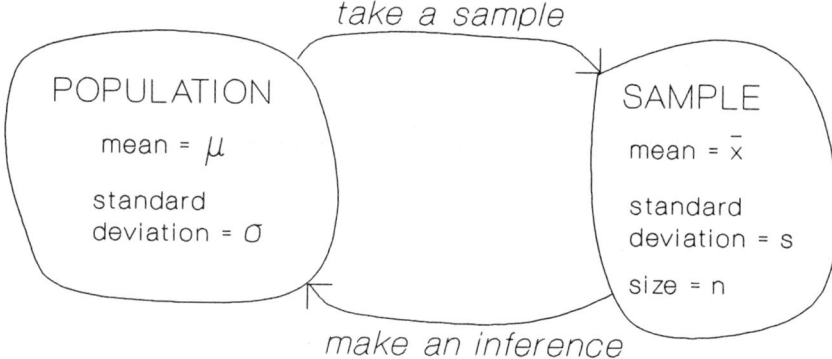

Figure 7.8 Sampling and inference

The most common problem is to estimate the population mean, μ. To do this we use Bayes' Rule.

The likelihood function is the probability distribution that describes how likely it is to observe a particular sample mean, $\bar{x}$, given a population mean, μ. You would expect that sample means tend to cluster around the population mean. This is just what happens and so the mean of the likelihood distribution is μ, but what is its variance and form?

The answer is given by the *Central Limit Theorem*. We draw upon two standard results. The first is that the sum of a number of independent random variables has a distribution that tends to be Normal as the number of variables in the sum increases, whatever the shape of the underlying distributions.

The second is that if

$$y = a_1 x_1 + a_2 x_2 + a_3 x_3 + \dots$$

where the x's are independent random variables and the a's are weights then the variance of the weighted sum is

$$\text{var}(y) = a_1^2 \text{var}(x_1) + a_2^2 \text{var}(x_2) + a_3^2 \text{var}(x_3) + \dots \qquad (7.1)$$

So for the distribution of the mean of a sample of size n comprising the observations $x_1, x_2, \dots, x_n$ we have

$$\bar{x} = \Sigma x_i / n$$

Each observation is drawn from the same population which has variance σ^2 so, using (7.1),

$$\text{var}(\bar{x}) = \Sigma \text{var}(x_i)|n^2 = \Sigma\sigma^2|n^2 = n\sigma^2|n^2$$

$$= \sigma^2|n \qquad (7.2)$$

The likelihood distribution, $f(x|\mu)$, is Normally distributed with mean μ and variance σ^2/n (Figure 7.9).

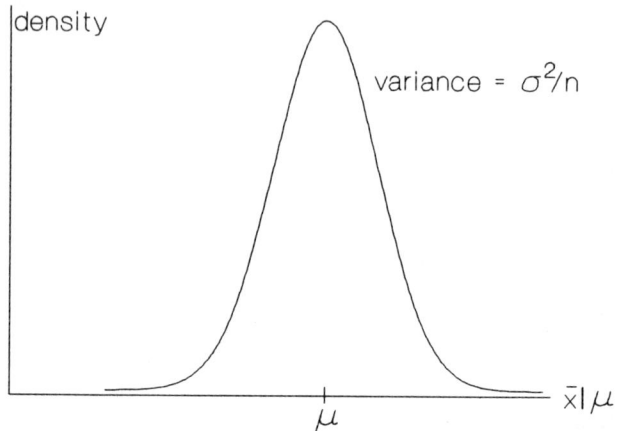

Figure 7.9 Likelihood distribution for sample means

Because of the frequency with which this result is used the standard distribution of this particular Normal distribution is often called the *standard error*:

$$\text{standard error} = \sigma/n^{0.5} \qquad (7.3)$$

The trouble is, of course, that we don't know σ^2. It might seem at first that we can use s^2, the variance of the sample, as a best guess at σ^2, the population variance. Unfortunately this is not the case since the sample variance is based upon deviations from the sample mean and not the population mean. This introduces a bias which can be easily corrected. The best estimate of the population variance from sample data is

$$\sigma^2_{EST} = [\Sigma(x-\bar{x})^2]|(n-1)$$
$$= s^2 [n|(n-1)] \qquad (7.4)$$

and we shall use this wherever we refer to the population variance.

The true population variance is not known and never will be in just about any real case. It can be shown that as a result it is inappropriate to use the Normal distribution of Figure 7.9 and that the correct likelihood distribution is the *t*

Sec. 7.3]   **Estimating the population mean**   167

distribution, which is sensitive to sample size as shown in Figure 7.10. As the sample size increases the *t* distribution more closely resembles the Normal, until with an infinite sample the two distributions are coincident. For samples of more than about 30 the Normal distribution may be used without introducing appreciable error.

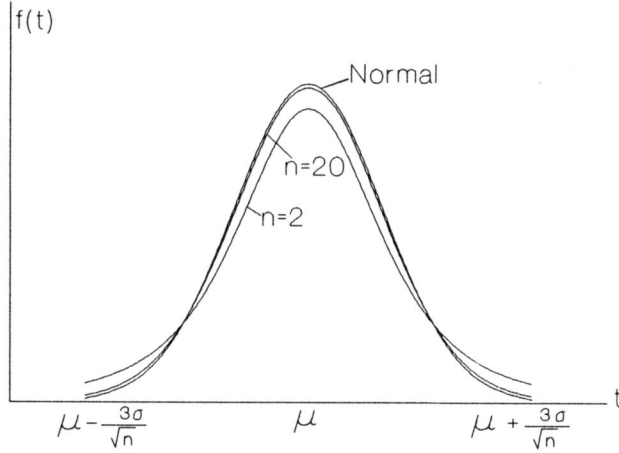

Figure 7.10 *t* distribution for different sample sizes, *n*

Suppose that a survey has been carried out to determine the frequency with which people in a town visit the theatre with the results shown in Table 7.2.

number of visits per year – x	number of responses $f(x)$	$xf(x)$	$x^2 f(x)$
0	36	0	0
1	31	31	31
2	12	24	48
3	5	15	45
4	1	4	16
5	0	0	0
6	2	12	72
	87	86	212

Table 7.2

We can calculate the following results:

mean = $\bar{x}$ = 86/87 = 0.989

From (1.19) the variance is

$$s^2 = [\Sigma(x-\bar{x})^2 f(x)]/n = [\Sigma x^2 f(x)]/n - (\Sigma xf(x)/n)^2$$

and so the sum of squares is

$$\Sigma(x-\bar{x})^2 f(x) = ns^2$$
$$= \Sigma x^2 f(x) - (\Sigma xf(x))^2/n$$
$$= 212 - 86^2/87 = 126.989$$

From (7.4)

$$\sigma^2_{EST} = 126.989/(87-1) = 1.477$$

From (7.2)

$$\text{var}(\bar{x}) = 1.477/87 = 0.017$$

and from (7.3)

$$\text{standard error} = 0.017^{0.5} = 0.130$$

So the likelihood distribution is Normally distributed with a mean of 0.989 and standard deviation of 0.130.

To make an estimate of μ use Bayes' Rule. Our prior estimate of μ is Normally distributed with mean m and variance v. The likelihood distribution is also Normal with mean μ and variance $\text{var}(\bar{x})$. The observed sample mean is $\bar{x}$. From (5.21) and (5.22) the revised estimate of μ, $(\mu|\bar{x})$, is a Normal distribution with mean

$$\mu_R = \frac{m/v + \bar{x}/\text{var}(\bar{x})}{1/v + 1/\text{var}(\bar{x})} \tag{7.5}$$

and variance

$$v_R = 1/[1/v + 1/\text{var}(\bar{x})] \tag{7.6}$$

As a special case suppose that we have no *a priori* feeling about the value of the mean number of visits per year and therefore use a diffuse prior with an infinite variance and so $1/v = m/v = 0$ and

$$\mu_R = \bar{x} \quad \text{and} \quad v_R = \text{var}(\bar{x}) \tag{7.7}$$

So the posterior distribution of $(\mu \mid \bar{x})$ has the same mean as the sample mean and the standard error found from the sample data (Figure 7.11).

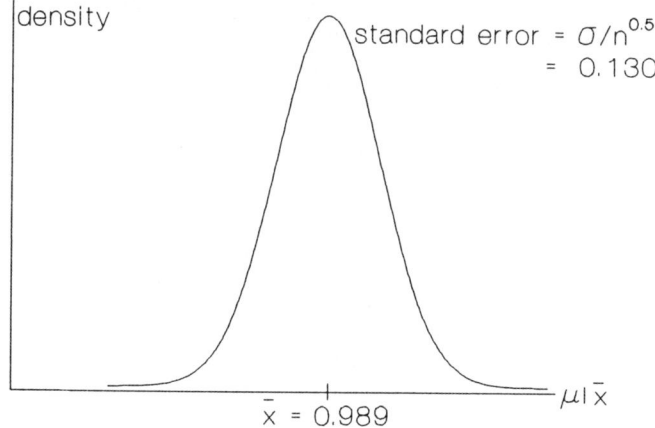

Figure 7.11 Bayes' posterior distribution for inference about the population mean

To find the 80% credible interval estimate of the population mean note from Table A5.3 that a range 1.28 standard deviations either side of the mean of a Normal distribution gives the required interval (find this by looking for $F(z) = 0.9$ in the body of the table). So

$$\mu = 0.989 \pm (1.28 \times 0.130)$$
$$= (0.823, 1.155)$$

7.4 HYPOTHESES

A hypothesis is an assertion that a particular system state is (or will be) the true state. For example:

10% of my colleagues read *The Times*

a certain new drug saves lives

the mean number of sweets in a branded pack is 50

In deciding whether or not to believe the hypothesis and therefore make the assertion I must bear in mind that whatever action I take I may be wrong. If I assert that the hypothesis is true and in fact it is then I make no error. If I assert that it is false and it is not I make an error that is called a Type 1 error. Table 7.3 shows the situation.

		system state	
		true	false
assertion	true	none	Type 2
	false	Type 1	none

Table 7.3
Types of error

If the loss associated with a Type 2 error is x times that associated with a Type 1 error we have, setting the magnitude of the Type 1 error arbitrarily to 1.0, the situation shown in Table 7.4,

		system state		E [loss]
		true	false	
assertion	true	0	x	$x(1-p)$
	false	1	0	p
		p	$1-p$	

Table 7.4
Losses

I believe p to be the probability that the hypothesised state is (or will be) the true state. I will assert it to be the true state if that action has the smaller expected loss

i.e.   if $x(1-p) < p$

$$p > x/(1+x) \qquad (7.8)$$

If the two losses are about the same magnitude then x is about 1 and I assert that the state is the true one provided p is greater than 0.5. This may be the case with a fairly routine estimation problem. Suppose, on the other hand, that we have a drug which provides improved relief for sufferers of hay fever. There is concern that the drug may have fatal side effects. The hypothesised system state with which we are concerned is therefore that there are no fatal side effects and that the drug is safe. The Type 1 error concerns the results of not releasing a safe drug: profit foregone by the company and relief of symptoms foregone by the patients. Type 2 errors are concerned with loss of life as a result of wrongly releasing a dangerous drug. We would expect x to be much greater than 1 and so would only assert that the drug was safe if p was almost 1.0. (You may care to review Exercise 6.3.)

Sec. 7.5]   **Discrimination**   171

This straightforward example gives the intuitively correct recommendation, but see how it has been derived by an explicit consideration of consequences via the expected losses.

7.5 DISCRIMINATION

It is often the case that we wish to compare a frequency or probability distribution describing some aspect of a real system to a theoretical distribution derived from a model. There are two main motivations for such a comparison. First, we may wish to describe the system's behaviour by a convenient model so that changes in the behaviour of the system may be simulated by changing the values of the model's parameters. For example, the Poisson distribution is likely to provide a good model of arrivals of customers at supermarket checkouts and this, in turn, would be useful in determining the best number of checkouts to provide in order to prevent excessive queuing. A key stage in this process is to *validate* the Poisson model by comparing actual arrival patterns with those predicted by the model.

Alternatively, we may wish to compare an observed distribution with some standard expected distribution as part of monitoring system behaviour.

Situations such as these may be conceptualised as tests of hypotheses. The model or standard describes a hypothesis about the system and we have to decide whether or not to believe that hypothesis.

In order to proceed we have to have some measure of the dissimilarity between the observed and hypothesised distributions. Here is an example.

Four identical machines in a factory are producing components at exactly the same rate. The number of defective components produced by each machine during a particular shift is counted with the results shown in Table 7.5. Is there any reason to believe that some machines are producing more defectives than others?

machine	number of defectives	proportion of defectives
A	12	0.27
B	17	0.38
C	10	0.22
D	6	0.13
	45	1.00

Table 7.5
Defectives produced by four machines

If we hypothesise that the machines are behaving differently from each other then in order to compare hypothesised with actual results we would need to specify exactly what we mean by different: A is producing twice as many defectives as B, and so on. This is, at best, difficult and is in any case likely to be arbitrary. We have no reason for preferring one of the infinite manifestations of "different" from any other and this renders the hypothesis, that the machines are behaving differently, essentially qualitative.

Although there are many ways in which things may differ there is only one way in which they can be the same. We therefore recast our problem so that instead of hypothesising a difference in behaviour and rejecting this hypothesis if the observed behaviours look sufficiently similar, we instead hypothesise identical behaviour and reject this hypothesis if the observed behaviours look sufficiently different from each other.

We now have a hypothesis that is sufficiently precise to permit the calculation of an expected distribution. If the hypothesis is correct then we would expect each machine to be producing the same number of defectives per shift; each would produce a quarter of all the defectives. Table 7.6 shows the results of applying this hypothesis.

machine	probabilities		frequencies	
	observed p_i	expected q_i	observed o_i	expected e_i
A	0.27	0.25	12	11.25
B	0.38	0.25	17	11.25
C	0.22	0.25	10	11.25
D	0.13	0.25	6	11.25
	1.00	1.00	45	45.00

Table 7.6
Observed and expected distributions

Why are the two distributions not identical? There are only two possible reasons:

the hypothesis of identical behaviour is incorrect

the hypothesis is correct and the differences are due to sampling fluctuations.

If the differences are not too large we may be inclined to believe that sampling fluctuations are the culprit and therefore that the hypothesis is not refuted by the data. The larger the differences the more we would be inclined to reject that hypothesis and believe that there are differences between the machines.

We can cast this as a hypothetical problem of the type depicted in Table 7.3. We have to decide whether or not to accept the hypothesis and in so doing may make errors shown in Table 7.7.

Sec. 7.5] **Discrimination** 173

	defective rates are:	
	the same	different
believe defective rates are the same		Type 2 error
believe defective rates are different	Type 1 error	

Table 7.7

How can we conveniently describe the dissimilarity between the observed distribution and that expected given the hypothesis?

A measure of the difference between the two probability distributions is provided by the relative entropy $I = \Sigma p_i \ln(p_i/q_i)$, introduced as equation (4.1). In this application p_i refers to the observed relative frequencies and q_i the hypothetical or modelled probabilities. From Table 7.8 we have $I = 0.067$.

machine	probabilities		$p_i \ln(p_i/q_i)$
	observed p_i	expected q_i	
A	0.27	0.25	0.021
B	0.38	0.25	0.159
C	0.22	0.25	−0.028
D	0.13	0.25	−0.085
	1.00	1.00	0.067

Table 7.8
Calculation of relative entropy, I

Using the same framework Kullback (1959) introduced the *minimum discrimination information statistic* as a means of measuring dissimilarity. It is reasonable to believe that when comparing an observed distribution with a theoretical or expected distribution the comparison is more forceful the more data that have been gathered, even though the pattern of the distributions as described by the expected probabilities and the observed relative frequencies are the same.

Just as the measure I permits the comparison of probability distributions so, alternatively, we may compare two frequency distributions. For a sample of size n (= 45 in this case) the observed frequencies are $o_i = np_i$ and the frequencies which the theory leads us to expect are $e_i = nq_i$. Using these in the equation for I gives a measure ψ (Greek "psi"):

$$\psi = \Sigma\, o_i \ln(o_i/e_i) \qquad (7.9)$$
$$= \Sigma\, np_i \ln[(np_i)/(nq_i)]$$
$$= n\Sigma\, p_i \ln(p_i/q_i) \qquad (7.10)$$

so $\psi = nI = 45 \times 0.067 = 3.02$

We now have a measure of the dissimilarity between the two frequency distributions. If the two distributions are identical then $\psi = 0$. The more dissimilar they are the larger ψ becomes. The greater the sample size, for a given value of I, the larger ψ becomes.

We now return to our decision problem and concentrate on Type 1 error: that we wrongly reject the hypothesis that the machines have the same propensity to produce defectives.

This hypothesis is to be rejected if the observed and expected distributions are sufficiently dissimilar; if ψ is sufficiently large. What does this mean?

Imagine taking a large number of samples, each of size n, from a system that exhibits the hypothesised behaviour. We would not expect each sample to show that each machine produced exactly the same number of defectives, and so, even though the hypothesis is known to be correct, the values of ψ would be greater than 0. If we were to plot these sample values of ψ we would obtain a distribution such as that shown in Figure 7.12.

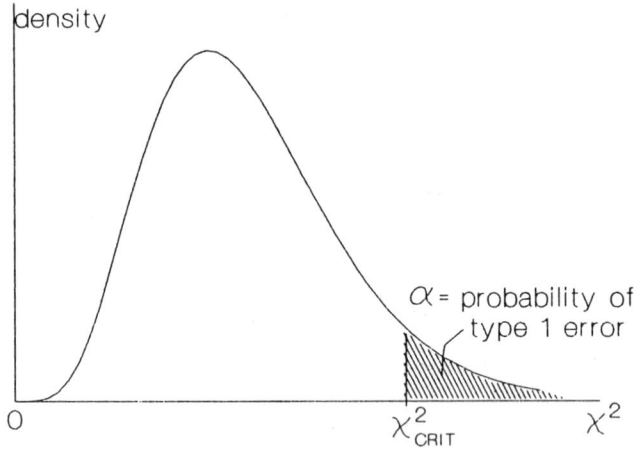

Figure 7.12 Likelihood distribution for $\chi^2 = G^2 = 2\psi$

You will see that the distribution given is for

$$G^2 = 2\psi = 2\Sigma\, o_i \ln(o_i/e_i) \qquad (7.11)$$

Sec. 7.5] **Discrimination** 175

The reason for this is one of convenience. The distribution of sample values of G^2 is well approximated by the theoretical distribution χ^2 ('chi-squared'), which is described in Appendix A5.6.

We have met this sort of thing before in Chapter 5 when discussing Bayes' Rule. It is just the likelihood function for the sample values of G^2 given the hypothesis of identical underlying behaviour.

We now decide that any value of G^2 greater than some critical value, χ^2_{CRIT}, is so large that we will reject the hypothesis whenever we find $G^2 > \chi^2_{CRIT}$. In adopting this decision rule the probability of making a Type 1 error is just the probability that a value of G^2 greater than the critical value has indeed been generated by a sample from a system with the hypothesised characteristic. This probability is shown in Figure 7.12 as α.

Rather than choosing the critical level χ^2_{CRIT} and then seeing what value of α is implied it makes sense first to set an acceptable value of α, the probability of making an error, and then to find the corresponding critical value to be used in the decision rule. Notice that α is a characteristic of the rule rather than of any particular decision made by applying the rule.

There is nothing in this methodology to help you arrive at an appropriate value for α. You will just have to think hard about the problem and decide what probability of error is acceptable. We faced a similar difficulty when discussing credibility levels for use in interval estimation and the comments made then apply equally here.

So, the procedure is clear: pick a value for α and then use the likelihood distribution for χ^2 to find χ^2_{CRIT}.

The χ^2 distribution is in fact a family of distributions, as shown in Figure 7.13. The parameter that determines the appropriate distribution to use in any particular case is called the *degrees of freedom*.

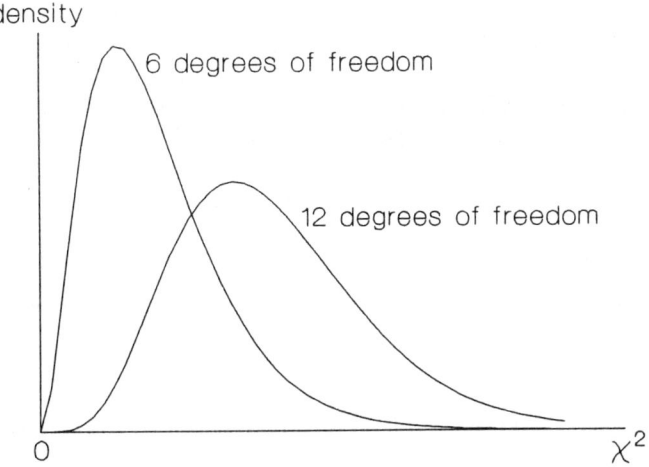

Figure 7.13 χ^2 distributions with different degrees of freedom

The degrees of freedom is a measure of the number of arbitrary assignments of expected values that it is possible to make *given the way that we have decided to model the system*. The model is determined by the hypothesis that we wish to test. Indeed, model and hypothesis may be considered synonymous.

Look at Table 7.8. How many arbitrary values of the expected probabilities q_i could I have written?

Arbitrary here means not completely determined yet respecting any natural requirements so that, for instance, I cannot "arbitrarily" choose a negative probability. In this case I could have written three values arbitrarily since the fourth must ensure that the sum is 1.0 and so is completely determined by the other three values. So

$$\text{degrees of freedom} = \text{no. of arbitrary assignments possible} \qquad (7.12)$$

Thinking now in terms of the modelling process we see that the model implied by the hypothesis of uniform behaviour is

$$\text{expected frequency, } e_i = \frac{\text{total number of defectives}}{\text{number of machines}} = \frac{45}{4}$$

This model has just one parameter that has to be estimated from the data: the total number of defectives. This leads to an alternative definition

$$\text{degrees of freedom} = \text{no. of classes} - \text{no. of parameters} \qquad (7.13)$$

The number of classes in Table 7.8 is just the number of machines, four, and the model has one parameter, so we have $4 - 1 = 3$ degrees of freedom, as before.

The concept of degrees of freedom is important. Any model or hypothesis that gives a good correspondence with observed system behaviour and yet is fairly parsimonious in that it does not need many parameters to be estimated from the data can be considered as useful or powerful. We could always design a model that gave perfect correspondence by adding more parameters and so making the model more complex. Each parameter would reduce by one the number of degrees of freedom until we reached zero at which point the hypothesis would just be a tautology and of no use at all.

At last we are able to consult the table in Appendix A5.6. If we are prepared to countenance a 10% chance of making an error then $\alpha = 0.1$ and, with 3 degrees of freedom, we find $\chi^2_{\text{CRIT}} = 6.25139$. The value calculated from the sample is

$$G^2 = (2 \times 3.02) = 6.04$$

Since this is less than the critical value, the hypothesis of uniform behaviour cannot be rejected.

7.6 FURTHER READING

Books which concentrate on problems of inference are provided by, for instance, Barnett (1973) and Silvey (1975); see also Dale (1991). Godambe and Sprott (1971) and Menges (1974) give interesting collections of papers as do Morrison and Henkel (1970). Most of the books on Bayesian statistics cited in Chapter 5 cover much of the ground, but see particularly Box and Tiao (1973) and also Iversen (1984) and Pollard (1986). Ferguson (1967) adopts a decision oriented approach similar to that found here.

Conditions for the satisfactory approximation of G^2 by χ^2 and appropriate references are given in Appendix A5.6.

Good discussions of the issues and methods of data collection and sampling are given by Moser and Kalton (1971) and, among others, by Cochran (1977), Jessen (1978), Sudman (1976) and Yates (1981).

7.7 EXAMPLES

Example 7.1

The first three columns of the table show 175 accidents that occurred in an ICI factory classified according to section.

section	no. of employees	no. of accidents	prob. (%)
By-products	22	10	1
Conversion	153	29	78
Day Gang	45	3	>99
Elect. and Inst.	89	8	>99
Fitters	103	29	6
Fitters (Labs.)	106	26	24
Laboratories	66	14	55
Laggers	18	6	17
Refinery	145	33	36
Riggers	21	5	47
Samplers	21	11	<1
U.S. plant	35	1	>99
	824	175	

(Source: Davies, 1961, p286)

Are there any sections which are particularly unsafe?

Solution

How best might the notion of "particularly unsafe" be formulated? We would say that a section was particularly unsafe if the number of accidents was surprisingly high, by which would be meant higher than expected (cf. Figure 2.1). The hypothesis that generates that expectation is that all sections are equally safe and so have the same accident rate. That rate is 175/824 = 0.212 accidents per employee.

Consider the first section, By-products. It seems reasonable to assume that accidents occur independently of each other and that all employees are equally accident prone. It follows that the probability of x accidents per year from a group of 22 employees is described by the Binomial distribution with $p = 0.212$ and $n = 22$. (There is an implicit assumption here that no employee has more than one accident per year.)

The probability that the number of accidents in any year is ten or more is 0.0095 or about 1%. This figure is shown in the last column of the table. I found these probabilities by setting up a spreadsheet to do the necessary calculations. Alternatively you may wish to use the tables given in Appendix A5. For sections with large numbers of employees Poisson or Normal approximations may be used. In the Conversion section, for example, use a Normal approximation with

$$\mu = np = 153 \times 0.212 = 32.436$$

and

$$\sigma = [np(1-p)]^{0.5} = (153 \times 0.212 \times 0.788)^{0.5} = 5.056$$

Since we are using a continuous approximation to a discrete variable we want prob($X \geq 28.5$) and so

$$z = (x - \mu)/\sigma = (28.5 - 32.436)/5.056 = -0.78$$

From the table in Appendix A5.3 find the required probability to be 0.7823 or 78%.

At what level of this probability ought we to be sufficiently surprised at the number of accidents to suspect that a section is particularly dangerous? This is exactly the question asked in any hypothesis test: what probability of wrongly rejecting the null hypothesis of uniform safety is appropriate? Let us say 5%, in which case By-products and Samplers ought to be inspected. We might want to have a look at the Fitters too.

Compare this approach with the problem of Table 7.8. Which do you prefer, and why?

Example 7.2

Researchers at the Moose Research Center measured the gestation period of captive moose cows. Here are the results:

gestation period – days	number observed
216	1
221	1
224	1
225	2
226	3
230	1
231	1
232	3
233	4
234	1
235	2
236	1
240	1

(Source: Schwartz and Hundertmark, 1993)

Is it plausible to believe that the durations of gestation periods are Normally distributed?

Solution

Just scanning the data indicates that the distribution is unimodal and roughly symmetrical. Draw a chart of the distribution if you need to be convinced of this.

Find the mean and standard deviation either directly via your calculator or spreadsheet or from the following table.

gestation x days	freq. f	rel. freq. p	px	px^2
216	1	0.045	9.818	2120.727
221	1	0.045	10.045	2220.045
224	1	0.045	10.192	2280.727
225	2	0.091	20.455	4602.273
226	3	0.136	30.818	5964.909
230	1	0.045	10.455	2404.545
231	1	0.045	10.500	2425.500
232	3	0.136	31.636	7339.636
233	4	0.182	42.364	9870.727
234	1	0.045	10.636	2488.909
235	2	0.091	21.364	5020.455
236	1	0.045	10.727	2531.636
240	1	0.045	10.909	2618.182
	22	1.000	229.909	52888.273

From this:

$\bar{x} = 229.909$

$sd(x) = s = (52888.273/22 - 229.909^2)^{0.5}$
$ = 5.485$

$\sigma_{EST} = 5.485(22/21)^{0.5} = 5.614$

Since the data are sparse they are grouped into classes of width 5 (<220, 200–224, etc.), as shown in the table below. Values of the Normal distribution function corresponding to the upper class limits are shown in the column "cum. prob.". Because the Normal distribution is being used to describe a discrete variable apply a continuity correction to the upper class limits so that, for instance, instead of prob($X \leq 119$) use prob($X \leq 119.5$) for the Normal approximation. These values are given in the column headed u.

The standardised Normal values are $z = (x - \bar{x})/\sigma_{EST}$. For the first class, for instance, $z = (119.5 - 229.09)/5.614 = -1.85$. Cumulative Normal probabilities can now be read directly from the table in Appendix A5.3 and expected class probabilities, e, found by subtracting successive cumulative probabilities. For instance,

prob($230 \leq X \leq 234$) = prob($X \leq 234$) – prob($X < 230$)
$ = 0.794 - 0.472 = 0.322$

Examples

The observed probabilities, p, are obtained directly from the data. The observed relative frequency in the second class is $2/22 = 0.091$.

x	u	z	cum. prob.	q	p	$p\ln(p/q)$
<220	119.50	−1.85	0.032	0.032	0.045	0.015
200–224	224.50	−0.96	0.169	0.136	0.091	−0.037
225–229	229.50	−0.07	0.472	0.304	0.227	−0.066
230–234	234.50	0.82	0.794	0.322	0.455	0.158
235–239	239.50	1.71	0.956	0.162	0.136	−0.024
>239			1.000	0.044	0.046	0.002
				1.000	1.000	0.048

Finally,

$$G^2 = 2 \times 22 \times 0.048 = 2.112$$

There are six classes and we have used three parameters (sum, mean and standard deviation) so that 3 degrees of freedom remain. With $\alpha = 0.05$ the critical χ^2 value is 7.81 which is greater than the calculated value of 2.112 and so we should not reject the hypothesis that the data are from a population that is Normally distributed.

Example 7.3

Suppose that before examining the data in the previous example we had believed that there was a 50% probability that the mean gestation period was 220 ± 2 days. What now would we believe the mean gestation period to be?

Solution

We have *a priori* information about mean and spread and so will use a maximum entropy prior, which is in this case a Normal distribution with mean 220 days. Spread is characterised by a 50% credible interval of ±2 days. From the table in Appendix A5.3 a 50% credible interval lies within ±0.67 standard deviations of the mean and so the standard deviation of our prior is $(2/0.67) = 3$ days.

The likelihood distribution is as defined by the Central Limit Theorem. It is Normal with a standard deviation of $5.614/22^{0.5} = 1.197$ days.

Use equations (7.5) and (7.6) to give the characteristics of the revised estimate:

$$\text{mean} = \frac{220/3^2 + 229.909/1.197^2}{1/3^2 + 1/1.197^2} = 228.55$$

variance = $1/(1/3^2 + 1/1.197^2) = 1.236$

standard deviation = $1.236^{0.5} = 1.113$

The precision of the data is about seven times that of the prior estimate and so the revised estimate is closer to the experimental result. The 50% credible interval estimate of the mean gestation period is now

$$228.55 \pm (0.67 \times 1.113) = (227.8, 229.3) \text{ days}$$

Example 7.4

The speeds (in mph) of ten vehicles on a motorway were

69.3   84.1   72.6   75.9   80.1   74.5   63.7   78.2   71.9   72.0

What is the 80% credible interval estimate of the mean speed of all vehicles using the motorway?

Solution

From the data:

$$n = 10, \quad \Sigma x = 742.30, \quad \Sigma x^2 = 55399.67$$

and so

$$\bar{x} = 742.30/10 = 74.23$$

$$s = (55399.67/10 - 74.23^2)^{0.5} = 5.466$$

$$\sigma_{EST} = 5.466(20/19)^{0.5} = 5.608$$

standard error = $5.608/10^{0.5} = 1.773$

This is a small sample so use the t distribution (Appendix A5.7) with $n - 1 = 9$ degrees of freedom to obtain

$$\begin{aligned} 80\% \text{ credible interval} &= 74.23 \pm (1.383 \times 1.773) \\ &= (71.78, 76.68) \end{aligned}$$

Example 7.5

Concrete pipe for a water supply is made up of standard 5 m long sections. In fact, all sections are not of exactly the same length due to variations in the manufacturing process. This variation is described by a Normal distribution such that 95% of all sections are within 1% of the nominal length. Give a 95% credible interval estimate of the length of a planned 100 m section.

Solution

From the table in Appendix A5.3 confirm that a 95% credible interval is given by $\mu \pm 1.96\sigma$. We know that 1.96σ is 1% of 5 m and so

$$1.96\sigma = 0.01 \times 5 \text{ m} = 5 \text{ cm}$$

and $\sigma = 5/1.96$ cm

The estimate of the length of 20 sections is Normally distributed with mean = 100 m and, from (7.1),

$$\text{variance} = 20\sigma^2 = 20(5/1.96)^2 = 130.15 \text{ cm}^2$$

and

$$\text{standard deviation} = 130.15^{0.5} = 11.41 \text{ cm}$$

The 95% credible interval estimate of the length of 20 sections is

$$100 \text{ m} \pm (1.96 \times 11.41 \text{ cm}) = 100 \text{ m} \pm 22.4 \text{ cm}$$

There is a 95% chance that the pipeline will be within 0.224% of the planned length.

Satisfy yourself of the more general result that the coefficient of variation (equation (1.12)) of the sum of n independent and identical variables is the coefficient of variation of each reduced by a factor equal to the reciprocal of the square root of n (in this case $1/20^{0.5} = 0.224$).

7.8 EXERCISES

7.1 The lives, measured in hours, of eight 60 watt light-bulbs were found to be

      1427   1591   1648   1532   1366   1484   1602 1555

Give a 90% confidence interval estimate of the mean life of such light-bulbs.

7.2 Weather forecasters never get plaudits for making correct forecasts but do get blamed if the forecasts are wrong. This is particularly so when severe storms are in the offing. The amount of blame attached to not forecasting storms that subsequently occur is much greater than that which results from forecasting storms that do not happen. Suppose that the first error is thought to be 50 times as serious as the second. If the forecaster obtains probabilistic estimates of imminent storms at what probabilities ought the forecaster to announce that storms will occur?

7.3 Using the data from Exercise 1.2 give an 85% credible interval estimate of the mean playing time of jazz CD's. By how much should the sample be increased in order to halve the width of the interval?

7.4 Trains arrive late at their final destinations. These delays follow a negative exponential distribution with mean 3.6 minutes. Give the 50% and 75% credible intervals for the delay to a particular train.

Give a 90% credible interval estimate of the mean delay suffered by a sample of 50 trains.

7.5 A company that installs central heating systems subcontracts a large proportion of its work. Inspectors are employed to check the work of subcontractors by selecting recently completed contracts at random and making a site visit. Here are some results of these inspections.

subcontractor	number of jobs inspected	jobs passed as acceptable
W. Tonks & Son	6	4
Acme Heating	13	8
StayWarm Contracts	18	15
A. Daly Installations	4	2

Subcontractors cease to be used if 15% or more of their jobs are unacceptable. Which of these four would you stop using?

7.6 The number of passengers arriving at an airport check-in desk in 20 successive 5 minute periods surveyed was

3, 0, 2, 2, 1, 4, 6, 1, 2, 4, 4, 0, 1, 2, 1, 5, 3, 2, 4, 1

Does this suggest a Poisson arrival pattern?

7.7 Ladislaus von Bortkewicz collected data showing the number of men killed by horse kicks in ten corps of the Prussian army for each of 20 years:

deaths/corps/year	0	1	2	3	4
frequency observed	109	65	22	3	1

(Source: Fisher, 1934, p56)

Suggest a model for this distribution and see how well it fits.

(I hope that the antiquity of these data appeals to you. Note that the methodology is valid for accidents on modern roads, in factories and the like. If you are interested in Bortkewicz's data read Quine and Seneta, 1987.)

7.8 A metal plate is coated on one side only. Two coats are used of nominal thicknesses 1 mm and 0.5 mm with tolerances, expressed as standard deviations, of 0.02 mm and 0.008 mm. The plate itself is nominally 1 cm thick with tolerance 0.1 mm. What proportion of finished coated plates would fail to fit into a slot exactly 11.6 mm wide?

7.9 A company forecasts its financial performance for the coming year. The forecasts are probabilistic:

	mean	standard deviation
Revenue (£m)	3.7	0.18
Costs (£m)	3.4	0.09

What is the probability that the company will make a loss?

7.10 Your company is in the clothing business; 70% of its employees are women and yet most managers are men. The Director of Human Resources has instituted a scheme whereby shortlists for jobs are monitored to see that the proportion of women on the shortlist is representative of the workforce. In a shortlist of ten, five are women. Ought the Director to be concerned? How few women would there need to be on the shortlist before it could be deemed to be unrepresentative?

8

Two Dimensions

We turn now to consider joint probability distributions and start by examining the associated entropies and how they may help us to interpret the relationship between the two variables. The transmission of information and Bayes' Rule are discussed as particularly relevant cases.

Relationships between entropies are sometimes given in this chapter without proofs. These may be found in the references on communication and information theory given in Chapter 2.

8.1 STRUCTURE IN CONTINGENCY TABLES

Frequency distributions describing the relationship between two variables are called *contingency tables*.

Some data were collected to show the disorders suffered by patients in an asylum. The results are shown in Table 8.1.

		sex of patient (Y)		
		women	men	
type of	Schizophrenia	155	492	647
disorder	Organic disorders	48	53	101
(X)	Mental retardation	138	256	394
	Other	19	18	37
		360	819	1179

Table 8.1
Characteristics of the patients in the Leros Asylum, Greece, 1988
(Source: Bouras et al., 1992)

The corresponding distributions of relative frequency, treated as probability, are given in Table 8.2. (Apparent discrepancies are due to rounding errors.)

Structure in contingency tables

		sex of patient (Y)		
		women	men	
type of	Schizophrenia	0.131	0.417	0.548
disorder	Organic disorders	0.041	0.045	0.086
(X)	Mental retardation	0.117	0.217	0.334
	Other	0.016	0.015	0.031
		0.305	0.695	1.000

Table 8.2
Joint and marginal probability distributions

We can describe these distributions in the usual way by calculating their entropies. For the joint distribution

$$H(X,Y) = -\sum_i \sum_j p_{ij} \ln(p_{ij}) \qquad (8.1)$$
$$= H(0.131, 0.041, \ldots, 0.015) = 1.614$$

where p_{ij} is the probability in row i and column j of Table 8.2. For the marginal distributions, using (2.2),

$$H(X) = H(0.549, 0.086, 0.334, 0.031) = 1.015$$
$$H(Y) = H(0.305, 0.695) = 0.615$$

As we saw in Chapter 1 the joint distribution is the result of the marginal distributions and, if the variables are not independent, of the conditional distributions too. Unsurprisingly, the entropies of these distributions are also related to each other.

Considering just the joint and marginal distributions, it can be proven that

$$H(X,Y) \leq H(X) + H(Y) \qquad (8.2)$$

This is shown graphically in Figure 8.1. (These useful diagrams were first used by Quastler, 1953, and then Attneave, 1959.) The significance of the inequality will become clear in what follows.

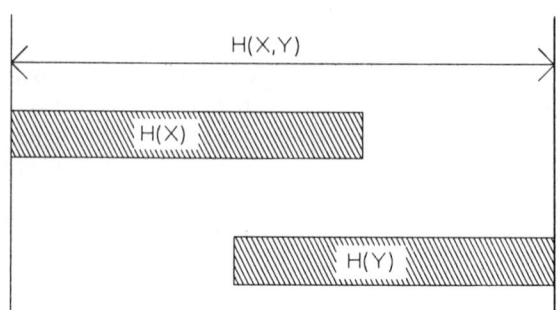

Figure 8.1 Joint and marginal entropies

Consider the problem of estimating which illness is being suffered by a particular patient. If we have no other information then we should choose according to the marginal distribution of X in Table 8.2. Our uncertainty in this situation is $H(X) = 1.015$.

Suppose now that we are able to discover the sex of the patient in question. How does this affect our uncertainty? Depending on whether the patient is male or female we choose according to the conditional probabilities shown in Table 8.3.

		sex of patient (Y)		
		women	men	
type of	Schizophrenia	0.431	0.601	0.548
disorder	Organic disorders	0.133	0.065	0.086
(X)	Mental retardation	0.383	0.313	0.334
	Other	0.053	0.022	0.031
		1.000	1.000	1.000

Table 8.3
Conditional probability distributions

If the patient is female then our uncertainty is

$$H(X \mid \text{woman}) = H(0.431, 0.133, 0.383, 0.053) = 1.154$$

While if the patient is male

$$H(X \mid \text{man}) = H(0.601, 0.065, 0.313, 0.022) = 0.931$$

The expected value of our uncertainty given that we can discover the sex of the patient, though we have not done so yet, is

$$H(X|Y) = (0.305 \times 1.154) + (0.695 \times 0.931) = 0.999$$

which is less than $H(X) = 1.015$. The extra knowledge of the sex of the patient is expected to reduce our uncertainty about the disorder being suffered. This is not unexpected, but let us see the mechanics. With p_{ij} the joint probability, $p(y_j)$ the marginal probability of the patient's sex and $p(x_i|y_j)$ the conditional probability of the disorder suffered we have

$$p_{ij} = p(x_i|y_j)p(y_j)$$

so

$$H(X,Y) = -\sum_i \sum_j p_{ij} \ln(p_{ij})$$

$$= -\sum_i \sum_j p(x_i|y_j) p(y_j) \ln(p(x_i|y_j) p(y_j))$$

$$= -\sum_i \sum_j p(x_i|y_j) p(y_j) [\ln(p(x_i|y_j)) + \ln(p(y_j))]$$

$$= -\sum_j p(y_j) \sum_i p(x_i|y_j) \ln(p(x_i|y_j))$$

$$- \sum_j p(y_j) \ln(p(y_j)) \sum_i p(x_i|y_j)$$

The first term is just $H(X|Y)$, the expected value of the conditional entropies. In the second term the sum of the conditional probability distribution is $\sum_i p(x_i|y_j) = 1$ so

$$H(X,Y) = H(X|Y) + H(Y) \tag{8.3}$$

This equation states that the uncertainty about the joint distribution is the sum of the expected conditional uncertainty and the marginal uncertainty. The general flavour is, naturally enough, similar to that of the probability equation (1.6):

prob(X and Y) = prob($X|Y$)prob(Y)

Similarly,

$$H(X,Y) = H(Y|X) + H(X) \tag{8.4}$$

The situation is shown in Figure 8.2. Note that the reduction in uncertainty about one variable due to knowledge of the other is

$$T(X,Y) = H(Y) - H(Y|X) = H(X) - H(X|Y) \qquad (8.5)$$

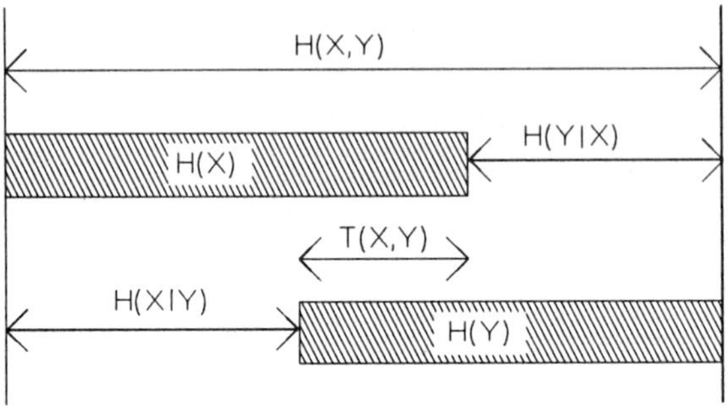

Figure 8.2 Joint information measures

$T(X,Y)$ is an important measure for it quantifies the strength of the relationship between the two variables. We can rearrange (8.5) to give

$$H(Y|X) = H(Y) - T(X,Y) \qquad (8.6)$$

so $\quad H(Y|X) \leq H(Y) \qquad (8.7)$

This tells us that knowledge about X is likely to reduce our uncertainty about Y. The worse that can happen is that the uncertainty will be left unchanged; it will never increase. $T(X,Y)$ is the measure of that reduction.

We can see from Figure 8.2 that

$$T(X,Y) = H(X) + H(Y) - H(X,Y) \qquad (8.8)$$

which is computationally convenient since it allows $T(X,Y)$ to be calculated directly from the joint and marginal entropies as shown in Table 8.4.

The expected conditional entropies may then be found by simple subtraction, for instance, from (8.3),

$$\begin{aligned} H(X|Y) &= H(X,Y) - H(Y) \\ &= 1.614 - 0.615 = 0.999 \end{aligned}$$

entropy	value
$H(X)$	1.015
$H(Y)$	0.615
$H(X,Y)$	1.614
$T(X,Y)$	0.016
$H(Y\|X)$	0.599
$H(X\|Y)$	0.999

Table 8.4

It may sometimes be more convenient to calculate entropy directly from a frequency distribution $f_1, f_2, \ldots$. If $\Sigma f = n$ then

$$-\Sigma p\ln(p) = -\Sigma(f/n)\ln(f/n)$$
$$= -(1/n)\,\Sigma[f\ln(f) - f\ln(n)]$$
$$= \ln(n) - [\Sigma f\ln(f)]/n \qquad (8.9)$$

From the column marginal frequencies of Table 8.1

$$H(Y) = \ln(1179) - [360\ln(360) + 819\ln(819)]/1179$$
$$= 0.615$$

as before.

8.2 CORRELATION AND INDEPENDENCE

$T(X,Y)$ measures the degree of *association* between the two variables. From Figure 8.2 and from (8.6) we can see that as $T(X,Y)$ increases the expected uncertainty in the conditional distributions decreases until, in the limit, it becomes zero. This means that once we know which value is taken by the marginal variable, X say, then we are certain of the value taken by the other variable, Y. Table 8.5 shows just such a distribution:

		y_1	y_2
X	x_1	0.6	0.0
	x_2	0.0	0.4

Table 8.5
Perfectly correlated variables

The variables are said to be *perfectly correlated*: we can predict one perfectly given the other. For perfectly correlated variables

$$H(X,Y) = H(X) = H(Y) = T(X,Y)$$
$$H(X|Y) = H(Y|X) = 0$$

Figure 8.3 shows this graphically.

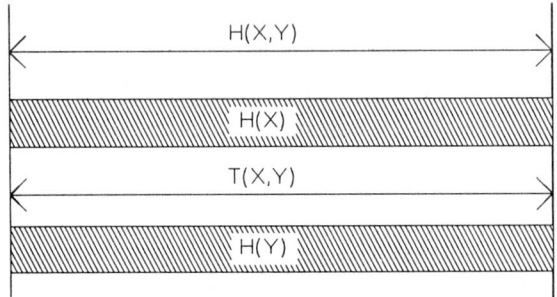

Figure 8.3   Perfect correlation

At the other extreme lies the case when information about X does nothing to reduce uncertainty about Y so that $H(Y|X) = H(Y)$ and consequently $T(X,Y) = 0$. When there is no association between variables we say that they are *independent* (see equation (1.8)) and

$$T(X,Y) = 0$$
$$H(X) = H(X|Y)$$
$$H(Y) = H(Y|X)$$
$$H(X,Y) = H(X) + H(Y)$$

as shown in Figure 8.4.

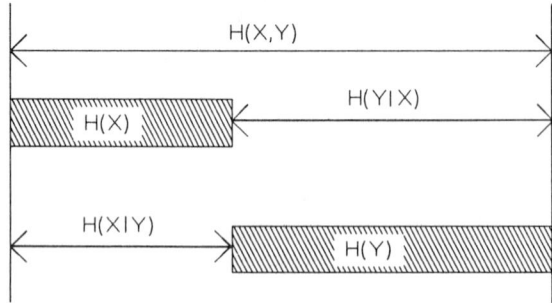

Figure 8.4   Independence

		y_1	y_2	
X	x_1	0.45	0.30	0.75
	x_2	0.15	0.10	0.25
		0.60	0.40	

Table 8.6
Independent variables

When variables are independent we make the same probabilistic estimates of X or Y whether we use marginal or conditional distributions, for they are identical. Table 8.6 shows such a case where, for instance,

$$\text{prob}(x_1|y_1) = \text{prob}(x_1|y_2) = \text{prob}(x_1) = 0.75$$

When two variables are independent the uncertainty of the joint distribution is highest. You can see from (8.8) that when $T(X,Y) = 0$ then

$$H_{max}(X,Y) = H(X) + H(Y) \tag{8.10}$$

8.3 TESTING FOR INDEPENDENCE

We have seen in Table 8.4 that for the Leros data $T(X,Y)$, or just T for convenience, took the value $T = 0.016$. How are we to interpret this value? If the variables were independent then T would be 0 while if they were perfectly correlated T would be $H(X) + H(Y) = 1.630$. The calculated value is only 1% of this maximum and so would seem to indicate that the degree of association between the sex of the patient and the type of disorder suffered is not particularly high. This may prompt the thought that there is in general no association between these two variables. Is this true?

We can formulate this as a problem that requires the testing of a hypothesis using the framework discussed in Chapter 7.

Consider the data in Table 8.1 to be a sample from a much larger population, say all patients in Greek asylums during the late 1980's. What hypothesis shall we test? The hypothesis of perfect correlation is immediately refuted since if in the population the data are perfectly correlated then so are the data in any sample, and we know that ours are not. We could hypothesise a population value of T less than the maximum but greater than 0. But which value, and why? The strategy we adopt is to assume no correlation and then see if the data support this assumption or not. If they do not then we say that the variables are related to some degree. So, we hypothesise that in this population $T = 0$ and will reject this hypothesis if the observed value of T exceeds some critical value.

We use a procedure similar to that used in Chapter 7 to test for correspondence between distributions. As was the case with the minimum discrimination information statistic, ψ, it can be shown that there is a simple transformation

$$Y^2 = 1.9712nT \tag{8.11}$$

whose sampling distribution is well approximated by the χ^2 distribution. In (8.11) n is the sample size and T is measured in nats.

In the present case

$$Y^2 = 1.9712 \times 1179 \times 0.016 = 37.184$$

To find the critical value of χ^2 we have to decide on the appropriate number of degrees of freedom. Using (7.12) we need to find the number of arbitrary assignments of values we could make in Table 8.1. Remember that the marginal totals are fixed and must be respected. A little thought will show that it is possible to make three arbitrary assignments. Table 8.7 shows just such an assignment. All other entries can be derived from these values.

155		647
48		101
138		394
		37
360	819	1179

Table 8.7

In general, given the hypothesis of independence in a contingency table,

$$\text{degrees of freedom} = (r-1)(c-1) \tag{8.12}$$

where r is the number of rows and c is the number of columns.

With, say, $\alpha = 0.01$ and 3 degrees of freedom you will see from the table in Appendix A5.6 that we have the rule

reject the hypothesis of independence if $Y^2 > \chi^2_{\text{CRIT}} = 11.3449$

and that this rule has a 1% chance of resulting in the incorrect rejection of the hypothesis.

The value found in our data is $Y^2 = 37.184$ and so we may reject the hypothesis and believe that in the population the distribution of disorders is different for men and women.

Note that our initial feeling was that $T = 0.016$ was quite a small value and, by implication, probably insignificant, but that it has proven not to be so. From (8.11) we can see that this is due to the size of the sample. Large samples mean that even quite slight evidence of association may be sufficient to reject the hypothesis of independence. Although the association within the population may still be considered slight, it is unlikely to be zero.

If we had had only a tenth of the data then Y^2 would have been reduced to a tenth of its value to give $Y^2 = 3.7184$ and we would have been unable to reject the hypothesis given the value of α chosen.

An alternative procedure is to compare the observed and expected distributions as we did in Chapter 7. We calculate ψ and then use (7.11) to obtain a value of G^2. This will hold whatever the source of the expected distribution: it is not limited to tests of independence. We shall now calculate G^2 for the Leros data using ψ.

If r_i and c_j are the marginal frequencies from Table 8.1 then the marginal probabilities are r_i/n and c_j/n. Assuming independence the joint probabilities are, using (1.8),

$$p_{ij} = (r_i/n)(c_j/n)$$

and the expected frequency is

$$e_{ij} = np_{ij} = n(r_i/n)(c_j/n) = r_i c_j /n \tag{8.13}$$

For example, for the first row and first column we have

$$e_{11} = (360 \times 647)/1179 = 197.56$$

The expected frequencies are shown in Table 8.8.

		sex of patient (Y)		
		women	men	
type of	Schizophrenia	197.56	449.44	647
disorder	Organic disorders	30.84	70.16	101
(X)	Mental retardation	120.31	273.69	394
	Other	11.30	25.70	37
		360	819	1179

Table 8.8
Expected frequencies assuming X and Y independent

Using (7.11) we now have

$$G^2 = 2[155\ln(155/197.56) + 492\ln(492/449.44) + \cdots]$$
$$= 37.132$$

which is quite close to the value $Y^2 = 37.184$ obtained above.

In modelling the expected frequencies using (8.13) we had to estimate the marginal probabilities and to do this used the marginal frequencies from the data in Table 8.1. These were the parameters of the model. How many were there? Provided that we know all the marginal totals but one, that last can be estimated from the rest. For instance, consider Table 8.9.

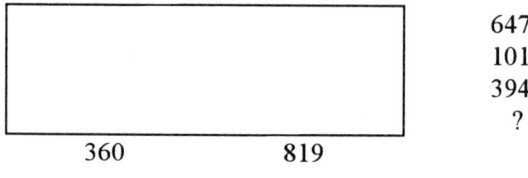

Table 8.9

The grand total is found from the column marginals and so the missing marginal total is

$$(360 + 819) - (647 + 101 + 394) = 37$$

Given the hypothesis of independence five parameters were needed to model the data. The table has eight classes and so, from (7.13),

degrees of freedom = 8 − 5 = 3

This is the same result as above but arrived at by a more satisfactory route that emphasises that the degrees of freedom is a result not just of the data but also of the way that we choose to model the data.

8.4 INFORMATION TRANSMISSION

Consider now the information transmission problem of Figure 2.5. In this section the word *information* is used for the entropy, H, as is common in the information theory literature.

The transmission of a message can be described by a joint probability distribution such as those discussed above. If s is the source and r the receiver then $H(s)$ and $H(r)$ represent the uncertainty at both ends of the transmission. Note that, in the case of noiseless transmission, $H(s) = H(r)$ means only that the *amount* of information received is the same as the amount sent, not that they are semantically identical. For instance, if a coin were tossed and the message giving the result always said heads when the result was tails and vice versa then $H(s) = H(r)$ but the identity of the side of the coin which showed is not known correctly at the destination. This once again emphasises the difference between the statistical and semantic concepts of information.

Suppose that an information source is capable of sending two signals, $S1$ and $S2$, rather like Morse code, and that because of noise in the system $S1$ was correctly received 90% of the time while $S2$ was received correctly only 80% of the time. This is shown in Table 8.10. (Notice the similarity with the Bayesian likelihoods of Table 5.1.)

		receiver – r		
		$S1$	$S2$	
source – s	$S1$	0.9	0.1	1.0
	$S2$	0.2	0.8	1.0

Table 8.10
Transmission characteristics – prob($r\,|\,s$)

We have no idea in advance of receiving a signal what sign will be sent and so prob($S1$) = prob($S2$) = 0.5 is, in Bayesian terminology, our non-informative prior. The situation is shown in Table 8.11 (cf. Table 5.4).

		receiver – r		
		$S1$	$S2$	
source – s	$S1$	0.45	0.05	0.5
	$S2$	0.10	0.40	0.5
		0.55	0.45	

Table 8.11
The information transmission system

The main entropies, in nats, are

$H(s) = H(0.5, 0.5) = 0.693$
$H(r) = H(0.55, 0.45) = 0.688$
$H(s,r) = H(0.45, 0.05, 0.10, 0.40) = 1.106$

The expected conditional entropy $H(s\,|\,r)$ measures the average uncertainty about the input once the output is known and is called the *equivocation* of the system.

equivocation = $H(s\,|\,r) = H(s,r) - H(r) = 0.418$

The amount of information transmitted, $T(s,r)$, is the difference between the information at the source and the noise induced equivocation of the system:

$T(s,r) = H(s) - H(s|r) = 0.275$

If the system is so noisy that the signal sent and that received are independent of each other then $H(s) = H(s|r)$ and the amount of information that can be transmitted is, as you would expect, zero. If, at the other extreme, the system is completely free of noise then transmission is perfect and the information transmitted is the same as that sent: $T(s,r) = H(s)$.

8.5 BAYES' RULE

Recall the weather forecasting problem of Chapter 5. Table 5.1 (shown here again as Table 8.12) describes the accuracy of a local weather forecaster.

A prior estimate that prob(wet) = 0.7 gives the result shown in Table 8.13.

		forecast – f		
		wet	dry	
actual state – s	wet	0.8	0.2	1.0
	dry	0.1	0.9	1.0

Table 8.12
The accuracy of the weather forecaster

		forecast – f		
		wet	dry	
actual state – s	wet	0.56	0.14	0.7
	dry	0.03	0.27	0.3
		0.59	0.41	

Table 8.13
The Bayesian analysis

The similarities with the information transmission problem discussed in the previous section are obvious. To emphasise this consider the statement of Bayes' Rule in (5.1). This can be rewritten using the notation of Table 8.13 as

$$\frac{\text{prob}(s|f)}{\text{prob}(s)} = \frac{\text{prob}(f|s)}{\text{prob}(f)}$$

Taking logarithms gives

$$\log[\text{prob}(s|f)] - \log[\text{prob}(s)] = \log[\text{prob}(f|s)] - \log[\text{prob}(f)]$$

Averaging over all possible states, s, and forecasts, f, and remembering to take the negative value of the logarithms, gives entropies

$$H(s) - H(s|f) = H(f) - H(f|s)$$

Just as the receipt of a signal over a (noisy) channel will reduce the uncertainty about which message is sent so a Bayesian revision will reduce the uncertainty of a probabilistic estimate. $H(s)$ is the entropy of the prior distribution and $H(s|f)$ is the entropy of the posterior distribution. Unless the data source is useless the posterior will be less uncertain than the prior so that $H(s|f) < H(s)$ and

Bayesian information gain = H(prior) – H(posterior) (8.14)

which is the uncertainty reduction "transmitted" by a Bayesian revision. We have met this idea before as the Bayesian learning of Figures 5.9 and 5.10.

From Table 8.13, the main entropies are

$H(s,f) = H(0.56, 0.14, 0.03, 0.27) = 1.059$
$H(s) = H(0.7, 0.3) = 0.611$
$H(f) = H(0.59, 0.41) = 0.677$
$H(s|f) = H(s,f) - H(f) = 0.382$
$H(f|s) = H(s,f) - H(s) = 0.448$

so,

information gain = $0.611 - 0.382 = 0.229$

$H(f)$ measures the uncertainty of the predictive distribution which estimates which forecast will in fact be made. Here, $H(0.59, 0.41) = 0.677$.

$H(f|s)$ measures the uncertainty of the likelihood distributions. If the forecasts are perfectly accurate (Table 8.14) $H(f|s)$ will be zero as will the posterior uncertainty, $H(s|f)$. The entropy of the predictive distribution will be equal to that of the prior: $H(s) = H(f)$.

		forecast – f		
		wet	dry	
actual	wet	p	0	p
state – s	dry	0	$1-p$	$1-p$
		p	$1-p$	

Table 8.14
A perfect forecaster

If the forecaster is useless (Table 8.15) then all the likelihood distributions will be uniform and $H(f \mid s)$ will be at a maximum. The posterior and prior distributions will be equal and the predictive distribution will be uniform, and so have maximum entropy.

		forecast – f		
		wet	dry	
actual	wet	$p/2$	$p/2$	p
state – s	dry	$(1-p)/2$	$(1-p)/2$	$1-p$
		$1/2$	$1/2$	

Table 8.15
A perfectly useless forecaster

8.6 FURTHER READING

There are many good introductory texts dealing with the analysis of contingency tables among which are those given by Fienberg (1991) and also Agresti (1990) and Upton (1978). An important general discussion of such tables is to be found in Goodman and Kruskal (1979) while Gokhale and Kullback (1978) proceed from an information theoretic point of view.

Derivations of equation (8.11) are given in, for instance, Miller (1955) and Tribus (1969, pp196–197). The result is referred to by Attneave (1959), Garner (1962), Garner and McGill (1956) and Rosenkrantz (1970). McGill (1954) discusses information transmission in the context of the perception of different noise levels.

8.7 EXAMPLES

Example 8.1

A survey was conducted (Mantle *et al.*, 1977) to explore the occurrence of backpain in pregnant women:

Number of previous children	Degree of backpain severity				Total
	0	1	2	3	
0	8	56	28	9	101
1	3	16	22	11	52
2 or more	2	8	10	7	27
Total	13	80	60	27	180

(Source: Chatfield, 1985)

Chatfield (1985) discusses these data and their collection and gives reasons for being wary of generalising from them. Based on what we have, however, is it reasonable to believe that backpain varies according to the number of children previously borne?

Solution

Calculate the table of frequencies that would be expected under the assumption of independence. Using (8.13) the expected number of women with backpain severity 0 and no previous children is

$$\frac{\text{(row total)} \times \text{(column total)}}{\text{grand total}} = \frac{101 \times 13}{180} = 7.29$$

Continuing this way gives:

Number of previous children	Degree of backpain severity				Total
	0	1	2	3	
0	7.29	44.88	33.67	15.15	101
1	3.76	23.11	17.33	7.80	52
2 or more	1.95	12.00	9.00	4.05	27
Total	13	80	60	27	180

Using (7.11),

$$G^2 = 2[8\ln(8/7.29) + 56\ln(56/44.88) + \ldots] = 14.87$$

With $(2-1)(4-1) = 3$ degrees of freedom and $\alpha = 0.05$, $\chi^2_{CRIT} = 7.815$. Since $G^2 > \chi^2_{CRIT}$ we reject the hypothesis of independence (with a 5% chance of doing so incorrectly) and say that there is a relationship between severity of backpain suffered by pregnant women and the number of children previously borne by them.

An alternative, and equivalent, measure of dissimilarity is Y^2. Using (8.9),

$$H(\text{backpain severity}) = \ln(180) - [13\ln(13) + 80\ln(80) + 60\ln(60) + 27\ln(27)]/180$$
$$= 1.201$$

Similarly,

$$H(\text{previous children}) = \ln(180) - [101\ln(101) + 52\ln(52) + 27\ln(27)]/180$$
$$= 0.968$$

$$H((\text{backpain severity}) \text{ and (previous children)})$$
$$= \ln(180) - [8\ln(8) + 56\ln(56) + \ldots + 7\ln(7)]/180$$
$$= 2.127$$

From (8.8)

$$T = 1.201 + 0.968 - 2.127 = 0.042$$

From (8.11)

$$Y^2 = 1.9712 \times 180 \times 0.042 = 14.90$$

which is close to G^2 and the same conclusion would have been reached.

Example 8.2

Here are 100 values that were generated by the uniform random number generator provided in the spreadsheet that I use. The values were generated by column and so were in the order 0, 5, 9, ...

0	0	0	6	0	0	5	9	8	3
5	7	4	7	2	1	8	4	8	9
9	4	5	0	3	3	3	1	9	3
0	8	3	5	1	2	4	3	8	2
0	9	4	5	7	1	8	8	9	2
4	8	7	1	6	9	2	0	1	3
4	2	4	8	0	7	4	5	2	5
1	4	5	2	2	7	8	8	1	4
4	8	4	1	6	3	2	1	6	5
9	9	6	7	2	8	4	0	3	4

Is the generator satisfactory?

Solution

The first requirement of a generator such as this is that all ten possible values occur with equal probability so that in a sample of 100 the expected frequency for each value, r, is 10. The table shows the necessary calculation for ψ.

Sec. 8.7] Examples

r	o	e	oln(o/e)
0	11	10	1.048
1	10	10	0.000
2	11	10	1.048
3	10	10	0.000
4	15	10	6.082
5	9	10	–0.948
6	5	10	–3.466
7	7	10	–2.497
8	13	10	3.411
9	9	10	–0.948
	100	100	3.731

$\psi = 3.731$ and so $G^2 = 2\psi = 7.46$. The only information needed from the data was the sample size and so there are $10 - 1 = 9$ degrees of freedom. With $\alpha = 0.05$ $\chi^2_{CRIT} = 16.92$. Since $G^2 < \chi^2_{CRIT}$ the hypothesis that all r values occur with equal frequency is not refuted.

So far so good. Another desirable property is that each value occurs independently of the occurrence of any other value or values. A fairly obvious way to start thinking about this is to consider each consecutive pair of values, (0,5), (5,9), (9,0) and so on. The second value of the pair should be independent of the first and so each should occur with probability 0.1 whatever the value of the first of the pair. Here is a tabulation of the 99 pairs:

first of pair	second of pair											
0	0	1	2	2	4	4	5	5	5	7	8		
1	0	2	3	3	4	6	7	7	8	9			
2	0	1	1	1	2	3	3	4	4	4	6		
3	1	2	2	3	4	4	5	8	8	9			
4	1	1	4	5	5	5	6	7	8	8	8	8	9	9
5	1	3	4	4	4	5	8	8	9				
6	0	2	3	6	7								
7	0	0	3	4	4	6	7						
8	0	1	2	2	2	2	3	5	8	9	9	9	9
9	0	0	0	1	3	4	7	8	8				

The straightforward approach is to form a contingency table and calculate G^2. There is some debate about the point at which the approximation of the distribution of G^2 by χ^2 ceases to be satisfactory (Appendix A5.7) but it seems safe to say that not too many cells should have small expected frequencies. The proposed contingency table would have 100 cells but only 99 observations and so is unlikely to provide the necessary preconditions. By grouping the r values (rows and columns) in pairs the number of cells is reduced to 25 and so we may proceed. Here is the contingency table; the rows show the values of the first of each of the 99 pairs and the columns show the second so that, for example, in five of the 99 pairs the first value was 2 or 3 and the second 0 or 1:

		\multicolumn{5}{c}{second of pair}	
		(0,1)	(2,3)	(4,5)	(6,7)	(8,9)	
	(0,1)	3	5	6	4	3	21
first	(2,3)	5	6	6	1	3	21
of	(4,5)	3	1	8	2	9	23
pair	(6,7)	3	3	2	4		12
	(8,9)	6	6	2	1	7	22
		20	21	24	12	22	99

The frequencies expected given the hypothesised independence are found in the usual way. For instance, in row 3 column 2 we put $(23 \times 21)/99 = 4.879$. Here is the complete table:

4.242	4.455	5.091	2.545	4.667	21
4.242	4.455	5.091	2.545	4.667	21
4.646	4.879	5.576	2.788	5.111	23
2.424	2.545	2.909	1.455	2.667	12
4.444	4.667	5.333	2.667	4.889	22
20	21	24	12	22	99

Finally compute the values of $o_{ij}\ln(o_{ij}/e_{ij})$:

−1.040	0.578	0.986	1.808	−1.325
0.822	1.787	0.986	−0.934	−1.325
−1.312	−1.585	2.888	−0.664	5.092
0.639	0.493	−0.749	4.046	0.000
1.801	1.508	−1.962	−0.981	2.513

14.067

So, $\psi = 14.067$ and $G^2 = 2\psi = 28.13$.

With $(5-1)(5-1) = 16$ degrees of freedom and $\alpha = 0.05$, $\chi^2_{\text{CRIT}} = 26.30$. G^2 exceeds this value and so the hypothesis is not supported. We conclude that the generator is suspect because of possible correlation between successive values.

There are many methods for testing random number generators. See, for example, Conolly (1981, Sec. 7.6).

Example 8.3

Here is a quotation from *The Independent on Sunday* (20 June 1993). The article from which it is taken was called "Who Pays the Piper?" by Stephen Castle and Nick Cohen:

> "In 1991, the last year for which reasonably full figures are available, Labour Research found that 42 of the top 100 companies in the country had given money to the Conservatives since 1980. Of these, 28 – or 67 per cent – had directors and senior executives who had been awarded honours. But, of the remaining 58 companies which made no donations, only 23 (40 per cent) had figured on honours lists."

What conclusions might reasonably be drawn about the relationship between donations and honours?

Solution

The data in the quotation may be tabulated as a contingency table:

	awarded honours	not awarded honours	
gave money	28	14	42
did not give money	23	35	58
	51	49	100

If we hypothesise that donations and honours are statistically independent then expected frequencies are found using (8.13). For example, the upper left cell of the table contains an expected frequency of

$$\frac{42 \times 51}{100} = 21.4$$

The four expected frequencies are:

	awarded honours	not awarded honours	
gave money	21.4	20.6	42
did not give money	29.6	28.4	58
	51	49	100

From (8.12) this problem is characterised by $(2-1)(2-1) = 1$ degree of freedom.

We now calculate the value of ψ:

$$\psi = \Sigma\, o_i \ln(o_i/e_i)$$

$$= 28\ln(28/21.4) + 14\ln(14/20.6) + 23\ln(23/29.6) + 35\ln(35/28.4)$$

$$= 7.527 - 5.407 - 5.802 + 7.314 = 3.632$$

so

$$G^2 = 2\psi = 2 \times 3.632 = 7.264$$

Now, with $\alpha = 0.05$, $\chi^2_{CRIT} = 3.842$ which is less than the calculated G^2 value of 7.264 and so we reject the null hypothesis and say that the data do not support the idea that political contributions and honours awarded are independent. We would therefore feel inclined to go along with the clear implication of the journalists Castle and Cohen, and Labour Research, that directors of companies giving money to the Conservatives are more likely to obtain honours than those from companies making no such contribution.

Are you happy with $\alpha = 0.05$? Remember that what it means is that in rejecting the proposition that contributions and honours are not linked there is a 1 in 20 chance that you are wrong.

It depends on who you are, of course, and on what are the penalties that you may suffer should the charge be shown to be false.

You may prefer a more cautious approach and be prepared to accept only a 1 in 200 chance of making an incorrect inference. In this case $\alpha = 0.005$ and so $\chi^2_{CRIT} = 7.879$. This is larger, though not by much, than the calculated value and so at this level of risk you would not reject the hypothesised independence and would not support the idea that contributions and honours are linked.

After some interpolation from the tables we can find that $\chi^2_{CRIT} = 7.264$, the calculated value, when $\alpha = 0.0075$. This represents a chance of error of 1 in 133.

In summary, if you are prepared to take a risk a little higher than 1 in 133 of being wrong then you should say that the proposition that there is no link between contributions and honours is not supported by the data. If you are more cautious then you will support the proposition.

If you decide against the hypothesis of independence ought you then to support allegations that honours have been sold? Not necessarily. An alternative behavioural explanation is that successful companies support the policies favoured by the Conservatives and that some will choose to express their support through financial contributions. Successful companies also contain the more able managers and directors (hence their success) who as individuals possess exactly those characteristics that attract honours.

Over to you.

Example 8.4

The *FT–SE 100* index (the *Footsie*) measures share prices on the London Stock Exchange. Here are the values of the index for the 44 trading days of September and October 1993:

3100.0	3085.1	3072.6	3057.3	3059.0	3038.6	3035.4	
3031.2	3037.0	3024.8	3028.0	2989.4	3003.9	3005.5	
3004.5	3001.6	3007.5	3001.3	3005.2	3026.3	3036.9	
3030.1	3037.5	3039.3	3067.7	3085.2	3100.8	3092.4	
3108.6	3102.2	3094.7	3080.9	3086.3	3120.8	3137.6	
3129.6	3156.3	3188.3	3199.0	3184.8	3165.3	3154.3	
3163.0	3170.0						

To an outsider it may seem reasonable to believe that the behaviour of the market ought not to be random and that, in consequence, knowledge of the movement of share prices between yesterday and today ought to be of help in determining the prices tomorrow. On the other hand, the *efficient market hypothesis* states that share prices reflect all publicly available information and so are of no use in making such forecasts. What might we conclude from the data?

Solution

There are 43 movements of *Footsie* during this period (fall, fall, fall, rise, fall, ...) and so 42 pairs of movements; let us call them yesterday's movement and today's movement. Here is a tabulation of those movements:

		today rise	today fall	
yesterday	rise	12	10	22
	fall	11	9	20
		23	19	42

There doesn't seem to be much evidence here of non-random behaviour. Let us see if some analysis supports this initial impression.

There are a number of ways that we can characterise this problem. Think of it as an information transmission process in which the current known behaviour of the market constitutes a message that we receive, r, about future behaviour. In information theoretic terms this future is the source, s. Now,

$$H(s,r) = H(12,10,11,9)$$
$$= \ln(42) - [12\ln(12) + 10\ln(10) + 11\ln(11) + 9\ln(9)]/42$$
$$= 1.381$$

$$H(r) = H(22,20)$$
$$= \ln(42) - [22\ln(22) + 20\ln(20)]/42 = 0.692$$

$$H(s) = H(23,19)$$
$$= \ln(42) - [23\ln(23) + 19\ln(19)]/42 = 0.689$$

The equivocation in the system is

$$\text{equivocation} = H(s\,|\,r) = H(s,r) - H(r)$$
$$= 1.381 - 0.692 = 0.689$$

which is the same as the unconditional entropy of the source and so the information transmitted is

$$T(s,r) = H(s) - H(s\,|\,r) = 0.0$$

Yesterday's *Footsie* movement contains no information about the direction of today's movement.

Alternatively, we could have found $Y^2 = G^2 = 0.001$, which tells the same story.

This example provides an interesting contrast with Example 8.2 since there seems to be less randomness in my random number generator than in the stock market, just as the *efficient market hypothesis* would lead us to believe. For an interesting view of market prediction, and a source of further reading, see Casti (1992, Ch. 4). The front page of the *Wall Street Journal* of 4 November 1993 may also be of interest.

8.8 EXERCISES

8.1 Revisit Example 1.3. What are you now able to say about the proposed independence?

8.2 The red-cockaded woodpecker makes its nest in a cavity in a tree. Some of these cavities are subsequently used by other creatures. Surveys to determine the contents of these cavities were undertaken in Georgia, USA. Here are some of the results:

	Year of survey	
Cavity contents	1990	1991
Southern flying squirrels	48	78
Birds	20	14
Black rat snakes	3	0
Grey squirrels	0	1
Racoons	1	0
Nesting or other material	204	163
Water	28	61

(Source: Loeb, 1993)

Has there been a significant change in the pattern of use of cavities vacated by the woodpeckers?

8.3 The number of vessels involved in collisions in different Japanese harbours in 1966–1968 were:

	Size of vessel (gross registered tons)			
Location	3000–20000	500–3000	100–500	20–100
Tokyo-Yokohama	122	295	1116	264
Nagoya	74	118	186	21
Osaka-Sakai	35	225	790	224
Kobe	53	123	434	220

(Source: Fuji and Yamanouchi, 1973)

Is there any relationship between the distribution of the sizes of ships involved in collisions and the harbour in which the collisions occur?

8.4 Among the concerns about industrial safety is that immigrant workers may not fully understand warning messages. A survey was carried out in 1985 in Edmonton, Canada, and is reported by Krahn *et al.* (1990). Immigrant workers of Portuguese, Vietnamese and Latin origins were shown the sign indicating Corrosion; 29.9% of the Portuguese correctly identified the sign. For the Vietnamese the figure was 54.1% and for the Latins 44.6%. The numbers interviewed in the three groups were 77, 74 and 74 respectively. Is there evidence in the data to support the proposition that recognition of this warning varies between immigrant groups?

8.5 In thinking about vehicles exceeding the speed limit on a particular stretch of road I decide that the proportion exceeding the limit is not less than 50% and my best (i.e. mean) guess is that it is 60%. I go to the road and of 30 vehicles seen 21 are speeding. By how much has my uncertainty, measured as entropy, changed?

Of a further 50 vehicles 41 are speeding. What is my uncertainty now?

Why did my uncertainty change as it did?

8.6 Binary numbers are like the decimal numbers with which we are familiar but each digit may have one of only two values, 0 and 1. Here are the first eight numbers in both systems:

decimal	0	1	2	3	4	5	6	7
binary	000	001	010	011	100	101	110	111

A simple device to display any of these eight values consists of three electric light-bulbs placed in a row. A lighted bulb indicates the binary digit 1 and an unlighted bulb indicates 0. Owing to a fault in the wiring when a bulb is switched on there is only a 90% chance that it will light. When switched off the light is always extinguished.

What is the equivocation in this system? How much information is transmitted?

If you see only the central bulb lit, what is your (probabilistic) estimate of the number being signalled?

8.7 Themes used in advertising found in women's magazines were examined for two American (*Glamour* and *Cosmopolitan*) and two British magazines (*Options* and *Woman's Journal*). The number of advertisements found in each of seven thematic categories was:

Theme	American	British
Beauty	47	6
Science	38	13
New and better	11	3
Traditional woman's role	12	10
Taste	21	26
Efficiency	0	11
Miscellaneous	12	10

(Source: Monk-Turner, 1990)

Was the distribution of themes used in America and Britain different?

8.8 In 1972 the US government set up Area Health Education Centers (AHECs) in which medical students could spend some time seeing unselected patients in order to obtain a better view of primary health care practice. A sample of 165 medical students was divided into three groups:

A   no AHEC exposure
B   AHEC exposure in a required ambulatory/primary care clerkship
C   AHEC exposure in a clinical elective

Seven years after graduation some had chosen primary care as their speciality while others had chosen other options:

		Student group		
		A	*B*	*C*
Primary care speciality?	Yes	26	21	16
	No	42	38	22

(Source: Brooks, 1992)

Has the AHEC programme affected choice of speciality?

9

Least Biased Estimates of Joint Distributions

Having considered the information characteristics of joint distributions we now apply the principle of entropy maximisation to show how these distributions may be estimated given certain *a priori* constraints, and how this may assist structural analyses.

9.1 MARGINAL DISTRIBUTIONS AS CONSTRAINTS

Imagine a brewery that sells three beers: *Windigut*, a keg beer; *Bigboy*, a strong ale; and *Litebru* for the health conscious drinker. It sells in four regions: North, South, East and West. As a competitor we estimate, based on our own sales, the proportions of beer sold by type and, separately, by region.

We now wish to estimate the joint distribution by type and region consistent with the above assessments. Our marginal probability distributions provide row and column constraints for the joint distribution, as shown in Table 9.1.

	N	S	E	W	
Windigut					0.5
Bigboy					0.3
Litebru					0.2
	0.3	0.2	0.1	0.4	

Table 9.1
Marginal probabilities

To find the least biased estimate of the joint probabilities p_{ij} (the probability in row i and column j) we use the methodology introduced in Chapters 3 and 4 and set out in detail in Appendix A1:

$$\text{maximise} \quad -\sum_i \sum_j p_{ij} \ln(p_{ij}) \tag{9.1}$$

Sec. 9.1]   **Marginal distributions as constraints**   213

subject to $\quad \sum_j p_{ij} = r_i \quad$ (9.2)

and $\quad \sum_i p_{ij} = c_j \quad$ (9.3)

where r_i and c_j are the row and column marginal probabilities.

It is shown in Appendix A1 that the appropriate estimate of a joint probability distribution is, most generally,

$$p_{ij} = a_i b_j q_{ij} \exp[\alpha g'(p_{ij}) + \beta h'(p_{ij}) + ...] \quad (9.4)$$

where:

p_{ij} is the required joint distribution
q_{ij} is an *a priori* estimate of p_{ij}
a_i is a multiplier associated with row i and constraint (9.2)
b_j is a multiplier associated with column j and constraint (9.3)

$g'(p_{ij})$ is the differential of constraint $g(p_{ij}) = k$
$\alpha, \beta, ...$ are the multipliers associated with $g(p_{ij}) = k$ etc.

This is the two-dimensional equivalent of equation (4.2). Look back and review that now so that you are comfortable with (9.4).

For our present problem we have just the row and column constraints and so (9.4) reduces to

$$p_{ij} = a_i b_j \quad (9.5)$$

Substituting this into (9.2) gives

$$\sum_j p_{ij} = \sum_j a_i b_j = a_i \sum_j b_j = r_i$$

so $\quad \sum_j b_j = r_i / a_i \quad$ (9.6)

and similarly,

$$\sum_i a_i = c_j / b_j \quad (9.7)$$

The sum of all joint probabilities is 1.0. Using this and (9.5) gives

$$\sum_i \sum_j p_{ij} = \sum_i \sum_j a_i b_j = \sum_i a_i \sum_j b_j = 1$$

And from (9.6) and (9.7)

$$(c_j/b_j)(r_i/a_i) = 1$$

so $\quad a_i b_j = r_i c_j = p_{ij}$ \hfill (9.8)

which means, in case you hadn't already guessed, that if the only *a priori* constraints are the marginal distributions the unbiased estimate of the joint distribution is obtained by assuming that the two variables are independent (Table 9.2).

	N	S	E	W	
Windigut	0.15	0.10	0.05	0.20	0.50
Bigboy	0.09	0.06	0.03	0.12	0.30
Litebru	0.06	0.04	0.02	0.08	0.20
	0.30	0.20	0.10	0.40	

Table 9.2
Estimate based only on marginal distributions and so assuming independence

9.2 USING A PREVIOUS ESTIMATE

While looking through some old records of beer sales we come upon some data, now a few years old, describing sales of our competitor's products. These are shown (a precision of just one figure was all that was given) in Table 9.3. The units are unknown.

	N	S	E	W	
Windigut	6	2	2	7	17
Bigboy	4	2	4	1	11
Litebru	1	5	9	3	18
	11	9	15	11	

Table 9.3
An old estimate of the distribution

We can incorporate this earlier estimate by treating it as our *a priori* distribution, q_{ij}, and so (9.4) gives

$$p_{ij} = a_i b_j q_{ij} \hfill (9.9)$$

Using a previous estimate

Remembering that the constants a_i and b_j come from the constraints imposed by the marginal distributions we should choose them to ensure that those constraints are satisfied.

The first row sum of the distribution in Table 9.3 is 17 but we want it to be 0.5. This can be achieved by multiplying every value in the first row by 0.5/17. Similarly the second row is multiplied by 0.3/11 and the third by 0.2/18 to give

	N	S	E	W
Windigut	0.18	0.06	0.06	0.21
Bigboy	0.11	0.05	0.11	0.03
Litebru	0.01	0.06	0.10	0.03
column sum:	0.30	0.17	0.27	0.27
required:	0.30	0.20	0.10	0.40

Table 9.4
Adjusted distribution with correct row margins

The row marginal totals are now correct but the column margins are not. Multiplying the columns by 0.3/0.3, 0.2/0.17, 0.10/0.27 and 0.40/0.27 gives

	N	S	E	W	sum	req'd
Windigut	0.18	0.07	0.02	0.31	0.58	0.50
Bigboy	0.11	0.06	0.04	0.04	0.26	0.30
Litebru	0.01	0.07	0.04	0.05	0.16	0.20

Table 9.5
Distribution with correct column margins

Now the row margins are incorrect and so we continue adjusting rows and columns in this way until we arrive at a satisfactorily accurate result. Continuing until marginal errors do not exceed 5% gives

	N	S	E	W	sum	req'd
Windigut	0.16	0.06	0.02	0.28	0.51	0.50
Bigboy	0.13	0.07	0.04	0.05	0.30	0.30
Litebru	0.01	0.07	0.04	0.06	0.19	0.20

Table 9.6
Final distribution with correct column margins and all row margins within 5%

The balancing or rows and columns in this way is sometimes called the *iterative proportional fitting procedure* (IPFP). Note that we could have saved the adjustment factors and so found values for the a_i's and b_j's but that, as here, this will not usually be required.

9.3 JOINT DISTRIBUTION WITH GIVEN MEAN

Reading in a freight transport journal an article written by the transport manager of our competitor brewery we learn that the average journey length for its vehicles is 125.8 miles. Can we make use of this information to estimate better the joint distribution of sales by product and region?

We know that the beers are brewed in three different breweries located around the country, each brewery producing just one type of beer. This enables us to make an estimate of the average length of journey to deliver each beer into a region. For example, we think that the mean journey length of delivery trucks from the Windigut plant to pubs and other outlets in the Northern region is 241 miles. Here are the rest of the journey distances:

	N	S	E	W
Windigut	241	126	58	73
Bigboy	197	106	124	62
Litebru	213	154	62	137

Table 9.7
Mean length of delivery – miles

For clarity of discussion we shall for the moment disregard the old distribution of Table 9.3, but shall return to it in the next section. So we now have, as well as the marginal constraints, a new constraint on delivery distance:

$$\sum_i \sum_j p_{ij} d_{ij} = \mu = 125.8 \text{ miles} \qquad (9.10)$$

where d_{ij} is the mean delivery distance for beer i into region j. From (9.4)

$$p_{ij} = a_i b_j \exp[\alpha g'(p_{ij})] \qquad (9.11)$$

with $g(p_{ij}) = \sum_i \sum_j p_{ij} d_{ij} = \mu$

so $g'(p_{ij}) = d_{ij}$

and, from (9.11),

$$p_{ij} = a_i b_j \exp(\alpha d_{ij}) \qquad (9.12)$$

Sec. 9.3] Joint distribution with given mean 217

The a's and b's are row and column factors as before. For some given value of α, which controls the mean distance, we proceed by first calculating $\exp(\alpha d_{ij})$ and then balancing that table, just as we did when using the old distribution as a first guess. For example, if we set $\alpha = -0.04$ we get as a first estimate Table 9.8 (scaled for legibility).

	N	S	E	W
Windigut	0.07	6.47	98.27	53.93
Bigboy	0.38	14.41	7.01	83.74
Litebru	0.20	2.11	83.74	4.17

Table 9.8
Initial estimate = $1000 \exp(-0.04 \times \text{distance})$

After balancing rows and columns we obtain the estimated joint distribution shown in Table 9.9. This gives a mean journey distance of 123.16 miles, which is too short. Setting $\alpha = 0$ is equivalent to assuming independence and so results in Table 9.2 and a mean journey distance of 132.64 miles, which is too long. Adjusting the parameter value by trial and error or by the method of successive bifurcation we find that $\alpha = -0.023$ gives the joint distribution we want (mean journey = 125.80 miles) and this is shown in Table 9.10.

	N	S	E	W
Windigut	0.06	0.11	0.06	0.27
Bigboy	0.11	0.07	0.00	0.12
Litebru	0.13	0.02	0.04	0.01

Table 9.9
Final distribution; $\alpha = -0.04$

	N	S	E	W
Windigut	0.09	0.10	0.06	0.24
Bigboy	0.10	0.07	0.01	0.13
Litebru	0.10	0.03	0.03	0.03

Table 9.10
Final distribution; $\alpha = -0.023$

9.4 MODIFYING A PREVIOUS ESTIMATE WITH A NEW MEAN

Since we want an estimate that uses all the data we find the least biased estimate that respects both the marginal constraints and the mean journey distance constraint and is also based on the earlier estimate of the distribution. From (9.4)

$$p_{ij} = a_i b_j q_{ij} \exp(\alpha d_{ij}) \qquad (9.13)$$

If we again assume $\alpha = -0.04$ then the initial, unscaled, estimate $q_{ij} \exp(-0.04 d_{ij})$ is found by multiplying the elements in Tables 9.3 and 9.8. This gives Table 9.11 which, when the rows and columns are correctly balanced, gives the estimated joint distribution shown in Table 9.12.

	N	S	E	W
Windigut	0.04	1.29	19.65	37.75
Bigboy	0.15	3.88	2.81	8.37
Litebru	0.02	1.06	75.37	1.25

Table 9.11
Initial estimate = (Table 9.3) × (Table 9.8) [scaled for legibility]

	N	S	E	W
Windigut	0.08	0.06	0.02	0.34
Bigboy	0.18	0.08	0.00	0.04
Litebru	0.05	0.06	0.08	0.01

Table 9.12
Final distribution; $\alpha = -0.04$

This gives a mean journey distance of 124.34 miles. Trying different values of α soon gives Table 9.13 which has the correct mean distance of 125.80 miles and is based on all the *a priori* data available.

	N	S	E	W
Windigut	0.09	0.06	0.02	0.33
Bigboy	0.17	0.08	0.00	0.05
Litebru	0.04	0.06	0.08	0.02

Table 9.13
Final distribution; $\alpha = -0.0288$

9.5 ESTIMATING MARGINAL DISTRIBUTIONS

Both marginal distributions may not be known. We can still make estimates of the joint distribution and can then extract the unknown marginal distribution(s) if it is of interest. We shall look at two examples using the beer sales data. Applications to different sets of constraints are straightforward.

We have already solved one of these problems as an intermediate step in a longer calculation. Suppose that we had the old estimate of the joint distribution (Table 9.3) and also knew the national sales of each of the three types of beer as before

Windigut	50%
Bigboy	30%
Litebru	20%

but had no knowledge of beer sales by region. Taking the relevant terms from (9.4) gives the least biased estimate of the joint distribution as

$$p_{ij} = a_i q_{ij} \tag{9.14}$$

This says we should take the old distribution, q_{ij}, and scale the rows so that the row marginal constraints, sales by type of beer, are met. This is exactly what we did to get Table 9.4 which is the estimate that we now need. As well as having the joint distribution the column sums provide an estimate of the column marginal distribution, the percentage breakdown of beer sales by region:

North	30%
South	17%
East	27%
West	27%

For the second illustration suppose that we have the earlier distribution, q_{ij}, and also the data about journey length but no *a priori* marginal distributions at all. The estimating model is

$$p_{ij} = k q_{ij} \exp(\alpha d_{ij}) \tag{9.15}$$

where k is a single scaling factor ensuring that the joint distribution sums to 1.0. With $\alpha = -0.04$, Table 9.11 shows values of $q_{ij} \exp(\alpha d_{ij})$. Scaling this to sum to 1.0 gives Table 9.14 and a mean journey length of 68.25 miles.

	N	S	E	W
Windigut	0.03	0.86	13.05	25.06
Bigboy	0.10	1.91	1.86	5.56
Litebru	0.01	0.70	50.03	0.84

Table 9.14
$p_{ij} = kq_{ij} \exp(-0.04d_{ij})$ [probabilities shown as percentages]

After some searching we find that $\alpha = -0.000265$ gives a distribution with the required mean journey length of 125.80 miles and the joint distribution in Table 9.15:

	N	S	E	W	
Windigut	12.65	4.35	4.43	15.43	36.86
Bigboy	8.53	4.37	8.70	2.21	23.81
Litebru	2.12	10.79	19.90	6.50	39.31
	23.30	19.51	33.03	24.14	

Table 9.15
$p_{ij} = kq_{ij} \exp(-0.000265d_{ij})$ [probabilities shown as percentages]

The row and column totals provide estimates of the marginal distributions so that, for example, 19.51% of all sales are made in the Southern region and 39.31% of beer sales are of Litebru.

9.6 THE STRUCTURAL COMPARISON OF TABLES

Massiah (1986) gives the results of a survey of Caribbean women showing the relationship between activity status, A, and union status, U. The categories are:

activity status, A:
    a_1 not employed
    a_2 home services/home production
    a_3 own business
    a_4 employed by others

union status, U:
    u_1 married
    u_2 common law
    u_3 visiting
    u_4 single
    u_5 none

Tables 9.16 and 9.17 show the results for two islands.

		u_1	u_2	u_3	u_4	u_5	
				U			
A	a_1	2.5	1.9	4.7	2.4	0.5	12.0
	a_2	20.9	4.7	3.4	8.5	1.0	38.5
	a_3	2.2	0.9	0.5	1.5	0.3	5.4
	a_4	14.1	4.4	13.4	11.0	1.2	44.1
		39.7	11.9	22.0	23.4	3.0	

Table 9.16
Joint distribution (%) in Barbados (sample size, $n = 590$)

		u_1	u_2	u_3	u_4	u_5	
				U			
A	a_1	1.7	1.3	4.0	2.4	0.4	9.8
	a_2	21.2	11.2	5.4	8.7	1.5	48.0
	a_3	5.2	2.2	0.6	1.5	0.2	9.7
	a_4	9.8	4.8	8.4	8.4	1.1	32.5
		37.9	19.5	18.4	21.0	3.2	

Table 9.17
Joint distribution (%) in St. Vincent (sample size, $n = 462$)

There are a number of similarities in the general shape but a true comparison of the underlying structural relationships is made difficult because of the different marginal distributions found in the samples. The row and column balancing procedure overcomes this, as suggested by Mosteller (1968).

Perhaps the most straightforward way to proceed is to take one table as an *a priori* estimate of the other and then use (9.9) and the subsequent balancing. Suppose we take the data from Barbados as an initial estimate, q_{ij}, of the distribution found in St. Vincent, p_{ij}. Balancing rows and columns of Table 9.16 to respect the margins of Table 9.17 gives

		U					
		u_1	u_2	u_3	u_4	u_5	
	a_1	1.7	2.4	3.6	1.7	0.4	9.8
A	a_2	23.4	9.4	4.2	9.6	1.3	48.0
	a_3	3.4	2.5	0.9	2.4	0.5	9.7
	a_4	9.3	5.2	9.7	7.3	0.9	32.5
		37.9	19.5	18.4	21.0	3.2	

Table 9.18
Joint distribution (%) in St. Vincent estimated from the Barbados data (Table 9.16)

First compare Tables 9.16 and 9.17 and then compare Tables 9.18 and 9.17. You will probably get a slightly different feeling for where the structural differences occur. In the comparison using the adjusted table most of the large interactions are seen to resemble each other more closely while some of the smaller interactions have become less similar.

To eliminate the marginal effect entirely and concentrate even more sharply on the interactions we could set both marginal distributions to be uniform:

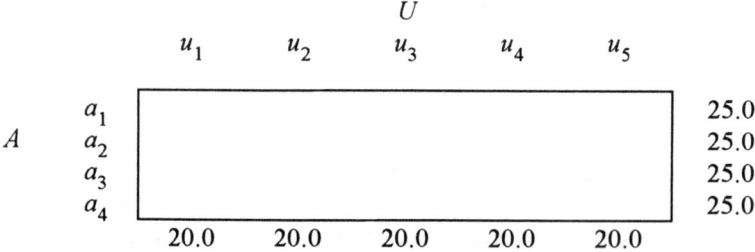

		U					
		u_1	u_2	u_3	u_4	u_5	
	a_1						25.0
A	a_2						25.0
	a_3						25.0
	a_4						25.0
		20.0	20.0	20.0	20.0	20.0	

Adjusting Tables 9.16 and 9.17 to these margins gives Tables 9.19 and 9.20.

		u_1	u_2	U u_3	u_4	u_5	
	a_1	2.5	5.3	8.3	3.8	5.1	25.0
A	a_2	8.2	5.2	2.3	5.3	4.0	25.0
	a_3	5.0	5.7	2.0	5.4	7.0	25.0
	a_4	4.4	3.9	7.4	5.5	3.9	25.0
		20.0	20.0	20.0	20.0	20.0	

Table 9.19
Joint distribution (%) in Barbados adjusted for uniform margins

		u_1	u_2	U u_3	u_4	u_5	
	a_1	2.0	3.1	9.0	5.2	5.8	25.0
A	a_2	6.0	6.3	2.9	4.5	5.2	25.0
	a_3	8.2	6.9	1.8	4.3	3.8	25.0
	a_4	3.8	3.7	6.3	6.0	5.2	25.0
		20.0	20.0	20.0	20.0	20.0	

Table 9.20
Joint distribution (%) in St. Vincent adjusted for uniform margins

Your feelings when comparing this pair of tables are likely to be quite different again. See, for instance, how the interactions between a_3 and u_5 vary between the two islands. The interactions between categories which initially had low marginal totals are given more prominence in this comparison.

Using this idea of structure enables you to discuss the differences in underlying interactions and differences in marginal distributions quite separately. For instance, we may say that although there is a large difference in the interaction between a_3 and u_5 the comparative rarity of women in the u_5 category may render this of limited practical importance.

Again, and perhaps more interestingly, the interactions between a_2 and u_1 appear to be quite similar in the original distributions but differences emerge when we focus on the underlying structures.

To see the differences in interpretation more clearly Tables 9.21 and 9.22 show the adjusted and unadjusted distributions for both islands displayed together. To aid clarity all figures have been rounded and adjusted to respect the integer marginal totals.

Table 9.21 depicts the structural similarities and differences much more clearly than does Table 9.22 and so helps an understanding of the data.

		u_1	u_2	U u_3	u_4	u_5	
	a_1	2	3	9	5	6	25
		3	5	8	4	5	
	a_2	6	6	3	5	5	25
A		8	5	3	5	4	
	a_3	8	7	2	4	4	25
		5	6	2	5	7	
	a_4	4	4	6	6	5	25
		4	4	7	6	4	
		20	20	20	20	20	

Table 9.21
Joint distributions (%) adjusted for uniform margins: top row of pair is St. Vincent, bottom is Barbados

		u_1	u_2	U u_3	u_4	u_5	
	a_1	2	2	4	2	0	9
		3	2	5	2	1	13
	a_2	21	11	5	9	2	48
A		21	5	3	9	1	39
	a_3	5	2	1	2	0	10
		2	1	1	1	0	5
	a_4	10	5	8	8	1	32
		14	4	13	11	1	43
		38	19	18	21	3	
		40	12	22	23	3	

Table 9.22
Unadjusted joint distributions (%): top row of pair is St. Vincent, bottom is Barbados

Remember the sense in which we may truly say that the distributions in Table 9.21 represent the underlying structures in the data. It is that the original tables may be obtained exactly from these structural distributions by simple row and column balancing to obtain the (different) marginal distributions found on the two islands.

9.7 MODELLING TWO-DIMENSIONAL CONTINGENCY TABLES

We now extend the application of maximum entropy estimates to the further analysis of the structure inherent in tabular data. The strategy is to see how few constraints are necessary to describe the data. A powerful model is one that provides a good description while maintaining a large number of degrees of freedom. In consequence we start with very simple models and then add constraints until all degrees of freedom have gone.

Staff turnover can often be a problem. Hiscott and Connop (1990) surveyed the nursing staff in an Ontario psychiatric hospital. Staff were described according to whether or not they had left their jobs at the hospital in the 3 year period 1984–1987, and also according to whether they worked long 12 hour shifts or the more regular 7.5 or 8 hour shifts. Table 9.23 shows some of the resulting tabulation.

What models are available for these data?

	long shift	regular shift	
stayed in job	33	45	78
left job	18	3	21
	51	48	99

Table 9.23
(Source: Hiscott and Connop, 1990)

The simplest model is obtained when all we have is the total number of observations. Working with probabilities, we wish to maximise

$$H = -\sum_i \sum_j p_{ij} \ln(p_{ij})$$

subject to

$$\sum_i \sum_j p_{ij} = 1$$

This results in the uniform distribution $p_{ij} = 1/n$, as we have already seen in (3.4). Scaling the probabilities to frequencies gives Table 9.24.

	long shift	regular shift	
stayed in job	24.75	24.75	49.5
left job	24.75	24.75	49.5
	49.5	49.5	99.0

Table 9.24
Total only model

The correspondence between this model and the data is

$$G^2 = 2[33\ln(33/24.75) + 45\ln(45/24.75) + 18\ln(18/24.75) + 3\ln(3/24.75)]$$
$$= 48.67$$

The model required only one parameter, the total, and so is characterised by $4 - 1 = 3$ degrees of freedom.

We may try to improve the correspondence between the model and the data by increasing the number of constraints. Suppose that we specified the column marginal distribution and so wished to maximise H subject to

$$\sum_i p_{ij} = c_j, \qquad j = 1,2,...,m$$

Satisfy yourself that this leads to

$$p_{ij} = \exp(-1 + \alpha_j) = a_j$$

Substituting in the row constraint equation gives

$$\sum_i p_{ij} = \sum_i a_j = ma_j = c_j$$

so

$$p_{ij} = a_j = c_j/m \qquad (9.16)$$

This means that the cells in each column contain equal values that sum to the marginal total. Table 9.25 gives the result:

	long shift	regular shift	
stayed in job	25.50	24.00	49.5
left job	25.50	24.00	49.5
	51.0	48.0	99.0

Table 9.25
Column totals model

Comparison with the data gives $G^2 = 48.58$ so incorporating knowledge of the column margins has not resulted in much of an improvement. This is because the column totals are barely different from those of the uniform marginal distribution of the previous total only model (Table 9.24). Two parameters were needed in the column total model, the two column totals (or, alternatively, the grand total and one column total), and so the number of degrees of freedom has been reduced to $4 - 2 = 2$.

Instead of column totals we could use the analogous row total model where only the row margin is specified. This gives Table 9.26:

	long shift	regular shift	
stayed in job	39.00	39.00	78.0
left job	10.50	10.50	21.0
	49.5	49.5	99.00

Table 9.26
Row totals model

This corresponds more closely to the data, $G^2 = 13.74$, so knowing only the row totals (and forming the appropriate maximum entropy estimate) offers some useful explanation of the data.

The natural next step is to see if a better explanation is obtained if both row and column margins are specified. We already know the model to use: it is given by equation (9.8). Table 9.27 gives the result. G^2 has been reduced to 13.65, only a slight improvement. The number of degrees of freedom is $4 - 3 = 1$.

	long shift	regular shift	
stayed in job	40.18	37.82	28.0
left job	10.82	10.18	21.0
	51.0	48.0	99.0

Table 9.27
Row and column totals model (independence)

We could, of course, give one more piece of information, that the frequency in the cell at row 1 column 1 is 33, for example, and achieve a perfect correspondence. G^2 would be zero but so too would the number of degrees of freedom: we would have produced a tautology. Such a model is said to be *saturated*.

Table 9.28 summarises the results obtained so far.

model	G^2	degrees of freedom	χ^2_{CRIT} ($\alpha = 0.05$)
total only	48.67	3	7.81
columns	48.58	2	5.99
rows	13.74	2	
rows and columns	13.65	1	3.84
saturated	0	0	

Table 9.28

What are we to conclude? The only substantial reduction in G^2 was obtained by the inclusion of given row totals as constraints. We may therefore say that the variations seen in the data are best explained by the difference between the number of nurses leaving their job at the hospital and those staying. And yet none of the models provides a satisfactory fit ($G^2 < \chi^2_{CRIT}$) and so the interaction effect of the saturated model is needed to explain the data: this interaction is significant.

9.8 MODELLING THREE-DIMENSIONAL CONTINGENCY TABLES

But there may be another effect at work. Table 9.29 depicts the same data but now disaggregated into a third dimension: whether or not the nurse had a child living at home.

Before moving on pause to check that you understand what has happened. We now have a three way table, a 2 × 2 × 2 cube. This apparently modest increase in problem size brings a large increase in complexity, in visualising the data. For instance, we could tabulate shift against children at home or not for both those that stayed in the job and those that left. This might look quite different to Table 9.29 (try it) but would still be the same data. We need to proceed with some care, using the same principles as those used above.

Sec. 9.8] Modelling three-dimensional contingency tables

	long shift	regular shift		
stayed in job	25	30	55	(a) children living
left job	5	1	6	at home
	30	31	61	

	long shift	regular shift		
stayed in job	8	15	23	(b) no children
left job	13	2	15	at home
	21	17	38	

Table 9.29
(Source: Hiscott and Connop, 1990)

As well as the one-dimensional marginal distributions with which we are familiar we now have three two-dimensional marginal distributions, each formed by collapsing (i.e. summing) the third dimension. Figure 9.1 shows these distributions with obvious abbreviations (C for children etc.). Make sure that you can construct these from Table 9.29. We shall call the one-dimensional marginal distributions [L], [S] and [C] and the two-dimensional distributions [LS], [LC] and [CS].

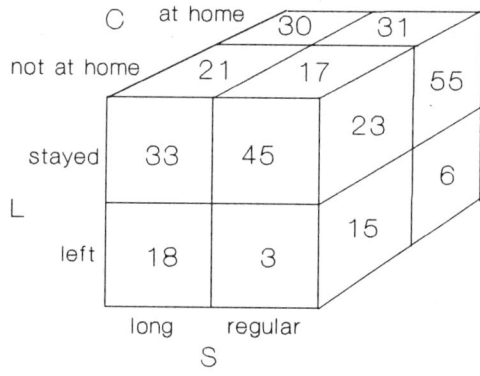

Figure 9.1 Three two-dimensional marginal totals

The eight small cubes that make up the large cube represent the eight frequencies of the 2 × 2 × 2 contingency table which we are about to model.
The first model is when all we know is the total, 99. In this case, as we have seen, all expected frequencies are the same and equal to 99/8 = 12.375. This is a natural extension of the two-dimensional result in the previous section. You may

derive the model yourself, if you wish. The other results in this section will be treated in the same way, without restating the full proofs provided in Appendix A1.

For this model

$$G^2 = 2[25\ln(25/12.375) + 30\ln(30/12.375) + ...]$$
$$= 66.98$$

Only one parameter, the total, was needed so the number of degrees of freedom is $8 - 1 = 7$.

We now consider the situation when we specify one one-dimensional and one two-dimensional distribution. The situation is shown in Figure 9.2. The distributions given are [C] and [LS]. If [C] has probabilities c_k and [LS] has probabilities g_{ij}, then we wish to find the maximum entropy estimate of the joint distribution p_{ijk} such that

$$\sum_k p_{ijk} = g_{ij}$$

and

$$\sum_i \sum_j p_{ijk} = c_k$$

which gives

$$p_{ijk} = c_k g_{ij} \qquad (9.17)$$

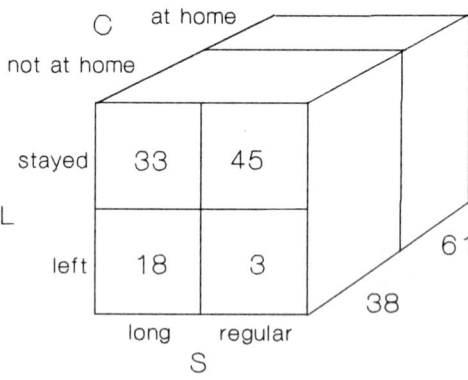

Figure 9.2 [LS][C] model

This model just apportions the joint marginal distribution g_{ij} according to the probabilities c_k. The frequency distribution corresponding to g_{ij} is [LS] and is shown in Table 9.23. From [C] $c_1 = 61/99$ and $c_2 = 38/99$. The estimated frequency corresponding to that in row 1 and column 1 of Table 9.29(a) is therefore $33(61/99) = 20.33$. The number of parameters needed by the model is all four values from the joint marginal and then either 38 or 61 and so the number of degrees of freedom is $8 - 5 = 3$. The results of applying this model and the two others of similar structure are given in Table 9.30. The cell references x_{111} etc. are given at the left and are followed by their definitions.

	L	S	C	data	[LS][C]	[LC][S]	[CS][L]
x_{111}	stayed	long	at home	25	20.33	28.33	23.64
x_{112}			not home	8	12.67	11.85	16.55
x_{121}		regular	at home	30	27.73	26.67	24.42
x_{122}			not home	15	17.27	11.15	13.39
x_{211}	left	long	at home	5	11.09	3.09	6.36
x_{212}			not home	13	6.91	7.73	4.46
x_{221}		regular	at home	1	1.85	2.91	6.58
x_{222}			not home	2	1.15	7.27	3.61
			G^2		12.92	14.45	26.22

Table 9.30
Observed and expected frequencies

The next set of models are those where two two-dimensional marginal distributions are given, as in Figure 9.3. The constraints are then

$$\sum_k p_{ijk} = g_{ij}$$

and $\sum_j p_{ijk} = h_{ij}$

which gives

$$p_{ijk} = a_{ik} b_{ij} \tag{9.18}$$

where the values of the a_{ik}'s and b_{ij}'s have to be found iteratively because of the interaction between them. This is similar to the situation in Tables 4.5–4.7 or the row and column balancing earlier in this chapter. Table 9.31 shows the calculation for the [LS][LC] model. In the first table we adjust to ensure that the [LS] margin is respected. Column (a) shows the data and (b) the starting estimate, a uniform distribution. Column (c) shows the values in the two cells whose sum must be 33, the number of nurses on long shift and who stayed in their jobs. The sum, at this stage in the calculation, is 2 and so both cells are multiplied by $33/2 = 16.5$ to give the new estimates in (d). Column (k) shows the new estimate of the table, corrected to meet the [LS] constraint, and these values are used as the starting point for the calculation in Table 9.31(b) wherein adjustments are made to ensure that the [LC] constraints are met. The values in column (k) of this table respect the values given in the [LS] margin and so no further iterations are required in this case.

	a	b	c	d	e	f	g	h	i	j	k
x_{111}	25	1.00	1.00	16.50							16.50
x_{112}	8	1.00	1.00	16.50							16.50
x_{121}	30	1.00			1.00	22.50					22.50
x_{122}	15	1.00			1.00	22.50					22.50
x_{211}	5	1.00					1.00	9.00			9.00
x_{212}	13	1.00					1.00	9.00			9.00
x_{221}	1	1.00							1.00	1.50	1.50
x_{222}	2	1.00							1.00	1.50	1.50
sum			2.00		2.00		2.00		2.00		99.00
constraint			33.00		45.00		18.00		3.00		
ratio			16.50		22.50		9.00		1.50		

(a)   adjusting for the [LS] margin

	a	b	c	d	e	f	g	h	i	j	k
x_{111}	25	16.50			16.50	23.27					23.27
x_{112}	8	16.50	16.50	9.73							9.73
x_{121}	30	22.50			22.50	31.73					31.73
x_{122}	15	22.50	22.50	13.27							13.27
x_{211}	5	9.00							9.00	5.14	5.14
x_{212}	13	9.00					9.00	12.86			12.86
x_{221}	1	1.50							1.50	0.86	0.86
x_{222}	2	1.50					1.50	2.14			2.14
sum			39.00		39.00		10.50		10.50		99.00
constraint			23.00		55.00		15.00		6.00		
ratio			0.59		1.41		1.43		0.57		

(b)   adjusting for the [LC] margin

Table 9.31
[LS][LC] model

Modelling three-dimensional contingency tables

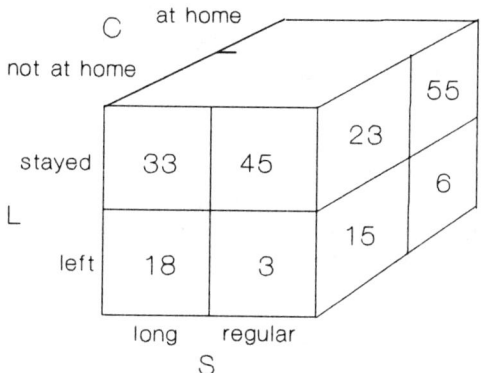

Figure 9.3 [LS][LC] model

How many parameters were required by this model? Given all four values in the [LS] marginal distribution we only needed two from the [LC] distribution, 23 and 15 say, for the others could be found by simple arithmetic (33 + 45 – 23 = 55). This model is characterised by 8 – 6 = 2 degrees of freedom. Table 9.32 shows the results from all three such models.

	L	S	C	data	[LS][LC]	[LC][SC]	[CS][LS]
x_{111}	stayed	long	at home	25	23.27	27.05	19.41
x_{112}			not home	8	9.73	12.71	13.59
x_{121}		regular	at home	30	31.73	27.95	29.06
x_{122}			not home	15	13.27	10.29	15.94
x_{211}	left	long	at home	5	5.14	2.95	10.59
x_{212}			not home	13	12.86	8.29	7.41
x_{221}		regular	at home	1	0.86	3.05	1.94
x_{222}			not home	2	2.14	6.71	1.06
			G^2		0.80	14.11	12.57

Table 9.32
Observed and expected frequencies

The final model, simpler than the last, is obtained by using the three one-dimensional margins as constraints: the [L][S][C] model. Satisfy yourself that the result is a simple extension of (9.8) and gives

$$p_{ijk} = a_i b_j c_k \tag{9.19}$$

where a_i, b_j and c_k are constants. Because of the interaction between them the calculations are performed iteratively and are shown in Table 9.33.

This model gives $G^2 = 26.57$ and 4 degrees of freedom. The results from all models are summarised in Table 9.34.

	a	b	c	d	e	f	g
x_{111}	25	1.00			1.00	15.25	15.25
x_{112}	8	1.00	1.00	9.50			9.50
x_{121}	30	1.00			1.00	15.25	15.25
x_{122}	15	1.00	1.00	9.50			9.50
x_{211}	5	1.00			1.00	15.25	15.25
x_{212}	13	1.00	1.00	9.50			9.50
x_{221}	1	1.00			1.00	15.25	15.25
x_{222}	2	1.00	1.00	9.50			9.50
sum			4.00		4.00		99.00
constraint			38.00		61.00		
ratio			9.50		15.25		

Table 9.33 (a)   adjusting for the [C] margin

	a	b	c	d	e	f	g
x_{111}	25	15.25	15.25	24.03			24.03
x_{112}	8	9.50	9.50	14.97			14.97
x_{121}	30	15.25	15.25	24.03			24.03
x_{122}	15	9.50	9.50	14.97			14.97
x_{211}	5	15.25			15.25	6.47	6.47
x_{212}	13	9.50			9.50	4.03	4.03
x_{221}	1	15.25			15.25	6.47	6.47
x_{222}	2	9.50			9.50	4.03	4.03
sum			49.50		49.50		99.00
constraint			78.00		21.00		
ratio			1.58		0.42		

Table 9.33 (b)   adjusting for the [L] margin

	a	b	c	d	e	f	g
x_{111}	25	24.03	24.03	24.76			24.76
x_{112}	8	14.97	14.97	15.42			15.42
x_{121}	30	24.03			24.03	23.30	23.30
x_{122}	15	14.97			14.97	14.52	14.52
x_{211}	5	6.47	6.47	6.67			6.67
x_{212}	13	4.03	4.03	4.15			4.15
x_{221}	1	6.47			6.47	6.27	6.27
x_{222}	2	4.03			4.03	3.91	3.91
sum			49.50		49.50		99.00
constraint			51.00		48.00		
ratio			1.03		0.97		

Table 9.33 (c) adjusting for the [S] margin

model	G^2	degrees of freedom	χ^2_{CRIT} ($\alpha = 0.05$)
total only	66.98	7	14.07
one factor model			
[L][S][C]	26.57	4	9.49
one and two factor models			
[LS][C]	12.92	3	7.81
[LC][S]	14.45	3	
[SC][L]	26.22	3	
two factor models			
[LS][LC]	0.80	2	5.99
[LS][SC]	12.57	2	
[LC][SC]	14.11	2	
saturated	0	0	

Table 9.34

What can we conclude? We see that none of the models with [SC] as a constraint offers significant improvements over the best models in the preceding class. Looking at the [SC] distribution (Figure 9.1) quickly shows why: there is very little correlation between C and S (check for yourself that hypothesising independence between C and S gives $G^2 = 0.35$). We find that the [LS][LC] model gives a very good fit to the data and so specification of the interaction effect is not needed. We could try fitting all two-dimensional constraints, the [LS][LC][SC] model, but, again, it seems not to be worthwhile in this case (but see Exercise 9.5). Example 9.1 shows how such estimates are made.

The ideas in this section may be extended to tables of more than two dimensions.

9.9 FURTHER READING

Much important work in the analysis of contingency tables is due to Goodman (see Goodman, 1984, for a collection). The models are frequently cast in the slightly different form of (A1.19) as *log-linear* models (Goodman, 1970). As such they are described in many texts. Fienberg (1991) is a good place to start. Gokhale and Kullback (1978) provide a text written from the viewpoint which this book shares.

The iterative proportional fitting procedure was first proposed by Deming and Stephan (1940). Mosteller (1968) provides a survey with some examples and Fienberg (1970) discusses the convergence of the process.

9.10 EXAMPLES

Example 9.1

What a nuisance it is when you drop something in a public place. Are you annoyed if others do not help you to pick up that which you have dropped?

Latane and Dabbs (1975) report a study in which an experimenter walked into a lift and then dropped some coins or pencils. The sex of whoever helped to pick up these objects was noted. Here is what happened in the cities of Columbus, Ohio and Atlanta, Georgia. What can we conclude from these data?

		experimenter		experimenter	
		male	female	male	female
helpful	female	715	588	389	180
bystander	male	815	505	448	165
		Columbus		Atlanta	

(Source: Latane and Dabbs, 1975)

Solution

The calculations follow those in the last section of this chapter. The calculation for three two-dimensional margins is shown in the following tables. The variable names are B for bystander, E for experimenter and C for city. Begin by adjusting the $[BE]$ margin. Starting with a uniform distribution pairs of cells are taken and adjusted so that constraints are met:

cell	data	start						adjusted
(111)	715	1.00	1.00					552.00
(112)	389	1.00	1.00					552.00
(121)	588	1.00		1.00				384.00
(122)	180	1.00		1.00				384.00
(211)	815	1.00			1.00			631.50
(212)	448	1.00			1.00			631.50
(221)	505	1.00				1.00		335.00
(222)	165	1.00				1.00		335.00
sum:			2.00	2.00	2.00	2.00		3805.00
constraint:			1104.00	768.00	1263.00	670.00		
ratio:			552.00	384.00	631.50	335.00		

For instance, in the first constraint the contents of cells (111) and (112) should sum to 1104.0 and they actually sum to 2.0. Both are multiplied by 552.0. The results of this and the other adjustments are shown in the last column, which then forms the starting point for the next series of adjustments to ensure that the $[BC]$ margin is respected:

cell	start					adjusted
(111)	552.00	552.00				768.44
(112)	552.00		552.00			335.56
(121)	384.00	384.00				534.56
(122)	384.00		384.00			233.44
(211)	631.50			631.50		862.47
(212)	631.50				631.50	400.53
(221)	335.00			335.00		457.53
(222)	335.00				335.00	212.47
sum:		936.00	936.00	966.50	966.50	3805.00
constraint:		1303.00	569.00	1320.00	613.00	
ratio:		1.39	0.61	1.37	0.63	

Finally the $[EC]$ margin is adjusted:

cell	start					adjusted
(111)	768.44	768.44				720.89
(112)	335.56		335.56			381.57
(121)	534.56			534.56		588.94
(122)	233.44				233.44	180.61
(211)	862.47	862.47				809.11
(212)	400.53		400.53			455.43
(221)	457.53			457.53		504.06
(222)	212.47				212.47	164.39
sum:		1630.91	736.09	992.09	445.91	3805.00
constraint:		1530.00	837.00	1093.00	345.00	
ratio:		0.94	1.14	1.10	0.77	

Now check the $[BE]$ margin again:

cell	data	start					adjusted
(111)	715	720.89	720.89				721.90
(112)	389	381.57	381.57				382.10
(121)	588	588.94		588.94			587.75
(122)	180	180.61		180.61			180.25
(211)	815	809.11			809.11		808.12
(212)	448	455.43			455.43		454.88
(221)	505	504.06				504.06	505.23
(222)	165	164.39				164.39	164.77
sum:			1102.46	769.55	1264.54	668.45	3805.00
constraint:			1104.00	768.00	1263.00	670.00	
ratio:			1.00	1.00	1.00	1.00	

All ratios are 1.00 and the differences in cell frequencies at the start and end of this iteration are slight. You may sometimes need to continue the iterations, but it is not necessary here. The results are:

model	G^2	df	χ^2_{CRIT} ($\alpha = 0.05$)
total only	861.55	7	14.07
[B][E][C]	71.86	4	9.49
[BE][C]	55.47	3	7.81
[BC][E]	71.09		
[EC][B]	16.69		
[BE][BC]	54.70	2	5.99
[BE][EC]	0.29		
[BC][EC]	15.92		
[BE][EC][BC]	0.13	1	3.84
saturated		0	0.00

We can see that large reductions in G^2 occur when the [EC] margin is used as a constraint and that a good fit between model and data is obtained using the [BE][EC] margins, with not much of an improvement when [BC] is added. The data may be well explained by the interactions between the sexes of the bystander and the experimenter, [BE]. The interaction between the sex of the experimenter and the city in which the experiment took place, [EC] also appears to have some effect. The first has a plausible behavioural interpretation.

Example 9.2

The table below shows some results given by Rudd (1987) showing the relationship between the subjects taken by male and female British university students and the occupation of their fathers.
Are there any underlying similarities between the distributions?

	Father's occupation					
	Doctor or Dentist		Scientist		Engineer	
Subject	m	f	m	f	m	f
Health (1)	39	27	15	13	9	9
Technology (2)	8	1	15	0	39	4
Science	23	22	42	38	29	27
Business (3)	20	23	13	23	13	24
Arts	8	26	11	23	9	24
Other	2	0	4	5	2	12
all subjects (%)	100	100	100	100	100	100
sample size	87	77	53	40	165	143

(Source: Rudd, 1987)

NOTES: The *m* and *f* at the column heads indicate male and female students
  (1)   includes medicine and dentistry
  (2)   includes engineering
  (3)   includes social, administrative and business studies

Solution

The data have been presented in the form of conditional distributions (each column sums to 100%) to show how the subjects chosen vary by sex of student and occupation of father. We may wish to go further and to ask if there is any strong effect due to the sex of the student. Notice, for instance, that although the proportions taking technology vary according to the father's occupation it is always more popular with male than with female students. To make the underlying comparison clear we need to recognise that the proportions choosing different subjects vary by father's occupation. One way of doing this is to adjust each table to have uniform margins; any differences will then be due entirely to interaction effects.
  The table below shows the situation for students whose fathers were doctors or dentists. Notice that because the published data were shown, for clarity, to only the nearest percentage point the second column sums to 99 rather than 100.

Sec. 9.10] **Examples** 241

Because the iterative process to be used is multiplicative any initial value of zero will remain at zero under all circumstances. This is undesirable, since the initial zero is based only on sample data and to say that this must mean that the number in that category in the population is certainly zero is not justified. To get over this problem set the cell containing the observed zero (female students taking other subjects) to some small value, here 0.01.

			sum	req'd	ratio
	39.00	27.00	66.00	16.67	0.25
	8.00	1.00	9.00	16.67	1.85
	23.00	22.00	45.00	16.67	0.37
	20.00	23.00	43.00	16.67	0.39
	8.00	26.00	34.00	16.67	0.49
	2.00	0.01	2.01	16.67	8.29
sum:	100.00	99.01			
req'd:	50.00	50.00			
ratio:	0.50	0.50			

We require that both columns sum to $100/2 = 50.00$ and that all rows sum to $100/6 = 16.67$. The marginal totals, required totals and ratios of the two are shown above.

Begin by balancing the rows. Both values in the first row are multiplied by 0.25 to give 9.85 and 6.82 rather than 39 and 27. Making this adjustment for all rows gives

	9.85	6.82
	14.81	1.85
	8.52	8.15
	7.75	8.91
	3.92	12.75
	16.58	0.08
sum:	61.44	38.56
req'd:	50.00	50.00
ratio:	0.81	1.30

where each row sums to 100/6, the required marginal total.

Now adjust the columns by multiplying all values in the first column by 0.81 and in the second by 1.30 to give:

		sum	req'd	ratio
8.01	8.84	16.86	16.67	0.99
12.06	2.40	14.46	16.67	1.15
6.93	10.57	17.50	16.67	0.95
6.31	11.56	17.87	16.67	0.93
3.19	16.53	19.72	16.67	0.85
13.50	0.11	13.60	16.67	1.23

Readjusting the rows and continuing the iterations finally gives the required result. This is shown below together with the similarly iterated tables for the children of scientists and engineers.

| | Father's occupation ||||||
| | Doctor or Dentist || Scientist || Engineer ||
Subject	*m*	*f*	*m*	*f*	*m*	*f*
Health	7	10	8	8	9	7
Technology	13	3	17	0	15	1
Science	6	11	8	9	10	7
Business	5	12	5	11	7	10
Arts	2	14	5	12	6	11
Other	17	0	7	10	3	14

These tables show just the interaction between the two variables, subject and sex, now that the differences due only to the marginal distributions have been eliminated. The similarities are easily seen as are the differences. One interesting variation is in the propensity of males and females to study science. The split is just about even for the children of scientists and skewed otherwise, but in different directions. This was not obvious in the data.

Example 9.3

People living in the small dormitory town of Groucho travel to work in nearby Chico or else in Harpo and travel by car or bus or train. Here are the journey times, in minutes:

		bus	train	car
from Groucho to:	Chico	50	40	45
	Harpo	60	55	40

60% of the residents of Groucho work in Chico and the mean time taken by all residents to get to work is 44.5 minutes. What proportion of workers travel by train?

Solution

We need a model similar to (9.12) but with only one set of marginal constraints:

$$p_{ij} = a_i \exp(\alpha d_{ij})$$

where p_{ij} is the percentage of workers and d_{ij} is the journey time. With $\alpha = 0$ we have $p_{ij} = a_i$:

		bus	train	car	
from Groucho to:	Chico	20.00	20.00	20.00	60.00
	Harpo	13.33	13.33	13.33	40.00
		33.33	33.33	33.33	

where the figures are in percentages of Groucho workers. Check that the mean journey time is 47.67 minutes.

To reduce journey time we need to reduce α. After some trial and error $\alpha = -0.08$ gives a mean journey time of 44.5 minutes. The calculation of the distribution of journeys began with values of $\exp(\alpha d_{ij})$. For bus journeys to Chico we have

$$\exp(-0.08 \times 50) = 0.0183$$

The first estimate of the distribution is

		bus	train	car	
from Groucho to:	Chico	0.0183	0.0408	0.0273	0.0864
	Harpo	0.0082	0.0123	0.0408	0.0613

Multiplying the values in the first row by 60/0.0864 and the second row by 40/0.0613 gives

		bus	train	car	
from Groucho to:	Chico	12.72	28.31	18.97	60.00
	Harpo	5.37	8.02	26.61	40.00
		18.09	36.33	45.58	

So, 36.33% of Groucho workers travel to work by train.

Example 9.4

The table below gives the distribution of families by dependent children and parental status in Alnwick, Northumberland, as determined by the 1991 10% census.

	married couple	cohabiting couple	lone parent
no children	358	35	0
dependent child(ren)	282	16	50
non-dependent child(ren) only	105	1	41

In nearby Berwick-upon-Tweed the 736 families surveyed had these marginal distributions:

no children	317
dependent child(ren)	278
non-dependent child(ren) only	141
married couple	624
cohabiting couple	29
lone parent	83

Estimate the joint distribution of the families in Berwick.

Solution

Use the Alnwick distribution as the staring point.

			sum	req'd	ratio
358	35	0	393.00	317	0.81
282	16	50	348.00	278	0.80
105	1	41	147.00	141	0.96

Begin by adjusting the row marginal totals. The numbers in the first row are multiplied by 0.81 (a small value replacing the zero) and so on to give

	288.76	28.23	0.01
	225.28	12.78	39.94
	100.71	0.96	39.33
sum:	614.75	41.97	79.28
req'd:	624	29	83
ratio:	1.02	0.69	1.05

Next multiply the columns by 1.02, 0.69 and 1.05 respectively:

			sum	req'd	ratio
293.11	19.51	0.01	312.62	317	1.01
228.67	8.83	41.82	279.31	278	1.00
102.23	0.66	41.17	144.07	141	0.98

Continuing the row and column balancing we eventually arrive at

297.34	19.65	0.01
227.07	8.71	42.22
99.58	0.64	40.77

If you are interested, the actual distribution for Berwick was

296	21	0
231	6	41
97	2	42

9.11 EXERCISES

9.1 Here is a description of the daily flow of goods vehicles in London:

		Destination		
		North London	South London	Outside London
	North London	375134	25354	18620
Origin	South London	26348	179976	8684
	Outside London	17989	9051	33168

(Source: Greater London Council, 1985)

Suppose that the number of origins and destinations might change to be

	Origins	Destinations
North London	440500	439000
South London	191500	200400
Outside London	70000	62600

What would the distribution of goods vehicle flows then be?

9.2 Use the data from Example 9.3. If it was known that 25% of Groucho residents went to work by bus and 45% took a train, what percentage of those working in Harpo went by car?

9.3 Pearson and Toby (1991), using data from the National Crime Survey carried out by the United States Bureau of the Census, investigated, among other things, the extent to which schoolchildren feared attack on their journeys to and from school depending on whether a street gang had been reported at the school and on the mode of transport used for the journey. Here are some of their results:

	Car		School bus		Public transport	
	gang	no gang	gang	no gang	gang	no gang
Fear of attack:						
Never	83.7	91.0	79.9	89.2	55.3	70.1
Sometimes	16.3	9.0	20.1	10.8	44.7	29.9
	100%	100%	100%	100%	100%	100%

(Source: Pearson and Toby, 1991)

Examine the interaction between fear of attack and the presence of gangs for the three transport modes.

9.4 Four small towns are connected by the road network shown in Figure 9.4. The numbers in the diagram show the journey time, in minutes, between towns. No public transport is in operation. A survey into Saturday shopping in the four towns gave

town	living in town but shopping elsewhere	shopping in town but living elsewhere
A	35	25
B	39	15
C	21	63
D	25	17

Responses from people shopping in the town in which they live have been omitted from this table. The survey also determined that the mean one-way journey time for these shopping trips was 19.7 minutes. All shoppers chose the quickest route for their journey.

How many people travel from A to B to shop?

Sec. 9.11] **Exercises** 247

What would your answer have been if you had not had the information about the mean journey time?

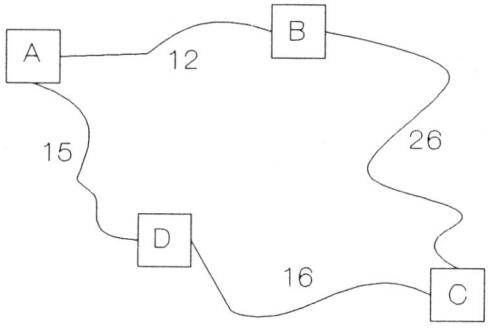

Figure 9.4

(This application, and those in Exercises 9.1 and 9.2, to such transportation and related problems is well known and extensively documented in, for instance, Wilson, 1970, 1974, and Batty, 1976.)

9.5 Fit the [LS][LC][SC] model to the data in Table 9.29. What distribution do you obtain and what value of G^2?

9.6 There are a number of reasons that might predispose some jurors to convict more readily than others. This is especially so in murder cases where the penalty is death. In the USA counsel may object to potential jurors on the grounds that they are biased in favour of or against the death penalty *per se*. Williams and McShane (1990) conducted a study of Texans in which they assessed whether or not they would be excluded by counsel and also asked whether they would convict in a hypothetical case described to them. Here are some of the results:

	excluded by counsel		not excluded	
	would convict	would not convict	would convict	would not convict
male	46	36	236	115
female	83	34	275	121

(Source: Williams and McShane, 1990)

Analyse the data and describe your results.

9.7 Hout *et al.* (1987) report a study by Heimer and Stinchcombe in which married couples in Tucson were asked to respond to the proposition "Sex is fun for me and my partner" by specifying how frequently this statement truly described their mutual activities. Here are the responses:

		Wife's response			
		never or occasionally (N)	fairly often (F)	very often (V)	almost always (A)
Husband's response	N	7	7	2	3
	F	2	8	3	7
	V	1	5	4	9
	A	2	8	9	14

(Source: Hout et al., 1987)

What do you conclude?

9.8 I guess that most of us would have difficulty in achieving a successful copulation if interrupted by the activities of our neighbours. This challenge is shared by some sharp-tailed grouse in south-western Manitoba:

	Copulations	
	Not disrupted	Disrupted
Successful	106	47
Unsuccessful	5	25

(Source: Gratson et al., 1991)

Find the importance of the marginal and interaction effects.

9.9 Using the data from Exercise 8.7 examine the interaction between the country of origin of the magazines and the advertising themes by eliminating the effects of differences in the marginal totals.

9.10 The following table shows the distribution of vehicle speeds on British motorways in 1992 for different types of vehicle. The columns show the percentage of vehicles of each type travelling at the indicated speeds.

speed (mph)	motor-cycles	cars	light goods	buses and coaches	goods vehicles
< 50	17	3	5	5	9
50–60	18	12	21	19	51
60–65	14	11	16	19	28
65–70	11	17	20	38	10
70–75	15	24	20	18	2
75–80	7	13	8	1	0
80–90	11	17	8	1	0
≥ 90	8	2	1	0	0

(Source: Transport Statistics Great Britain: 1993 Edition)

Suppose that on a certain road the composition of traffic is

    motorcycles   5%
    cars   57
    light goods   12
    buses and coaches   8
    goods vehicles   18

and that 40% of all vehicles travel at no more than 60 mph.

(a) What proportion of all vehicles exceed the speed limit of 70 mph?

(b) What proportion of cars travel at between 60 mph and 80 mph?

10

Correlation and Regression

In the previous two chapters we have seen how to deal with two-dimensional relationships presented as tables. The variables were either *nominal* (e.g. sex) or *ordinal* (e.g. small, medium, large) and so could only be described categorically. The third type is that of *cardinal* or *real* variables where it is meaningful to talk about the ratios of differences. Distance is such a variable. It is meaningful for me to say not only that Edinburgh is farther from Durham than is Newcastle but also how much farther. Real variables are what we understand as ordinary numbers.

Real variables may be grouped into categories but to do so would involve a loss of precision that we would wish to avoid. In this chapter we introduce some ways of describing and modelling the relationship between two real variables.

10.1 COVARIANCE AND CORRELATION

	t.v. sets per 1000 people (y)	net disposal income ('000 ECU per head) (x)
Belgium	303	8.1
Denmark	369	10.6
France	375	9.2
Germany	335	10.3
Greece	257	3.8
Ireland	205	5.0
Italy	243	6.3
Netherlands	310	9.3
Portugal	151	2.3
Spain	258	4.0
United Kingdom	328	8.0

(Source: Eurostat Review 1976–1985)

Table 10.1
Television ownership and income, 1983

Sec. 10.1] Covariance and correlation

Table 10.1 shows data for some European countries. Does anything occur to us when inspecting this table? We perhaps bring to the problem the general notion that in richer countries there may be a tendency for more people to own televisions. Table 10.2 shows more clearly that this is the case.

	t.v. sets per 1000 people (y)	net disposal income ('000 ECU per head) (x)
Portugal	151	2.3
Ireland	205	5.0
Italy	243	6.3
Greece	257	3.8
Spain	258	4.0
Belgium	303	8.1
Netherlands	310	9.3
United Kingdom	328	8.0
Germany	335	10.3
Denmark	369	10.6
France	375	9.2

Table 10.2

Here is a lesson about presenting data for two or more variables: beware of irrelevant ordering. Table 10.1 presented the data in alphabetical order of the countries. This is fine for some purposes but see how much clearer for our present purpose is the picture in Table 10.2 wherein the data are ordered according to television ownership.

The data may also be depicted as a *scattergram* (Figure 10.1). This shows quite clearly that there is a strong relationship between the two variables and that in this case the relationship appears to be linear, but more of that later.

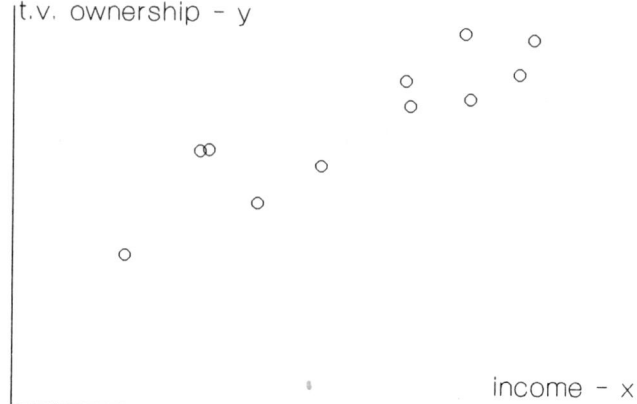

Figure 10.1 Scattergram

Just as we had simple numerical descriptors such as variance to measure the spread of a single variable we now require a measure of the joint spread of two variables: a single figure summary of the scattergram.

	y	x	$(y-\bar{y})$	$(x-\bar{x})$	$(y-\bar{y})^2$	$(x-\bar{x})^2$	$(x-\bar{x})(y-\bar{y})$
	151	2.3	−133.91	−4.69	17931.89	22.00	628.04
	205	5.0	−79.91	−1.99	6385.61	3.96	159.02
	243	6.3	−41/91	−0.69	1756.45	0.48	28.92
	257	3.8	−27.91	−3.19	778.97	10.18	89.03
	258	4.0	−26.91	−2.99	724.15	8.94	80.46
	303	8.1	18.09	1.11	327.25	1.23	20.08
	310	9.3	25.09	2.31	629.51	5.34	57.96
	328	8.0	43.09	1.01	1856.75	1.02	43.52
	335	10.3	50.09	3.31	2509.01	10.96	165.80
	369	10.6	84.09	3.61	7071.13	13.03	303.56
	375	9.2	90.09	2.21	8116.21	4.88	199.10
sum:	3134	76.9	0	0	48086.91	82.01	1775.49
mean:	284.91	6.99	0	0	4371.54	7.46	161.41

Table 10.3

Table 10.3 shows some basic calculations with which, with the exception of the last column, we are already familiar. We see that

Sec. 10.1] Covariance and correlation

for variable y: mean $= \bar{y}$ $= 284.91$
variance $= \text{var}(y) = 4371.54$
standard deviation $= \text{sd}(y) = 66.12$

for variable x: mean $= \bar{x}$ $= 6.99$
variance $= \text{var}(x) = 7.46$
standard deviation $= \text{sd}(x) = 2.73$

These are descriptors of each of the two variables considered separately. Measures of spread, variance and standard deviation, are based on deviations from the mean. This idea can be extended to two dimensions as shown in Figure 10.2.

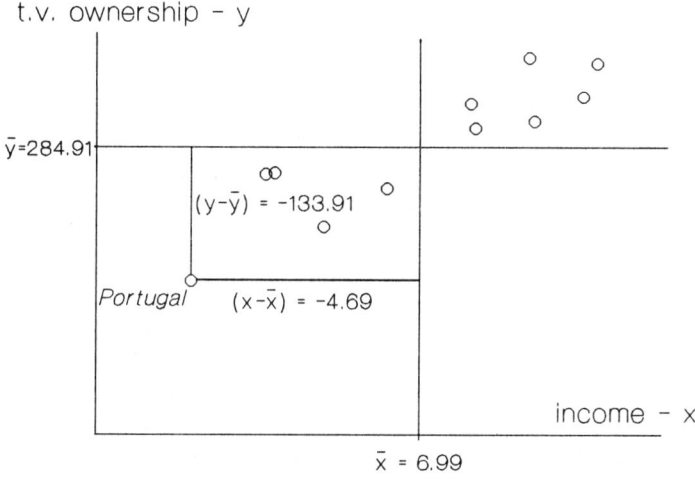

Figure 10.2

We can think of any point as having coordinates measured not from the axes $x = 0$ and $y = 0$, which is how we customarily think, but rather from the axes representing the means $x = \bar{x}$ and $y = \bar{y}$. For instance, consider the point representing Portugal, which is shown in the graph. We have previously thought of this as being described by

$x = 2.3$
$y = 151$

but now see that it may also be represented as

$(x - \bar{x}) = -4.69$

$(y - \bar{y}) = -133.91$

To measure how far away this point is from both means it is sensible to take the product

$$(x - \bar{x})(y - \bar{y}) = (-4.69)(-133.91) = 628.04$$

These products are shown in the last column of Table 10.3. The product has a small value when the point either is close to one of the axes drawn through the means or is close to their intersection. The further away from the axes, the larger the value.

The mean of these products is called the *covariance*:

$$\text{cov}(x, y) = \Sigma(x - \bar{x})(y - \bar{y}) / n \qquad (10.1)$$

where n is the number of points. You can see from Table 10.3 that in this case $\text{cov}(x,y) = 161.41$.

What does this mean? Look again at Figure 10.2. The axes drawn through the means divide the two-dimensional space into four quadrants. In the upper right and lower left quadrants the deviations from mean values both have the same sign: both positive and both negative respectively. This means that the product in (10.1) is positive for all points that lie in either of these quadrants, as is the case with our data. In the other two quadrants the product is negative. We may therefore think of three extreme cases: when the data lie predominantly in the upper right and lower left quadrants then covariance will be positive, when the data lie predominantly in the other two quadrants then covariance will be negative, and when the data are distributed equally in all four quadrants then covariance will tend towards zero.

So, covariance measures pattern. Our data have a fairly strong upward sloping pattern: high values of x are associated with high values of y and vice versa. Consequently we would expect a fairly high positive value for the covariance. But the value we find depends on the units in which the variables are measured. If income had been given in ECU rather than thousands of ECU then $\text{cov}(x,y)$ would have been 161410 rather than 161.41 though the pattern of the data in the scattergram would have been unaltered. To overcome this we divide by the product of the standard deviations to give the *correlation coefficient*

$$r = \frac{\text{cov}(x, y)}{\text{sd}(x)\text{sd}(y)} \qquad (10.2)$$

$$= \frac{161.41}{2.73 \times 66.12} = 0.89$$

This is a dimensionless number and so is unaffected by the units in which the variables are measured. The correlation coefficient is also bounded. When $r = 1$ the data are perfectly linear and upward sloping and we describe them as perfectly positively correlated. When $r = -1$ the data are again perfectly linearly correlated but this time negatively. If the data show no relationship between the variables then $r = 0$ (Figure 10.3).

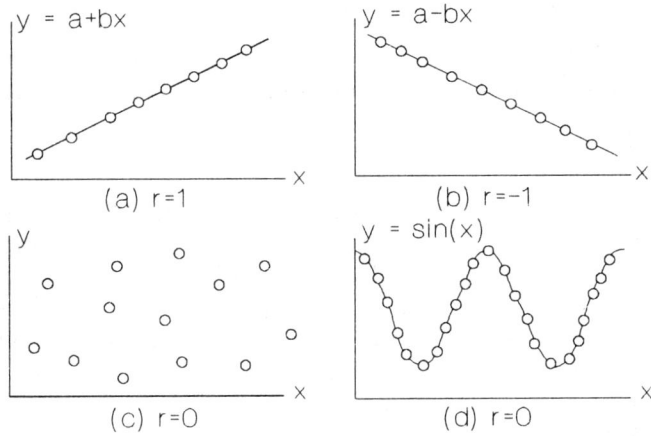

Figure 10.3 Some values of r

For our data $r = 0.89$ is fairly close to the maximum value of $r = 1$, indicating a strong positive correlation. This agrees with the impression gained from the scattergram.

The correlation coefficient provides a convenient summary of the association between two variables. Always remember, however, that it is a measure of *linear* association. Other quite strong patterns may exist and yet result in only a low value of r (Figure 10.3d). Plot the data and inspect the scattergram before calculating the correlation coefficient.

10.2 FITTING A LINEAR MODEL

If the data seem to be reasonably linear then it is natural to wish to have as a summary a straight line that best describes the relationship between the variables.

There is another reason why we may wish to determine a linear model. Suppose that we need to estimate the level of television ownership in some country not shown in our data. With no other knowledge of that country we would have to say that our best guess is characterised by the data we have on y. Rather than just referring to the data it would be convenient to be able to assume that this sample of 11 countries comes from a population that is Normally distributed. Table 10.4 compares our observed data with what would be expected from a Normal distribution with the same mean and variance. You might like to try generating this last column yourself and find $G^2 = 1.1$ so that the assumption of Normality is not refuted ($\chi^2_{CRIT} = 6.0$).

t.v. sets/1000 y	observed frequency	frequency if Normal
$y \leq 200$	1	1.1
$200 < y \leq 250$	2	2.2
$250 < y \leq 300$	2	3.2
$300 < y \leq 350$	4	2.7
$350 < y$	2	1.8

Table 10.4

So our estimate of television ownership for this other (for any other) country is characterised by a Normal distribution with mean 284.91 and standard deviation 66.12. We may quote appropriate credible intervals if we wish.

Suppose now that the average income, x, is given for this other country. We would naturally wish to use our knowledge of the linear relationship between y and x to improve our estimate of y. We wish to estimate y conditional upon x, $y|x$, rather than make the unconditional estimate $y = \bar{y}$. In doing this we would hope to reduce the uncertainty (variance or entropy) of our estimate.

Figure 10.4 shows the situation. The linear model is

$$y = a + bx + e \tag{10.3}$$

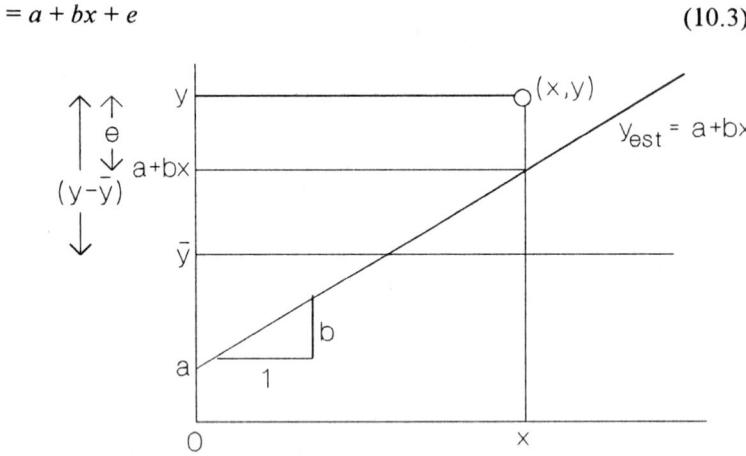

Figure 10.4 The linear model

Sec. 10.2]   Fitting a linear model

In this equation b is the slope of the line, a is the intercept (the value of y at $x = 0$) and e is the error in using the linear model to estimate y,

so $e = y - (a + bx)$ (10.4)

or $e = y - y_{est}$ (10.5)

where $y_{est} = a + bx$ (10.6)

is the estimated value of y.

We want to choose the parameters a and b so that the effects of the *residual error*, e, are somehow minimised. In Chapter 7 we saw that an average was set by solving a decision problem that involved minimising expected loss. In particular, if the loss was proportional to the square of the difference between the estimated and actual values (Figure 7.2) then the optimal solution was to take the mean as the average value.

Using exactly the same criterion here we wish to minimise

$$S = \Sigma_i e_i^2 \qquad (10.7)$$

to give a line of *best fit* as an optimal estimator of y given x. Provided that we restrict any subsequent forecasting estimates to interpolations (not going outside the range of the existing data) we may reasonably hope to have found the best forecasting model too.

In Appendix A8 it is shown that the optimal parameter values are given by

$b = \text{cov}(x,y)/\text{var}(x)$ (10.8)

and $a = \bar{y} - b\bar{x}$ (10.9)

which in our example gives

$b = 161.41/7.46 = 21.64$

and $a = 284.91 - (21.64 \times 6.99) = 133.65$

giving $y_{est} = 133.65 + 21.64x$ (10.10)

The results are shown in Figure 10.5 and Table 10.5. This line of best fit is called the *regression* line of y on x, and the method that we have used is called *least squares regression*.

x	y	y_{est}	e	e^2
2.3	151	183.42	−32.42	1051.19
5.0	205	241.85	−36.85	1357.92
6.3	243	269.98	−26.98	728.03
3.8	257	215.88	41.12	1690.69
4.0	258	220.21	37.79	1428.08
8.1	303	308.93	−5.93	35.21
9.3	310	334.90	−24.90	620.11
8.0	328	306.77	21.23	450.71
10.3	335	356.54	−21.54	464.06
10.6	369	363.03	5.97	35.59
9.2	375	332.74	42.26	1786.08
mean: 6.99	284.91	284.93	0	877.06

Table 10.5
Regression results

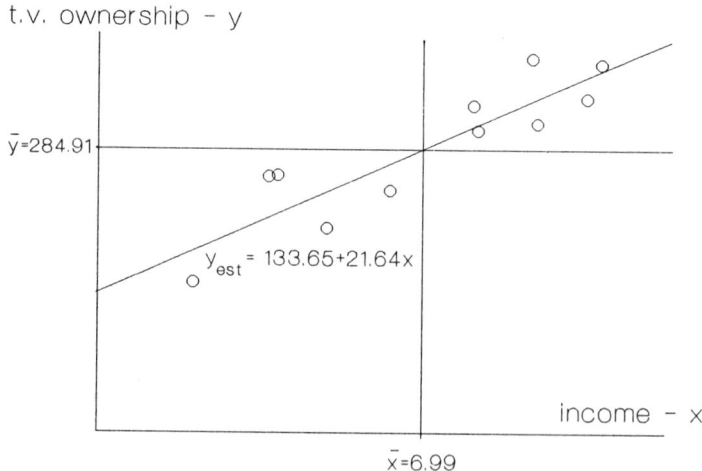

Figure 10.5 The fitted model

10.3 MODEL PERFORMANCE

To see how well the model performs as an estimator look at the column of e values in Table 10.5. The mean error is zero. The model tends to neither overestimate nor underestimate: it is *unbiased*. In addition, there seems to be no obvious pattern in the errors and so we are able to say that the underlying error of the estimate is constant and so independent of the value being estimated. Errors with these characteristics are said to constitute *white noise*.

Assume that the distribution of errors in the population may be taken as Normal with

$\bar{e} = 0$
$\text{var}(e) = 877.06$
$\text{sd}(e) = 29.62$

Our estimate of y for some other country has been greatly improved, as Table 10.6 shows.

Characteristics of Normal estimate	no information about x	with linear model
mean	284.91	$133.65 + 21.64x$
variance	4371.54	877.06
entropy	5.61	4.81

Table 10.6
Estimates of y with and without linear model

The last line of the table is from the standard result (e.g. Shannon and Weaver, 1949) that the entropy of a Normal distribution is

$$H = \log[(2\pi e)^{0.5} \sigma] = \log(4.133\sigma) \quad (10.11)$$

$$= \ln(\sigma) + 1.419$$

Variance is just the expected loss with a quadratic loss function (Appendix A6). One way to measure the performance of the model is to calculate the reduction in this expected loss as a result of using the linear model:

reduction in expected loss = $\text{var}(y) - \text{var}(e)$
$= 4371.54 - 877.06$
$= 3494.48$

It is more usual to express this reduction as a proportion of the initial expected loss. This ratio is then called the *coefficient of determination*, R^2, where

R^2 = proportionate reduction in expected loss
$= [\text{var}(y) - \text{var}(e)]/\text{var}(y)$
$= 1 - \text{var}(e)/\text{var}(y) \quad (10.12)$
$= 1 - 877.06/4371.54 = 0.8$

You may also think of this as the proportion of the variance of y accounted for by the model.

Unsurprisingly, the extent to which the model reduces expected loss is intimately related to the degree to which the data are correlated. If correlation were perfect then there would be no residual error: $\text{var}(e) = 0$ and $R^2 = 1$, a perfect model. In Appendix A8 it is shown that

$$r^2 = R^2 \qquad (10.13)$$

10.4 ENTROPY REDUCTION AND CORRELATION

In Chapter 8 we saw how the association of two variables in a contingency table could be measured by the reduction in entropy,

$$T(X,Y) = H(Y) - H(Y|X) \qquad (10.14)$$

We have a similar situation here. $H(Y)$ is just the entropy of y and measures the uncertainty of estimates made without the linear model. $H(Y|X)$ is the entropy of the residual uncertainty as indicated by the errors, e, and measures the (reduced) uncertainty of estimates made using the model. From Table 10.6

$$T(X,Y) = 5.61 - 4.81 = 0.80$$

(This is only coincidentally the same as the value of R^2 in this example.) Writing this generally, and assuming Normality, we have, from (10.11),

$$T(X,Y) = [\ln(\text{sd}(y)) + 1.419] - [\ln(\text{sd}(e)) + 1.419]$$

$$= \ln[\text{sd}(y)] - \ln[\text{sd}(e)]$$

so, $\quad T(X,Y) = \ln[\text{sd}(y)/\text{sd}(e)] \qquad (10.15)$

But, from (10.12) and (10.13),

$$r^2 = 1 - \text{var}(e)/\text{var}(y)$$

and $\quad \text{sd}(y)/\text{sd}(e) = 1/[\text{var}(e)/\text{var}(y)]^{0.5}$

$$= (1 - r^2)^{-0.5}$$

so $\quad T(X,Y) = -0.5\ln(1 - r^2) \qquad (10.16)$

$$= -0.5\ln(0.2) = 0.80$$

as before, Y^2 and r^2 are quite distinct measures of association, though with obvious similarities of purpose. We see through (8.11) and (10.16) that the common unifying concept is that of uncertainty reduction as measured by the change in entropy, $T(X,Y)$. You may like to think of this as the amount of information transmitted by one variable about the other via either a regression equation or a contingency table.

The discussion in this section follows Attneave (1959, Appendix 1).

10.5 INFERENCE AND FORECAST

We may wish to consider the data as constituting a sample from some larger population. Remember that the model permits estimates of y conditional upon some known value of x. The conditional variance of y is

$$s_y^2 = \Sigma e_i^2 / (n-2) \tag{10.17}$$

where $(n-2)$ is the number of degrees of freedom (the model requires that two parameters, a and b, be estimated from the data). In the current example

$$s_y^2 = 877.06 / 9 = 97.45$$

Just as in Chapter 7 we were able to make estimates of the population mean μ based upon a sample mean $\overline{x}$, so here we can estimate the population slope β based upon that calculated from the sample, b. The estimate is assumed to be Normally distributed (t for small samples) with mean b and variance

$$\text{var}(\beta) = s_y^2 / \Sigma (x_i - \overline{x})^2 \tag{10.18}$$

This result is proved in Appendix A8. In the example

$$\text{var}(\beta) = 97.45/82.01 = 1.19$$

and sd(β) = 1.09

A 50% credible interval estimate for β is

$$\beta = 21.64 \pm (0.70 \times 1.09) = (20.88, 22.40)$$

(0.70 is the appropriate t value). This assumes no *a priori* estimate of β and so a non-informative Bayes prior.

We may now wish to interpolate a value of y for some value of x. It is shown in the appendix that the regression line passes through the centroid of the data, $(\bar{x}, \bar{y})$, and that we may write

$$y_{\text{est}} = \bar{y} + b(x - \bar{x}) \tag{10.19}$$

so that the required value of y given x is

$$y|x = \bar{y} + b(x - \bar{x}) + e \tag{10.20}$$

We have assumed a model in which the variable Y, television ownership, varies according to X, income level, and so all variation is associated with Y. Table 10.7 shows the sources of variation leading to uncertainty about $y|x$ in (10.20).

source of uncertainty	variance
the residual error, e	s_y^2
the mean, $\bar{y}$	s_y^2/n
the slope, b	$s_y^2(x-\bar{x})^2/\Sigma(x_i-\bar{x})^2$

Table 10.7
Sources of variance in regression estimates (for derivation see Appendix A8)

The effect of uncertainty about the mean will become small as n becomes large, as will the effect of uncertainty about the slope. This last will increase a little the further away from the mean, $\bar{x}$, a prediction is required.

Combining these uncertainties gives

$$\text{var}(y|x) = s_y^2[1 + 1/n + (x-\bar{x})^2/\Sigma(x_i-\bar{x})^2] \tag{10.21}$$

$$= 97.45[1 + 1/11 + (x-6.99)^2/82.01]$$

$$= 97.45[1.09 + (x-6.99)^2/82.01] \tag{10.22}$$

For a country with a per capita disposable income of 7000 ECU the estimate of the number of television sets owned per 1000 people is Normally distributed with mean

$$133.65 + (21.64 \times 7) = 285.13$$

and variance

$$97.45[1.09 + (7 - 6.99)^2/82.01] = 106.22$$

and so standard deviation 10.31.

A 50% credible interval estimate, given a non-informative prior, is

$$y = 285.13 \pm (0.70 \times 10.31) = (277.91, 292.35)$$

We have here constructed a credible interval estimate of the television ownership in a particular country with 7000 ECU disposable income. If, instead, we had been interested in the mean television ownership of all countries with this income (albeit that this may be an imaginary rather than a real group) then the variability of ownership between such countries would be irrelevant in determining the variance of the estimate. Instead of (10.21) we have

$$\text{var}(\bar{y}|x) = s_y^2[1/n + (x - \bar{x})^2/\Sigma(x_i - \bar{x})^2] \qquad (10.23)$$

$$= 97.45[0.09 + (7 - 6.99)^2/82.01]$$

$$= 8.77$$

and $\text{sd}(\bar{y}|x) = 2.96$

The 50% credible interval estimate is now

$$285.13 \pm (0.70 \times 2.96) = (283.06, 287.20)$$

10.6 EXAMPLES

Example 10.1

In the issue of *The Independent on Sunday* published on 28 July 1991 there appeared some data showing, among other things, the percentage rises in top director's salary and in earnings per share for the previous year for the 110 biggest British companies. Here is a sample of 20 results chosen at random. The companies are listed in order of decreasing market capitalisation.

company	previous year's change (%)	
	top salary	earnings per share
Shell Transport	16.8	−6.5
ICI	−12.8	−32.7
Barclays Bank	17.9	16.0
Prudential	43.2	−53.2
Argyll Group	23.6	25.3
Racal Electronic	12.9	−8.1
Reckitt & Colman	17.4	11.5
Enterprise Oil	54.0	13.9
Land Securities	14.3	−4.4
Legal & General	11.1	−43.7
Rank	9.6	−4.1
Forte	−7.9	2.2
Redland	17.7	−12.2
MEPC	22.5	16.6
Willis Corroon	46.4	16.2
Sedgwick Group	2.0	26.8
Ranks, Hovis	13.8	−20.9
Harrisons & Crosfield	2.9	−25.6
Bowater	7.0	−8.0
Wolseley	19.9	−2.1

What conclusions may be drawn about the relationship between the changes in top pay and in earnings per share?

Solution

The data are shown in the scattergram (Figure 10.6).

Examples

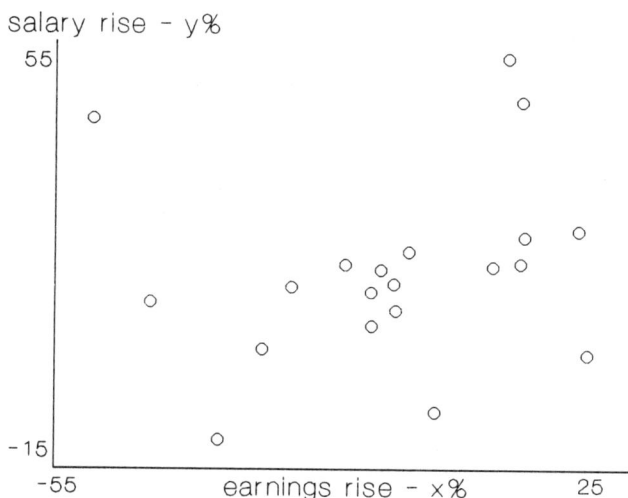

Figure 10.6

From Table 10.8 some simple summaries are easily obtained:

	% change in salary y	earnings x	$(y-\bar{y})$	$(x-\bar{x})$	$(y-\bar{y})^2$	$(x-\bar{x})^2$	$(x-\bar{x})(y-\bar{y})$
	16.8	−6.5	0.185	−1.850	0.034	3.423	−0.342
	−12.8	−32.7	−29.415	−28.050	865.242	786.803	825.091
	17.9	16.0	1.285	20.650	1.651	426.423	26.535
	43.2	−53.2	26.585	−48.550	706.762	2357.103	−1290.702
	23.6	25.3	6.985	29.950	48.790	897.003	209.201
	12.9	−8.1	−3.715	−3.450	13.801	11.903	12.817
	17.4	11.5	0.785	16.150	0.616	260.823	12.678
	54.0	13.9	37.385	18.550	1397.638	344.103	693.492
	14.3	−4.4	−2.315	0.250	5.359	0.063	−0.579
	11.1	−43.7	−5.515	−39.050	30.415	1524.903	215.361
	9.6	−4.1	−7.015	0.550	49.210	0.303	−3.858
	−7.9	2.2	−24.515	6.850	600.985	46.923	−167.928
	17.7	−12.2	1.085	−7.550	1.177	57.003	−8.192
	22.5	16.6	5.885	21.250	34.633	451.563	125.056
	46.4	16.2	29.785	20.850	887.146	434.723	621.017
	2.0	26.8	−14.615	31.450	213.598	989.103	−459.642
	13.8	−20.9	−2.815	−16.250	7.924	264.063	45.744
	2.9	−25.6	−3.715	−20.950	1788.101	438.903	287.329
	7.0	−8.0	−9.615	−3.350	92.448	11.223	32.210
	19.9	−2.1	3.285	2.550	10.791	6.503	8.377
sum:	332.300	−93.000	0.000	0.000	5156.326	9312.850	1183.665
mean:	16.615	−4.650	0.000	0.000	257.816	465.643	59.183

Table 10.8

for % top salary rise (y): $\bar{y} = 16.615$
$\text{var}(y) = 257.816$
$\text{sd}(y) = 257.816^{0.5} = 16.057$

for % earnings rise (x): $\bar{x} = -4.650$
$\text{var}(x) = 465.643$
$\text{sd}(x) = 465.643^{0.5} = 21.579$

Considering the relationship between the two variables:

from (10.1),

$$\text{cov}(x,y) = 59.183$$

from (10.2),

$$r = \frac{\text{cov}(x,y)}{\text{sd}(x)\text{sd}(y)} = \frac{56.183}{21.579 \times 16.057} = 0.162$$

The entropy reduction is, from (10.16),

$$T = -0.5\ln(1 - r^2)$$
$$= -0.5\ln(1 - 0.162^2) = 0.013$$

For line $y_{est} = a + bx$ use (10.8) and (10.9) to give

$$b = \frac{\text{cov}(x,y)}{\text{var}(x)} = \frac{56.183}{465.643}$$

$$= 0.127$$

$$a = \bar{y} - b\bar{x}$$
$$= 16.615 + (4.65 \times 0.127)$$
$$= 17.206$$

and so

salary rise = 17.206 + 0.127(earnings rise)

Sec. 10.6]   Examples

Note that both the slope, b, and the correlation coefficient, r, are small, indicating that there is not much of a relationship between the two variables.

Table 10.9 shows the estimated y values and the residuals

$$e = y - y_{est}$$

	y	x	y_{est}	e	e^2	$e/\text{sd}(y_{est})$
	16.8	−6.5	16.380	0.420	0.177	0.025
	−12.8	−32.7	13.050	−25.850	668.214	−1.455
	17.9	16.0	19.240	−1.340	1.795	−0.077
	43.2	−53.2	10.444	32.756	1072.937	**1.721**
	23.6	25.3	20.422	3.178	10.102	0.178
	12.9	−8.1	16.177	−3.277	10.735	−0.192
	17.4	11.5	18.668	−1.268	1.607	−0.073
	54.0	13.9	18.973	35.027	1226.911	**2.015**
	14.3	−4.4	16.647	−2.347	5.507	−0.137
	11.1	−43.7	11.652	−0.552	0.304	−0.030
	9.6	−4.1	16.685	−7.085	50.196	−0.415
	−7.9	2.2	17.486	−25.386	644.431	−1.482
	17.7	−12.2	15.655	2.045	4.180	0.119
	22.5	16.6	19.316	3.184	10.139	0.182
	46.4	16.2	19.265	27.135	736.30	1.554
	2.0	26.8	20.612	−18.612	346.418	−1.038
	13.8	−20.9	14.550	−0.750	0.562	−0.043
	2.9	−25.6	13.952	−11.052	122.152	−0.633
	7.0	−8.0	16.189	−9.189	84.442	−0.537
	19.9	−2.1	16.939	2.961	8.767	0.173
sum:				0.000	5005.881	

Table 10.9

The characteristics of the residuals are

$$\bar{e} = 0.000$$

from (10.17),

$$\text{var}(e) = s_y^2 = 5005.881/18 = 278.105$$

$$\text{sd}(e) = s_y = 278.105^{0.5} = 16.676$$

We may wish to identify those companies for which the top salary rise was unusually high or unusually low given the earnings growth. We may further decide that "unusually" means that there is no more than a 10% chance of such extreme high or low values being observed. Assuming that residuals are Normally distributed unusual values are those more than 1.734 standard deviations from the predicted value. (The 1.734 is obtained from t tables with $n - 2 = 18$ degrees of freedom.) The appropriate standard deviation to use is the square root of the variance of the estimated value of y, which is, from (10.21),

$$\text{var}(y_{est}) = s_y^2[1 + 1/n + (x - \bar{x})^2 / \Sigma(x_i - \bar{x})^2]$$

$$= 278.105[1.05 + (x + 4.65)^2 / 9312.85]$$

For the first company, Shell Transport, $x = -6.5$ and so

$$\text{var}(y_{est}) = 278.105[1.05 + (-1.85)^2 / 932.85]$$

$$= 292.112$$

$$\text{sd}(y_{est}) = 292.112^{0.5} = 17.091$$

$$e/\text{sd}(y_{est}) = 0.420/17.091 = 0.025$$

This and the remaining values are shown in the last column of the table. Figure 10.7 shows the regression line. The two outside lines on the graph show the 90% credible interval, $y_{est} \pm 1.734\text{sd}(y_{est})$.

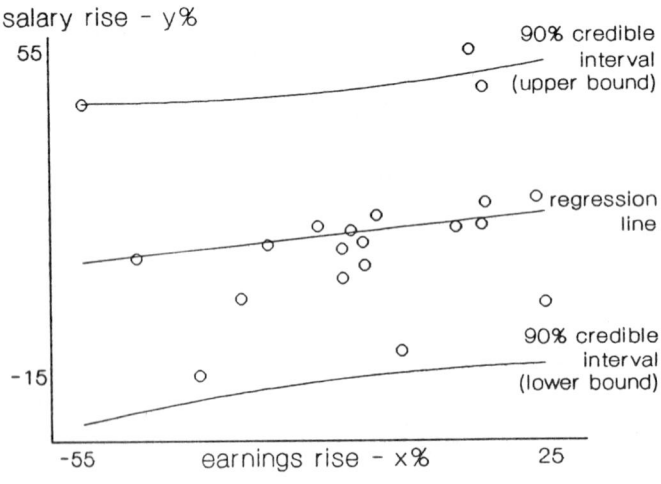

Figure 10.7

From this analysis it seems that only Enterprise Oil and, possibly, Prudential gave unusually high top salary increases. No company gave unusually low increases.

Remember that the increases have been compared with the average behaviour of all companies. We would have got a somewhat different picture if we had compared against some other model, for instance that top pay rises ought to be the same as the rises in earnings per share (Figure 10.8).

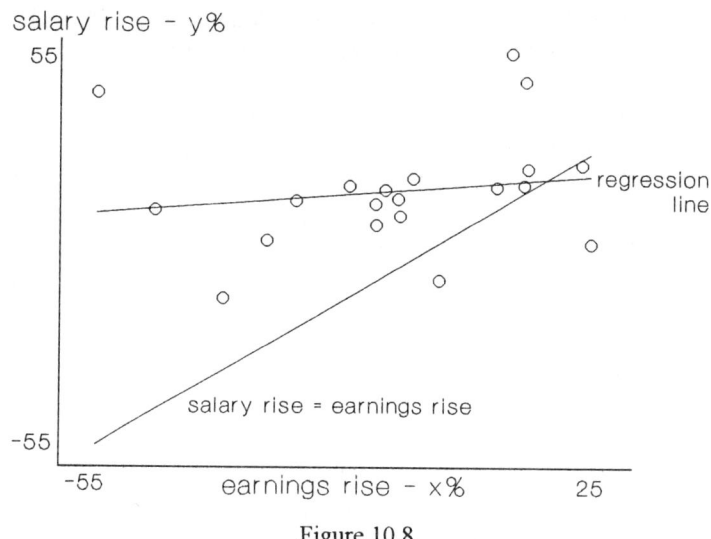

Figure 10.8

Example 10.2

Using the data in the previous example give credible interval estimates of the slope and intercept of the linear model.

Solution

From the sample of companies we have the regression line $y = a + bx$. In the population of all companies the regression line is $y = \alpha + \beta x$. We need appropriate estimates for α and β.

From (10.18)

$$\text{sd}(\beta) = [s_y^2 / \Sigma(x_i - \bar{x})^2]^{0.5}$$

$$= [278.105 / 9312.850]^{0.5} = 0.173$$

The t value for a 90% credible interval is 1.734 and so the interval estimate for the population slope is

$$\text{slope} = \beta = 0.127 \pm (1.734 \times 0.173)$$

$$= (-0.173, 0.427)$$

Note that the interval includes $\beta = 0$ and so we could not credibly deny the proposition that in the population of companies there is no relationship between changes in top salary and in earnings per share. This confirms the earlier observation of a low r value.

Note also that the interval does not include the value $\beta = 1$ that we would expect if the salary changes were equal to the changes in earnings per share.

The intercept α is just the mean value of y when $x = 0$ and so, from (10.23),

$$\text{sd}(\alpha) = [s_y^2 (1/n + \bar{x}^2 / \Sigma(x_i - \bar{x})^2)]^{0.5}$$

$$= [278.105(0.05 + (-4.650)^2 / 9312.850)]^{0.5}$$

$$= 3.815$$

The 90% credible interval estimate is

$$\text{intercept} = \alpha = 17.206 \pm (1.734 \times 3.815)$$
$$= (10.591, 23.821)$$

This shows that even when the rise in earnings per share was zero the mean top salary increase was between 11% and 24%.

Example 10.3

In Great Britain the consumption of butter is falling:

year	consumption (oz/person/week)
1974	5.61
5	5.63
6	5.16
7	4.70
8	4.55
9	4.45
1980	4.05
1	3.69
2	3.17
3	3.27
4	2.87
5	2.83
6	2.27
7	2.14
8	2.00
9	1.75
1990	1.61
1	1.54

(Source: Annual Abstract of Statistics, 1982 and 1993)

It is necessary to make a forecast of consumption in 1994. You have heard people describe the changes in terms of annual percentage decline.

Solution

You will often hear or read of changes described as percentages. The motive for and usefulness of this is clear enough: to give an easy appreciation of the magnitude of the change in relation to the base value. However, it is easy to forget that a constant percentage change implies a curvilinear relationship (exponential change over time).

The data are shown in Figure 10.9. From this scattergram it seems obvious to try a linear model.

model 1: $y = a + bt$

where y is the consumption and t is the year.

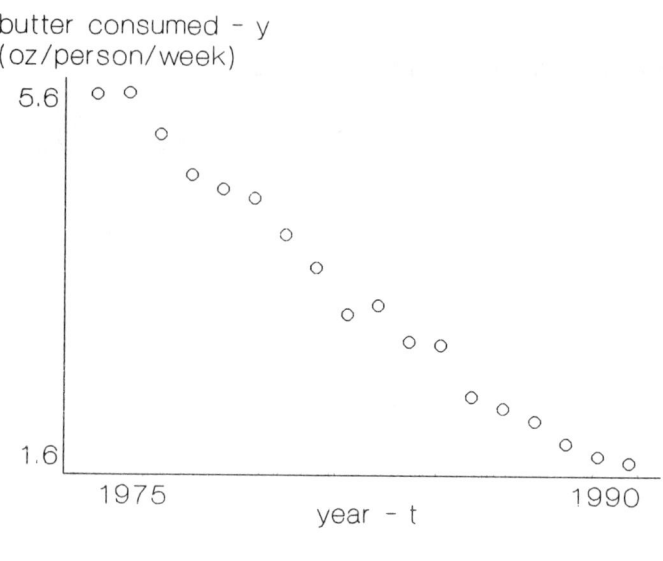

Figure 10.9

But what about the percentage change model? If the (constant) rate of change is $c\%$ per year then we have

model 2: $$y = A(1 + c/100)^t \tag{10.24}$$

where A is the consumption at $t = 0$ (a clear case of not worrying about the physical interpretation of the constant). This is the sort of formula that describes compound interest, radioactive decay and many other phenomena. But it is not linear and we only know how to deal with linear regression. Taking logarithms of both sides gives

$$\ln(y) = \ln(A) + t\ln(1 + c/100) \tag{10.25}$$

This relationship between $\ln(y)$ and t is linear with slope $\ln(1 + c/100)$ and intercept $\ln(a)$.

Now we know that we can deal with both models begin by estimating the parameters for linear model 1. Table 10.10 shows the calculations.

y	t	$(t-\bar{t})$	$(y-\bar{y})$	$(t-\bar{t})^2$	$(y-\bar{y})^2$	$(t-\bar{t})(y-\bar{y})$
5.610	1974	−8.5	2.205	72.250	4.862	−18.743
5.630	1975	−7.5	2.225	56.250	4.951	−16.688
5.160	1976	−6.5	1.755	42.250	3.080	−11.408
.	.	.	.	.	.	.
.	.	.	.	.	.	.
.	.	.	.	.	.	.
1.610	1990	7.5	−1.795	56.250	3.222	−13.462
1.540	1991	8.5	−1.865	72.250	3.478	−15.852
61.290	35685	0.0	0.000	484.500	32.462	−124.465

Table 10.10

From the table

$\bar{t}$ = 35685/18 = 1982.5
var(t) = 484.5/18 = 26.917
sd(t) = $26.917^{0.5}$ = 5.188

$\bar{y}$ = 61.290/18 = 3.405
var(y) = 32.462/18 = 1.803
sd(y) = $1.803^{0.5}$ = 1.343

cov(y,t) = −124.465/18 = −6.915
r = −6.915/(5.188 × 1.343) = −0.992

slope b = −6.915/26.917 = −0.257
intercept a = 3.405 − (−0.257 × 1982.5) = 512.908

Table 10.11 shows the calculations to estimate model 2, putting $L = \ln(y)$ for ease of notation.

$L = \ln(y)$	t	$(t-\bar{t})$	$(L-\bar{L})$	$(t-\bar{t})^2$	$(L-\bar{L})^2$	$(t-\bar{t})(L-\bar{L})$
1.725	1974	−8.5	0.584	72.250	0.341	−4.965
1.728	1975	−7.5	0.588	56.250	0.345	−4.407
1.641	1976	−6.5	0.500	42.250	0.250	−3.253
.	.	.	.	.	.	.
.	.	.	.	.	.	.
.	.	.	.	.	.	.
0.476	1990	7.5	−0.664	56.250	0.441	−4.982
0.432	1991	8.5	−0.709	72.250	0.502	−6.024
20.528	35685	0.0	0.000	484.500	3.209	−39.113

Table 10.11

The mean and variance of t are unaltered. From the table

$\bar{L}$ = 20.528/18 = 1.140
var(L) = 3.209/18 = 0.178
sd(L) = $0.178^{0.5}$ = 0.422

cov(L,t) = −39.113/18 = −2.173
r = −2.173/(5.188 × 0.422) = −0.993

slope b = −2.173/26.917 = −0.081
intercept a = 1.140 − (−0.081 × 1982.5) = 161.722

The two models are

butter consumption = y = 512.908 − 0.257(year)

and $\ln(y)$ = 161.722 − 0.081(year)

or, in the original form of (10.24),

$\exp(\ln(y))$ = $\exp(161.722 - 0.081(\text{year}))$
= $\exp(161.722)\exp(-0.081)^{\text{year}}$

so $y = (1.717 \times 10^{70})(0.922)^{\text{year}}$

Comparing with (10.24)

$(1 + c/100) = 0.922$

so $c/100 = 0.922 - 1 = -0.078$

and $c = -7.8\%$

In the linear model butter consumption declines at a constant rate of 0.257 oz/person/year and in the exponential model the decline is 7.8% per year. Both models have the same correlation coefficient so which one should be used?

Figure 10.10 shows the two models.

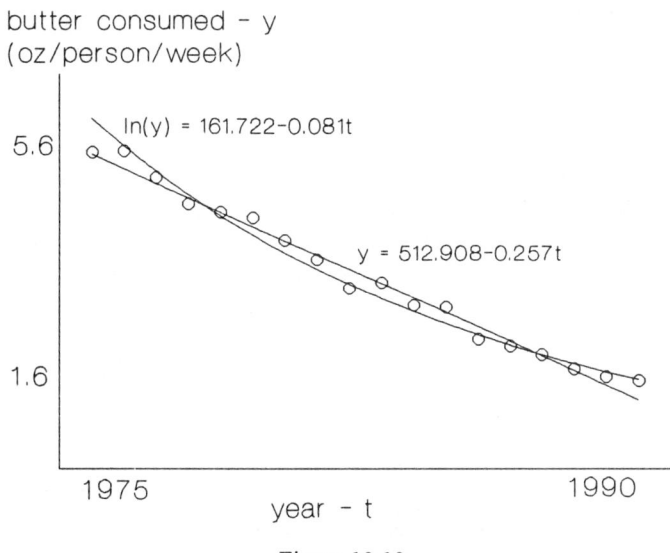

Figure 10.10

Looking at the linear model 1 over the whole time span there appears to be no pattern to the residuals. In contrast, the exponential model 2 results in positive residuals at either end of the range and negative residuals in the central portion. The residuals are not randomly placed and so this model is inferior.

You may object on two grounds, one empirical and one theoretical. Empirically you may observe that the exponential model seems to fit the last few observations pretty well and that the series does appear to be flattening out. Perhaps, but remember that as humans we are all too keen to see patterns where sometimes they don't exist. And is this flattening out no more than the behaviour seen in the late 1970's? On theoretical grounds you may say that the series cannot be entirely linear since if it were consumption would become negative at some point (during 1995 in fact; show this for yourself). Exponential decline means that the consumption will approach but never become zero.

The difficulty arises because we are extrapolating beyond the range of the collected data. Forecasting inevitably involves doing this. Here are the two forecasts for 1994:

linear model 1: $y = 512.908 - (0.257 \times 1994)$
                   $= 0.450$ oz/person

exponential model 2: $y = \exp(161.722 - (0.081 \times 1994))$
                        $= \exp(0.208)$
                        $= 1.231$ oz/person

Which would you choose?

Forecasting is a large subject in its own right. If you are interested try, for instance, Granger (1980) or Hanke and Reitsch (1989) or, more compendiously, Makridakis *et al.* (1983).

Example 10.4

Suppose that an expert on food and diet has compared current dietary fashions in a number of countries for 1992 and based on research expresses an opinion that the consumption of butter in Great Britain in 1994 will be 0.6 oz/person. Being of a statistical cast of mind the expert further acknowledges that there must inevitably be some uncertainty about the estimate and, further, that this may be characterised by a standard deviation of 0.08.

How would this modify the estimate obtained using the linear model in the previous example?

Solution

It seems safe to treat the two estimates, from the regression model and from the expert, as independent since the first is based on longitudinal data for one country and the second is based on a cross-sectional comparison between countries at one recent point in time.

Given only the mean and standard deviation of the expert's estimate the maximum entropy distribution is Normal. Regression residuals are assumed Normal too and so the estimates may be combined using (5.21) and (5.22).

Assuming the residuals from the regression model to be Normally distributed with mean 0.450 (above) and, from (10.21), variance

$$\text{var}(y_{\text{est}}) = s_y^2 \, [1 + 1/18 + (1994 - 1982.5)^2/484.500]$$

$$= 1.329 s_y^2$$

From (10.12) and (10.13)

$$s_y^2 = \text{var}(e) = (1 - r^2)\text{var}(y) \qquad (10.26)$$

$$= 1.803(1 - (-0.992)^2)$$
$$= 0.029$$

and so

$$\text{var}(y_{est}) = 1.329 \times 0.029 = 0.039$$

The combined estimate is Normally distributed with, from (5.22),

$$\text{variance} = 1/(1/0.039 + 1/0.08^2) = 0.024$$

so   standard deviation $= 0.024^{0.5} = 0.155$

and, from (5.21),

$$\text{mean} = (0.45/0.039 + 0.6/0.08^2)/(1/0.039 + 1/0.08^2)$$
$$= 0.507$$

A 95% credible interval forecast is

$$\text{butter consumption} = 0.507 \pm (1.96 \times 0.155)$$
$$= (0.203, 0.811) \text{ oz/person}$$

Figure 10.11 shows the combination of the estimates.

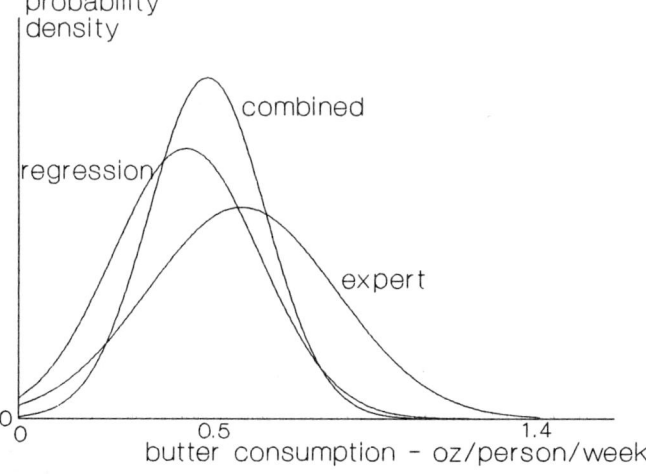

Figure 10.11 Combination of estimates

Example 10.5

The strength of beer is measured by its original gravity (OG). Here are some data gathered from a pub in Durham in November 1993:

Beer	OG	Price of a pint (£)
Theakston Best Bitter	1038	1.39
McEwan's 80/-	1042	1.44
Theakston XB	1044	1.49
Younger's No. 3	1043	1.44
Marston's Pedigree	1043	1.54
Charles Wells' Eagle Best Bitter	1035	1.47
Moorhouse's Pendle Witches' Brew	1050	1.57
Banks and Taylor's Dragon Slayer	1045	1.52
Shepherd Neame Bishop's Finger	1053	1.57
Everard's Tiger Best Bitter	1041	1.47
Timothy Taylor's Landlord	1042	1.47

I intend to produce a beer, *Old Scrotum*, with an OG of 1049. How much should I charge to be in line with the market? How much is an increase of 1 unit of OG worth in revenue?

Solution

The data are shown in Figure 10.12.

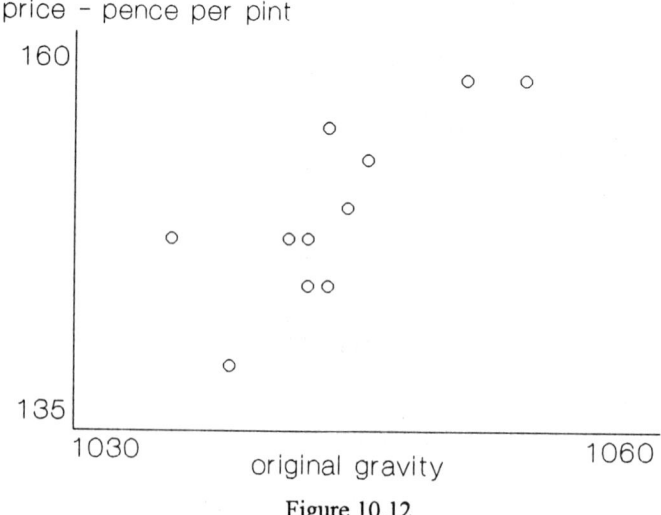

Figure 10.12

Examples

The table below shows the necessary calculations. y is the price of a pint, in pence, and x is the OG. The variances given in the bottom row are the variances of the data, not estimated population variances.

	y	x	$(x-\bar{x})(y-\bar{y})$	y_{est}	e	e^2
	139	1038	51.769	144.154	−5.154	26.562
	144	1042	6.132	147.692	−3.692	13.633
	149	1044	0.132	149.462	−0.462	0.213
	144	1043	1.314	148.577	−4.577	20.948
	154	1043	−1.413	148.577	5.423	29.410
	147	1035	15.041	141.500	5.500	30.250
	157	1050	55.041	154.769	2.231	4.976
	152	1045	5.496	150.346	1.654	2.735
	157	1053	79.587	157.423	−0.423	0.179
	147	1041	4.132	146.808	0.192	0.037
	147	1042	2.314	147.692	−0.692	0.479
mean:	148.818	1043.273	19.959	148.818	0.000	11.766
variance:	29.421	22.562				

As a preliminary, note that

$$r^2 = \frac{\text{cov}(x,y)^2}{\text{var}(x)\,\text{var}(y)} = \frac{19.959^2}{22.562 \times 29.421} = 0.600$$

Next calculate the parameters of the regression line.

slope: $\quad b = \dfrac{\text{cov}(x,y)}{\text{var}(x)} = \dfrac{19.959}{22.562} = 0.885$

intercept: $\quad a = \bar{y} - b\bar{x}$
$= 148.818 - (0.885 \times 1043.273)$
$= -774.077$

The performance of the model is described by looking at the residuals and, in particular, finding the variance of the residuals

$$s_y^2 = \frac{\Sigma e^2}{(n-2)} = 11.766 \times \frac{11}{9} = 14.380$$

so $\quad s_y = (14.380)^{0.5} = 3.792$

We may, equivalently, find the coefficient of determination:

R^2 = proportion of var(y) accounted for by the model

$R^2 = 1 - 11.766/29.421 = 0.600$

(Remember that for simple regression, $R^2 = r^2$.)

Now, think about the questions that were posed.

A pint of *Old Scrotum* ought to sell for

$-774.077 + (0.885 \times 1049) = 154$ pence

An increase of 1 unit of OG is worth 0.885 pence per pint in increased revenue. But this is based only on sample data gathered in one pub. What might be the population slope, β?
We have

$$\text{var}(\beta) = s_y^2 / \Sigma(x - \bar{x})^2 = s_y^2 / n \, \text{var}(x)$$

$$= 14.380/(11 \times 22.562) = 0.058$$

and $\text{sd}(\beta) = (0.058)^{0.5} = 0.241$

Using the t distribution with $11 - 2 = 9$ degrees of freedom the 95% confidence interval for β is

$\beta = 0.885 \pm (2.262 \times 0.241) = (0.340, 1.430)$

where the 2.262 is the appropriate t value.
So, the increase in revenue per unit increase in OG could be anywhere between 0.34 and 1.43 pence per pint.

Computer solution

Although the calculations in this chapter have been shown in some detail it will usually be more convenient to use the regression capability possessed by most spreadsheets. Here is the output of the regression part of a spreadsheet. (This is from *SmartWareII*, but others are pretty similar.)

MULTIPLE LINEAR REGRESSION

Dependent Variable: PRICE

Variable	Mean	Parameter Estimate	Standard Error	t for HO: parameter = 0
Intercept		−774.077	251.132	−3.082
OG	1043.273	0.885	0.241	3.675

Source	DF	Sum of Squares	Mean Square	F-Value
Model	*1.000*	*194.213*	*194.213*	*13.505*
Error	*9.000*	*129.423*	*14.380*	
total	*10.000*	*323.636*		

Dependent Mean	148.818
Root Mean Square Error	3.792
Coefficient of Variation	2.548
R-Square	0.600
Adjusted R-Square	0.556

The results are headed MULTIPLE LINEAR REGRESSION because the software can deal with equations with more than one independent variable (more than one X) although we have only used it for simple regression in this case.

The results shown in italics are used for *analysis of variance*, which is not covered in this book.

At the top are the results already calculated above. Check that you can recognise them. The standard deviation (here called standard error) of the estimate of α, the population intercept, is given by treating the intercept as just $x = 0$ and then applying (10.23).

$$\text{var}(\alpha) = s_y^2 [1/n + (0 - \overline{x})^2 / \Sigma(x_i - \overline{x})^2]$$

$$= 14.380[1/11 + 1043.273^2/(11 \times 22.562)]$$

$$= 63065.748$$

so $\text{sd}(\alpha) = 251.129$

(which is the same as found in the computer solution, given rounding error).

The t values are to test the hypothesis that α and β are 0 and are found by dividing the estimate by the standard deviation of the estimate (e.g. 0.885/0.241). Since the absolute values of both exceed 2.262 both hypotheses may be rejected at the 95% level: α and β are significantly different from 0.

At the bottom, the Root Mean Square Error is just s_y.

The Coefficient of Variation shows s_y expressed as a percentage of the mean value of y:

$$\frac{3.792}{148.818} \times 100\% = 2.548\%$$

Finally, adjusted R^2 is R^2 calculated to take account of degrees of freedom. Remember that R^2 is the proportion of the variance of y accounted for by the regression. We have used

$$R^2 = 1 - \frac{\Sigma(y - y_{est})^2 / n}{\Sigma(y - \bar{y})^2 / n} = 0.600$$

But it would have been more correct to divide by the appropriate number of degrees of freedom rather than dividing by the sample size, n. Without the regression there are $n - 1$ degrees of freedom. With the regression with m explanatory variables there are $n - m - 1$ degrees of freedom and so

$$\text{adjusted } R^2 = 1 - \frac{\Sigma(y - y_{est})^2 / (n - m - 1)}{\Sigma(y - \bar{y})^2 / n}$$

$$= 1 - (1 - R^2) \frac{(n - 1)}{(n - m - 1)}$$

In the current example we have just one explanatory variable and so with $n = 11$ and $m = 1$

$$\text{adjusted } R^2 = 1 - (1 - 0.600) \frac{(11 - 1)}{(11 - 2)}$$

$$= 1 - 0.400(10 / 9) = 0.556$$

When n is large and m is small compared to n the correction is fairly trivial.

10.7 EXERCISES

10.1 An experiment was conducted to show the relationship between the yield and the number of tillers used in growing the Milfor 6(2) variety of rice. Here are the results:

Grain Yield (kg/ha)	Tillers (no./m^2)
4862	160
5244	175
5128	192
5052	195
5298	238
5410	240
5234	252
5608	282

(Source: Gomez and Gomez, 1976, p132)

Find the covariance, correlation coefficient and entropy change that characterise these data. Give an 80% credible interval forecast for the yield if the number of tillers/m^2 is 260.

10.2 Triaxial tests were carried out on samples of Erksak sand to determine the relationship between the peak friction angle and dilation rate at peak. The results obtained for dense samples (relative density = 70%) were:

peak friction angle (deg)	maximum rate of dilatancy
40.7	0.822
40.9	0.887
41.7	0.923
41.8	0.929
36.4	0.325
35.9	0.310
39.5	0.299
35.8	0.261
40.0	0.706
37.8	0.434

(Source: Vaid and Sasitharan, 1992)

Fit a linear model to estimate friction angle from dilatancy and give 90% credible interval estimates of the slope and intercept of the model.

10.3 The data below show measurements of diastolic and systolic blood pressure (in mmHg) for 32 middle aged male patients at a general practice.

Systolic	Diastolic	Systolic	Diastolic
152	71	124	77
105	61	146	96
167	120	156	94
133	89	144	81
186	138	103	75
98	67	131	87
155	99	163	90
136	74	129	66
170	112	160	85
142	86	142	82
115	76	201	119
129	83	158	92
113	70	149	84
157	98	132	78
146	88	175	103
142	79	118	68

(Source: Woodward and Lesley, 1988, p312)

Find the entropy reduction, T, due to the correlation between the two variables.

If it is known that a patient has a systolic blood pressure of 120 mmHg give a 70% credible interval estimate of his diastolic blood pressure. Give a similar estimate for the mean diastolic blood pressure of all men with this systolic pressure.

10.4 The following data show the relationship between density and Janaka hardness for 36 species of timber.

Janaka hardness	Density (lb/ft³)	Janaka hardness	Density (lb/ft³)
2700	67.4	1100	40.7
2890	68.8	3260	66.00
2740	69.1	1940	59.8
2020	57.3	1210	39.4
427	24.8	2310	59.2
704	32.7	2010	51.5
1710	51.5	1070	38.8
914	38.5	1880	53.4
1400	46.9	979	35.6
517	28.4	587	30.3
549	28.4	413	27.3
1160	40.3	989	39.9
648	29.0	1980	56.0
1820	56.5	1010	40.6
1270	42.9	3140	69.1
1130	40.7	1980	57.6
484	24.7	1180	45.8
1760	48.2	1020	39.3

(Source: Williams, 1959, p43)

It is suggested that a relationship of the form

$$\text{hardness} = k(\text{density})^b$$

may be appropriate. Obtain an estimate of the exponent b. Plot the data and the fitted curve.

10.5 In eleventh century England taxes were levied by the Crown on the lords of manors. The amount of tax to be paid was assessed and expressed in fiscal acres. The Doomsday Book gives data showing these manorial assessments and also the annual value (income) of the manor, in shillings. The following table shows such data for 45 manors in Wiltshire.

tax	value	tax	value	tax	value
1200	180	300	50	384	60
60	60	240	40	1200	160
720	120	300	25	1200	200
120	15	1335	300	360	60
480	60	1320	280	1200	200
600	120	60	10	750	160
840	220	480	60	720	100
120	15	1200	80	1200	140
720	240	600	70	210	15
360	60	270	50	120	13
2400	240	30	10	30	7
1200	200	300	30	60	12
150	20	120	10	600	80
240	40	1440	200	120	20
120	15	120	30	600	100

(Source: McDonald and Snooks, 1985)

It would be a whimsical monarch who levied taxes independently of ability to pay. Find the association between the variables as measured by T, the information reduction.

Now group the data into a contingency table and calculate T again. Why are the two values for T not the same?

10.6 Sulphate of ammonia is a crystalline substance sold in sacks or boxes. For the automatic filling process to proceed easily it is important that the crystals flow smoothly. The table gives data from 48 samples showing the relationship between flow rate (F g/s) and the initial moisture content of the crystals (M, in units of 0.01%).

F	M	F	M	F	M
5.00	21	5.10	14	5.00	21
4.81	20	5.05	14	2.43	24
4.46	16	4.27	20	0.00	37
4.81	18	4.90	12	4.10	21
4.46	16	4.55	11	3.70	28
3.85	18	5.32	10	3.36	29
3.21	12	4.39	10	3.79	23
3.25	12	4.85	16	3.40	32
4.55	13	4.59	17	1.51	26
4.85	13	5.00	17	0.00	28
4.00	17	3.82	17	1.72	21
3.62	24	3.68	15	2.33	22
5.15	11	5.15	17	2.38	34
3.76	10	2.94	21	3.68	29
4.90	17	3.18	23	4.20	17
4.13	14	2.28	22	5.00	11

(Source: Davies, 1961, p221)

It is suggested that a model of the form $F = a + bM$ may be used to describe the relationship.

(a) Find values for a and b and for the correlation coefficient.

(b) Give a 75% credible interval estimate for the slope, β.

(c) A knowledgeable colleague gives an estimate of β, based on years of experience, as best described as β having a mean of -0.1 and a standard deviation of 0.05. What is your credible interval for β if this opinion is taken into account?

(d) How would your answer to (c) change if the colleague's estimate had been characterised by a range of between -0.05 and -0.14 and a mean of -0.1?

10.7 In Example 8.4 it was concluded that knowledge of yesterday's *Footsie* movement was no help in estimating today's movement. Considering all the data, is it possible to estimate the first *Footsie* value in November? If it is, do it.

10.8 Refer back to Example 10.5. Is there any evidence that any of the beers and their brand names are being strongly exploited or woefully underexploited?

10.9 The mean household size in the recent past has been:

year	mean household size (people/household)
1971	2.91
3	2.83
5	2.78
7	2.71
9	2.67
1981	2.70
3	2.64
5	2.56
6	2.55
7	2.55
9	2.55
1990	2.46
1	2.48

(Source: General Household Survey)

Using this table and that in Exercise 4.1 estimate what the distribution of household sizes will be in 1995.

References

Agresti, A. (1990) *Categorical Data Analysis*, New York, Wiley.

Arkes, H.R. and Hammond, K.R. (eds.) (1986) *Judgement and Decision Making: An Interdisciplinary Reader*, Cambridge, Cambridge University Press.

Attneave, F. (1959) *Applications of Information Theory to Psychology: A Summary of Basic Concepts, Methods and Results*, New York, Henry Holt and Company.

Aune, B. (1991) *Knowledge of the External World*, London, Routledge.

Aykac, A. and Brumat, C. (eds.) (1977) *New Developments in the Applications of Bayesian Methods*, Amsterdam, North-Holland.

Barnett, V. (1973) *Comparative Statistical Inference*, Chichester, Wiley.

Batty, M. (1976) *Urban Modelling: Algorithms, Calibrations, Predictions*, Cambridge, Cambridge University Press.

Bayes, T. (1763) An essay towards solving a problem in the doctrine of chances, *Philos. Trans. R. Soc.*, **56**, 370–418. [This is reprinted, with a biographical note by G.A. Barnard, in *Biometrika*, **45**, 293–315. See also Press 1989.]

Bell, D.E., Keeney, R.L. and Raiffa, H. (eds.) (1977) *Conflicting Objectives in Decisions*, Chichester, Wiley.

Berger, J.O. (1985) *Statistical Decision Theory and Bayesian Analysis (2nd edn.)*, New York, Springer-Verlag.

Bernardo, J.M., DeGroot, M.H., Lindley, D.V. and Smith, A.F.M. (eds.) (1980) *Bayesian Statistics: Proceedings of the First International Meeting Held in Valencia (Spain)*, Valencia, University Press.

Bernardo, J.M., DeGroot, M.H., Lindley, D.V. and Smith, A.F.M. (eds.) (1985) *Bayesian Statistics 2*, Amsterdam, North-Holland.

Bernardo, J.M., DeGroot, M.H., Lindley, D.V. and Smith, A.F.M. (eds.) (1988) *Bayesian Statistics 3*, Oxford, Clarendon Press.

Bouras, N., Webb, Y., Clifford, P., Papadatos, Y. and Zouri, M. (1992) A needs survey among patients in Leros asylum, *Br. J. Psychiatr.*, **161**, 75–79.

References

Box, G.E.P. and Tiao, G.C. (1973) *Bayesian Inference in Statistical Analysis*, Reading, Mass., Addison-Wesley.

Brillouin, L. (1962) *Science and Information Theory (2nd. edn.)*, New York, Academic Press.

Brooks, C.H. (1992) Do area health education programs produce primary care specialists? Results of a longitudinal study, *Int. J. Health Servi.*, **22**(3), 567–578.

Bunn, D.W. (1984) *Applied Decision Analysis*, New York, McGraw-Hill.

Casti, J.L. (1992) *Searching for Certainty: What Science Can Know About the Future*, London, Scribners.

Chapman, J-A.W. (1976) A comparison of the χ^2, $-2\log R$ and multinomial probability criteria for significance tests when expected frequencies are small, *J. Am. Stat. Assoc.*, **71**, 854–863.

Chatfield, C. (1985) The initial examination of data, *J. R. Stat. Soc. A*, **148**(3), 214–253.

Cherry, C. (1966) *On Human Communication (2nd edn.)*, Cambridge, Mass., MIT Press.

Chung, K.L. (1974) *Elementary Probability Theory with Stochastic Processes*, New York, Springer-Verlag.

Clemen, R.T. (1986) *Making Hard Decisions: An Introduction to Decision Analysis*, Boston, PWS-Kent Publishing Co.

Cochran, W.G. (1952) The χ^2 test and goodness of fit, *Ann. Math. Stat.*, **23**(3), 315–345.

Cochran, W.G. (1977) *Sampling Techniques (3rd edn.)*, New York, Wiley.

Conolly, B. (1981) *Techniques in Operational Research Volume 2: Models, Search and Randomisation*, Chichester, Ellis Horwood.

References

Covello, V.T., Flamm, W.G., Rodricks, J.V. and Tardiff, R.G. (eds.) (1983) *The Analysis of Actual Versus Perceived Risks*, New York, Plenum Press.

Coveney, P. and Highfield, R. (1990) *The Arrow of Time: A Voyage Through Science to Solve Time's Greatest Mystery*, London, W.H. Allen.

Cover, T.M. and Thomas, J.A. (1991) *Elements of Information Theory*, New York, Wiley.

Coyle, R.G. (1972) *Decision Analysis*, London, Nelson.

Crum, R.L., Laughhann, D.J. and Payne, J.W. (1981) Risk Preference: Empirical Evidence and its Implications for *Capital Budgeting:* In Derkinderen, F.G.J. and Crum, R.L. (eds.) *Risk, Capital Costs and Project Financial Decisions*, Boston, Martinus Nijhoff.

Dale, A.I. (1991) *A History of Inverse Probability from Thomas Bayes to Karl Pearson*, New York, Springer-Verlag.

Davies, O.L. (ed.) (1961) *Statistical Methods in Research and Production (3rd edn.)*, London, Oliver and Boyd.

de Finetti, B. (1937) Foresight: its logical laws, its subjective sources, *Annales de l'Institut Henri Poincaré (vol. 7)*. [A translation is given in Kyburg and Smokler, 1964.]

de Finetti, B. (1972) *Probability, Induction and Statistics: The Art of Guessing*, New York, Wiley.

de Finetti, B. (1974) *Theory of Probability: A Critical Introductory Treatment (vol. 1)*, New York, Wiley.

DeGroot, M.H. (1970) *Optimal Statistical Decisions*, New York, McGraw-Hill.

Delucchi, K.L. (1983) The use and misuse of chi-square: Lewis and Burke revisited, *Psychol. Bull.*, **94**(1), 166–176.

Deming, W.E. and Stephan, F.F. (1940) On a least squares adjustment of a sampled frequency table when the expected marginal totals are known, *Ann. Math. Stat.*, **11**, 427–444.

Denbigh, K.G. and Denbigh, J.S. (1985) *Entropy in Relation to Incomplete Knowledge*, Cambridge, Cambridge University Press.

Domotor, Z. (1970) *Qualitative Information and Entropy Structures:* In Hintikka, J. and Suppes, P. (eds.) *Information and Inference,* Dordrecht, D. Reidel.

Dretske, F. (1981) *Knowledge and the Flow of Information,* Oxford, Basil Blackwell.

Dreze, J.H. (ed.) (1974) *Allocation under Uncertainty: Equilibrium and Optimality,* London, Macmillan.

Earman, J. (1992) *Bayes or Bust?: A Critical Examination of Bayesian Confirmation Theory,* Cambridge, Mass., MIT Press.

Easton, A. (1973) *Complex Managerial Decisions Involving Multiple Objectives,* New York, Wiley.

Edwards, W. and Tversky, A. (eds.) (1967) *Decision Making,* Harmondsworth, Penguin Books.

Eells, E. (1982) *Rational Decision and Causality,* Cambridge, Cambridge University Press.

Ehrenberg, A.S.C. (1982) *A Primer in Data Reduction,* Chichester, Wiley.

Employment Department (1992) *Labour Market Quarterly Report,* London, Employment Dept., May 1992.

Fast, J.D. (1962) *Entropy,* Eindhoven, Philips Technical Library.

Feinstein, A. (1958) *Foundations of Information Theory,* New York, McGraw-Hill.

Ferguson, T.S. (1967) *Mathematical Statistics: A Decision Theoretic Approach,* New York, Academic Press.

Fienberg, S.E. (1970) An iterative procedure for estimation in contingency tables, *Ann. Math. Stat.,* **41**, 907–917.

Fienberg, S.E. (1991) *The Analysis of Cross-Classified Categorical Data,* Cambridge, Mass., MIT Press.

Fischhoff, B., Lichtenstein, S., Slovic, P., Derby, S.L. and Keeney, R.L. (1981) *Acceptable Risk,* Cambridge, Cambridge University Press.

Fishburn, P.C. (1970) *Utility Theory for Decision Making,* New York, Wiley.

References

Fishburn, P.C. (1981) Subjective expected utility: a review of normative theories, *Theory Decis.*, **13**, 139–199.

Fishburn, P.C. (1988) *Non-Linear Preference and Utility Theory*, Brighton, Wheatsheaf Books.

Fisher, R.A. (1934) *Statistical Methods for Research Workers (5th edn.)*, Edinburgh, Oliver and Boyd.

Fletcher, C., Higginbotham, R. and Norris, P. (1993) The interrelationships of managers' work time and personal time, *Pers. Rev.*, **22**(2), 56–64.

French, S. (1986) *Decision Theory: an Introduction to the Mathematics of Rationality*, Chichester, Ellis Horwood.

French, S. (1989) *Readings in Decision Analysis*, London, Chapman and Hall.

Fuji, Y. and Yamanouchi, H. (1973) The distribution of collisions in Japan and methods of estimating collision damage: In *Marine Traffic Engineering: Proceedings of a Conference*, London, Royal Institution of Naval Architects and Royal Institute of Navigation.

Garner, W.R. (1962) *Uncertainty and Structure as Psychological Concepts*, New York, Wiley.

Garner, W.R. and McGill, W.J. (1956) The relation between information and variance analyses, *Psychometrika*, **21**, 219–228.

Godambe, V.P. and Sprott, D.A. (eds.) (1971) *Foundations of Statistical Inference*, Toronto, Holt, Rinehart and Winston of Canada.

Gokhale, D.V. and Kullback, S. (1978) *The Information in Contingency Tables*, New York, Marcel Dekker.

Goldman, A.I. (1986) *Epistemology and Cognition*, Cambridge, Mass., Harvard University Press.

Goldman, S. (1953) *Information Theory*, London, Constable.

Gomez, K.A. and Gomez, A.A (1976) *Statistical Procedures for Agricultural Research with Emphasis on Rice*, Manila, The International Rice Research Institute.

References

Good, I.J. (1983) *Good Thinking: The Foundations of Probability and its Applications*, Minneapolis, University of Minnesota Press.

Goodman, L.A. (1970) The multivariate analysis of qualitative data: interactions among multiple classifications, *J. Am. Stat. Assoc.*, **65**, 225–256.

Goodman, L.A. (1984) *The Analysis of Cross-Classified Data Having Ordered Categories*, Cambridge, Mass., Harvard University Press.

Goodman, L.A. and Kruskal, W.H. (1979) *Measures of Association for Cross Classifications*, New York, Springer-Verlag. [Brings together four papers originally published in the *J. Am. Stat. Assoc.*, **49**, 732–764, **54**, 123–163, **58**, 310–364, **67**, 415–421.]

Goodwin, P. and Wright, G. (1991) *Decision Analysis for Management Judgement*, Chichester, Wiley.

Granger, C.W.J. (1980) *Forecasting in Business and Economics*, New York, Academic Press.

Gratson, M.W., Gratson, G.K. and Bergerud, A.T. (1991) Male dominance and copulation disruption do not explain variance in male mating success on sharp-tailed grouse (tympanuchus phasianellus) leks, *Behaviour*, **118**, 187–213.

Greater London Council (1985) *GLTS 81: Transport Data for London*, London, Greater London Council.

Guiasu, S. (1977) *Information Theory with Applications*, New York, McGraw-Hill.

Hacking, I. (1975) *The Emergence of Probability*, Cambridge, Cambridge University Press.

Haight, F.A. (1967) *Handbook of the Poisson Distribution*, New York, Wiley.

Hanke, J.E. and Reitsch, A.G. (1989) *Business Forecasting (3rd edn.)*, Boston, Allyn and Bacon.

Hartigan, J. (1983) *Bayes Theory*, New York, Springer-Verlag.

Henery, R.J. (1985) On the average probability of losing bets on horses with given starting price odds, *J. R. Stat. Soc. A*, **148**(4), 342–349.

Hertz, D.B. and Thomas, H. (1983) *Risk Analysis and its Applications*, Chichester, Wiley.

Hertz, D.B. and Thomas, H. (1984) *Practical Risk Analysis: An Approach Through Case Histories*, Chichester, Wiley.

Hey, J.D. (1983) *Data in Doubt: An Introduction to Bayesian Statistical Inference for Economists*, Oxford, Martin Robertson.

Hintikka, J. (1970) On Semantic Information: In Hintikka, J. and Suppes, P. (eds.) *Information and Inference*, Dordrecht, D. Reidel.

Hiscott, R.D. and Connop, P.J. (1990) Job turnover among nursing professionals: impact of shift length and kinship responsibilities, *Sociol. Soc. Res.*, **75**(1), 32–37.

Hobson, A. and Cheng, B. (1973) A comparison of the Shannon and Kullback information measures, *J. Stat. Phys.*, **7**, 301–310.

Hogarth, R.M. (1975) Cognitive processes and the assessment of subjective probability distributions, *J. Am. Stat. Assoc.*, **70**, 271–294 (reprinted in Aykac and Brumat, 1977).

Holmes, D.I. (1985) The analysis of literary style – a review, *J. R. Stat. Soc. A*, **148**(4), 328–341.

Horwich, P. (1981) *Probability and Evidence*, Cambridge, Cambridge University Press.

Hout, M., Duncan, O.D. and Sobel, M.E. (1987) Association and heterogeneity: structural models of similarities and differences, *Sociol. Method*, **17**, 145–184.

Hull, J., Moore, P.G. and Thomas, H. (1973) Utility and its measurement, *J. R. Stat. Soc. A,* **136**(2), 226–247.

Humphreys, P., Svenson, O. and Vari, A. (eds.) (1983) *Analysing and Aiding Decision Processes*, Amsterdam, North-Holland.

Hutton, C., Shaw, G., and Pearson, R. (1809) *The Philosophical Transactions of the Royal Society of London From Their Commencement in 1665 to the Year 1800; Abridged*, **23**, p41.

Iversen, G.R. (1984) *Bayesian Statistical Inference*, Beverly Hills, Sage.

Jamison, D. (1970) Bayesian Information Usage: In Hintikka, J. and Suppes, P. (eds.) *Information and Inference*, Dordrecht, D. Reidel.

Jaynes, E.T. (1957) Information theory and statistical mechanics, *Phys. Rev.*, **106**, 620–630 and **108**, 171–190 (reprinted in Jaynes, 1983).

Jaynes, E.T. (1968) Prior probabilities, *IEEE Trans. Systems Science and Cybernetics*, **SSC-4**, 227–241 (reprinted in Jaynes, 1983).

Jaynes, E.T. (1983) *Papers on Probability, Statistics and Statistical Physics (ed. R.D. Rosenkrantz)*, Dordrecht, D. Reidel.

Jaynes, E.T. (1985) Some random observations, *Synthese*, **63**, 115–138.

Jeffreys, H. (1939) *Theory of Probability*, Oxford, Clarendon Press.

Jessen, R.J. (1978) *Statistical Survey Techniques*, New York, Wiley.

Jessop, A. (1990) *Decision and Forecasting Models: With Transport Applications*, Chichester, Ellis Horwood.

Jones, D.S. (1979) *Elementary Information Theory*, Oxford, Clarendon Press.

Jones, J.M. (1977) *Introduction to Decision Theory*, Homewood, Ill., Richard D. Irwin.

Journal of the Institute of Statisticians (1983) **32** (1 & 2).

Kahneman, D., Slovic, P. and Tversky, A. (eds.) (1982) *Judgement Under Uncertainty: Heuristics and Biases*, Cambridge, Cambridge University Press.

Kaufman, M. and Thomas, H. (eds.) (1977) *Modern Decision Analysis: Selected Readings*, Harmondsworth, Penguin Books.

Keeney, R.L. (1972) Utility functions for multi-attributed consequences, *Manage. Sci.*, **18**, 276–287.

Keynes, J.M. (1973) *The Collected Writings of John Maynard Keynes Volume VIII: A Treatise on Probability*, London, Macmillan. [The treatise was originally published in 1921.]

Khinchin, A.I. (1957) *Mathematical Foundations of Information Theory*, New York, Dover.

Krahn, H., Fernandes, A. and Adebayo, A. (1990) English language ability and industrial safety among immigrants, *Sociol. Soc. Res.*, **75**(1), 17–26.

Kullback, S. (1959) *Information Theory and Statistics*, New York, Wiley.

Kullback, S. and Leibler, R.A. (1951) On information and sufficiency, *Ann. Math. Stat.*, **22**, 79–86.

Kyburg, H.E. (1961) *Probability and the Logic of Rational Belief*, Middletown, Conn., Wesleyan University Press.

Kyburg, H.E. and Smokler, H.E. (eds.) (1964) *Studies in Subjective Probability*, New York, Wiley.

Latane, B. and Dabbs, J.M. (1975) Sex, group size and helping in three cities, *Sociometry*, **38**(2), 180–194.

LaValle, I.H. (1978) *Fundamentals of Decision Analysis*, New York, Holt, Rinehart and Winston.

Lee, P.M. (1989) *Bayesian Statistics: an Introduction*, New York, Oxford University Press.

Levine, R.D. and Tribus, M. (eds.) (1979) *The Maximum Entropy Formalism*, Cambridge, Mass., MIT Press.

Lindley, D.V. (1965) *Introduction to Probability and Statistics (2 vols)*, Cambridge, Cambridge University Press.

Lindley, D.V. (1971) *Making Decisions*, London, Wiley.

Lindley, D.V. (1972) *Bayesian Statistics, A Review*, Philadelphia, Society for Industrial and Applied Mathematics.

Lindley, D.V., Tversky, A. and Brown, R.V. (1979) On the reconciliation of probability assessments (with discussion), *J. R. Stat. Soc. A*, **142**, 146–180.

Loeb, S.C. (1993) Use and selection of red-cockaded woodpecker cavities by southern flying squirrels, *J. Wildl. Manage.*, **57**(2), 329–335.

Makridakis, S., Wheelwright, S.C. and McGee, V.E. (1983) *Forecasting: Methods and Applications (2nd edn.)*, New York, Wiley.

Mansuripur, M. (1987) *Introduction to Information Theory*, Englewood Cliffs, N.J., Prentice Hall.

Mantle, M.J., Greenwood, R.M. and Curry, H.L.F. (1977) Backache in pregnancy, *Rheumatol. Rehabil.*, **16**, 95–101.

Maritz, J.S. (1970) *Empirical Bayes Methods*, London, Methuen.

Massiah, J. (1986) Work in the lives of Caribbean women, *Soc. Econ. Stud.*, **35**(2), 177–239.

McCarthy, R.G. (ed.) (1959) *Drinking and Intoxication: Selected Readings in Social Attitudes and Controls*, New Haven, Conn., College and Universities Press.

McDonald, J. and Snooks, G.D. (1985) Statistical analysis of Domesday Book (1086), *J. R. Stat. Soc. A*, **148**(2), 147–160.

McGill, W.J. (1954) Multivariate information transmission, *Psychometrika*, **19**(2), 97–116.

Menges, G. (ed.) (1974) *Information, Inference and Decision*, Dordrecht, D. Reidel.

Miller, G.A. (1955) Note on the Bias of Information Estimates: In Quastler, 1955a.

Miller, G.A. (1956) The magic number seven, plus or minus two, *Psychal. Rev.*, **63**, 81–97.

Moles, A. (1966) *Information Theory and Esthetic Perception (trans. J.E. Cohen)*, Urbana, University of Illinois Press.

Monk-Turner, E. (1990) Comparing advertisements in British and American women's magazines: 1988–89, *Sociol. Soc. Res.*, **75**(1), 53–56.

Moore, P.G. (1972) *Risk in Business Decision*, London, Longman.

Moore, P.G. (1983) *The Business of Risk*, Cambridge, Cambridge University Press.

Moore, P.G. and Thomas, H. (1976) *The Anatomy of Decisions*, Harmondsworth, Penguin Books.

References

Moore, P.G., Thomas, H., Bunn, D.W. and Hampton, J. (eds.) (1976) *Case Studies in Decision Analysis*, Harmondsworth, Penguin Books.

Morris, P.A. (1974) Decision analysis expert use, *Manage. Sci.*, **20**, 1233–1241.

Morris, P.A. (1977) Combining expert judgements: a Bayesian approach, *Manage. Sci.*, **23**, 679–693.

Morrison, D.E. and Henkel, R.E. (1970) *The Significance Test Controversy: A Reader*, Chicago, Aldine.

Moser, C.A. and Kalton, G. (1971) *Survey Methods in Social Investigation*, London, Heinemann.

Mosteller, F. (1968) Association and estimation in contingency tables, *J. Am. Stat. Assoc.*, **63**, 1–28.

Mosteller, F. and Wallace, D.L. (1984) *Applied Bayesian and Classical Inference: The Case of The Federalist Papers*, New York, Springer-Verlag.

Munro, J. (1979) Uncertainty and fuzziness in engineering decision-making: In Wirasinghe, S.C. and Jordaan, I.J. (eds.) *Proceedings of the First Canadian Seminar on Systems Theory for the Civil Engineer*, University of Calgary, Dept. of Civil Engineering.

Nisbett, R. and Ross, L. (1980) *Human Inference: Strategies and Shortcomings of Social Judgement*, Englewood Cliffs, N.J., Prentice Hall.

Nozick, R. (1969) Newcomb's problem and two principles of choice: In Rescher, N. *et al.* (eds.) *Essays in Honor of Carl G. Hempel*, Dordrecht, D. Reidel.

O'Brien, D.P. and Darnell, A.C., (1982) *Authorship Puzzles in the History of Economics: A Statistical Approach*, London, Macmillan.

Olmsted, J. (1988) Observations on Evolution: In Weber, B.H., Depew, D.J and Smith, J.D. (eds.) *Entropy, Information and Evolution: New Perspectives on Physical and Biological Evolution,* Cambridge, Mass., MIT Press.

Owen, D.B. (ed.) (1976) *On the History of Statistics and Probability*, New York, Marcel Dekker.

Pearson, F.S. and Toby, J. (1991) Fear of school-related predatory crime, *Sociol. Soc. Res.*, **75**(3), 117–125.

References

Perutz, M.F. (1993) An intellectual bumblebee, *The New York Review of Books*, **40**(16), 17–20 (7 October).

Peters, J. (1975) Entropy and information: conformities and controversies: In Kubat, L. and Zeman, J. (eds.) *Entropy and Information in Science and Philosophy*, Prague, Academia.

Philips, L.D. (1973) *Bayesian Statistics for Social Scientists*, London, Nelson.

Pollard, W.E. (1986) *Bayesian Statistics for Evaluation Research: an Introduction*, Beverly Hills, Sage.

Press, S.J. (1989) *Bayesian Statistics: Principles, Models, and Applications*, New York, Wiley.

Quastler, H. (ed.) (1953) *Essays on the Use of Information Theory in Biology*, Urbana, University of Illinois Press.

Quastler, H. (ed.) (1955a) *Information Theory in Psychology: Problems and Methods*, Glencoe, Ill., The Free Press.

Quastler, H. (1955b) Information Theory Terms and Their Psychological Correlates: In Quastler, 1955a.

Quine, M.P. and Seneta, E. (1987) Bortkewicz's data and the law of small numbers, *Int. Stat. Rev.*, **5**, 173–181.

Raiffa, H. (1968) *Decision Analysis*, Reading, Mass., Addison-Wesley.

Raiffa, H. and Schlaifer, R. (1961) *Applied Statistical Decision Theory*, Cambridge, Mass., Harvard University Press.

Ramsey, F.P. (1926) Truth and Probability, reprinted in Kyburg and Smokler, 1964.

Rescher, N. (1985) *Pascal's Wager: A Study of Practical Reasoning in Philosophical Theology*, Notre Dame, Ind. University of Notre Dame Press.

Rosenkrantz, R.D. (1970) Experimentation as Communication with Nature: In Hintikka, J. and Suppes, P. (eds.) *Information and Inference*, Dordrecht, D. Reidel.

Rosenkrantz, R.D. (1977) *Inference, Method and Decision: Towards A Bayesian Philosophy of Science*, Dordrecht, D. Reidel.

References

Rosenkrantz, R.D. (1981) *Foundations and Applications of Inductive Probability*, Atascadero, Calif., Ridgeview Publishing Company.

Rudd, E. (1987) The educational qualifications and social class of the parents of undergraduates entering British universities in 1984, *J. R. Stat. Soc. A.*, **150**(4), 346–372.

Savage, L.J. (1954) *The Foundations of Statistics*, New York, Wiley.

Savage, L.J. (1962) *The Foundations of Statistical Inference*, London, Methuen.

Savage, L.J. (1971) Elicitation of personal probabilities and expectations, *J. Am. Stat. Assoc.*, **66**, 783–801.

Schlaiffer, R. (1959) *Probability and Statistics for Business Decisions*, New York, McGraw-Hill.

Schlaiffer, R. (1961) *Introduction to Statistics for Business Decisions*, New York, McGraw-Hill.

Schmitt, S.A. (1969) *Measuring Uncertainty: An Elementary Introduction to Bayesian Statistics*, Reading, Mass., Addison-Wesley.

Schoemaker, P.J.H. (1980) *Experiments on Decisions Under Risk: The Expected Utility Hypothesis*, Boston, Martinus Nijhoff.

Scholz, R.W. (ed.) (1983) *Decision Making Under Uncertainty: Cognitive Decision Research, Social Interaction, Development and Epistomology*, Amsterdam, North-Holland.

Schrader-Frechette, K.S. (1985) *Risk Analysis and Scientific Method: Methodological and Ethical Problems with Evaluating Societal Hazards*, Dordrecht, D. Reidel.

Schwartz, C.C. and Hundertmark, K.J. (1993) Reproductive characteristics of Alaskan moose, *J. Wildl. Manage.*, **57**(3), 454–468.

Shannon, C.E. and Weaver, W. (1949) *The Mathematical Theory of Communication*, Urbana, University of Illinois Press. [Contains Shannon's original 1948 paper from the *Bell Systems Technical Journal*.]

Shimony, A. (1985) The status of the principle of maximum entropy, *Synthese*, **63**, 35–53.

References

Silvey, S.D. (1975) *Statistical Inference*, London, Chapman and Hall.

Skyrms, B. (1985) Maximum entropy as a special case of conditionalization, *Synthese*, **63**, 55–74.

Smith, C.R. and Erickson, G.J. (eds.) (1987) *Maximum-Entropy and Bayesian Spectral Analysis and Estimation Problems*, Dordrecht, D. Reidel.

Smith, C.R. and Grandy, W.T. (eds.) (1985) *Maximum-Entropy and Bayesian Methods in Inverse Problems*, Dordrecht, D. Reidel.

Smith, J.Q. (1988) *Decision Analysis: A Bayesian Approach*, London, Chapman and Hall.

Sorensen, R.A. (1985) The iterated versions of Newcomb's paradox and the prisoner's dilemma, *Synthese*, **63**, 157–166.

Stigler, S.M. (1986) *The History of Statistics: The Measurement of Uncertainty Before 1900*, Cambridge, Mass., The Bellknap Press of Harvard University Press.

Strong, N. and Walker, M. (1987) *Information and Capital Markets*, Oxford, Basil Blackwell.

Sudman, S. (1976) *Applied Sampling*, New York, Academic Press.

Swalm, R.O. (1966) Utility theory – insights into risk taking, *Harvard Bus. Rev.*, **44**, 123–136.

Tribus, M. (1969) *Rational Descriptions, Decisions and Designs*, New York, Pergamon Press.

Tukey, J.W. (1977) *Exploratory Data Analysis*, Reading, Mass., Addison-Wesley.

Tversky, A. and Kahneman, D. (1974) Judgement under uncertainty: heuristics and biases, *Science*, **185**, 1124–1131.

Upton, G.J.G. (1978) *The Analysis of Cross-tabulated Data*, Chichester, Wiley.

Vaid, Y.P. and Sasitharan, S. (1992) The strength and dilatancy of sand, *Can. Geotech. J.*, **29**, 552–526.

References

von Neumann, J. and Morgenstern, O. (1947) *Theory of Games and Economic Behaviour*, Princeton, Princeton University Press.

Waller, R.A. and Covello, V.T. (eds.) (1984) *Low-Probability High-Consequence Risk Analysis: Issues, Methods, and Case Studies*, New York, Plenum Press.

Watanabe, S. (1969) *Knowing and Guessing: A Quantitative Study of Inference and Information*, New York, Wiley.

Watson, S.R. and Buede, D.M. (1987) *Decision Synthesis: The Principles and Practice of Decision Analysis*, Cambridge, Cambridge University Press.

Wendt, D. and Vlek, C. (eds.) (1975) *Utility, Probability and Human Decision Making*, Dordrecht, D. Reidel.

Wiener, N. (1948) *Cybernetics: or Control and Communication in the Animal and the Machine*, Cambridge, Mass., MIT Press.

Wijeratne, A. and Wirasinghe, S.C. (1985) Estimation of the number of fire stations and their allocated areas to minimise fire service and property damage costs, *Civ. Eng. Syst.*, **3** (1), 2–6.

Williams, E.J. (1959) *Regression Analysis*, New York, Wiley.

Williams, F.P. and McShane, M.D. (1990) Inclinations of prospective jurors in capital cases, *Sociol. Soc. Res.*, **74**(2), 85–94.

Wilson, A.G. (1970) *Entropy in Urban and Regional Modelling*, London, Pion.

Wilson, A.G. (1974) *Urban and Regional Models in Geography and Planning*, London, Wiley.

Winkler, R.L. (1967a) The assessment of prior distributions in Bayesian analysis, *J. Am. Stat. Assoc.*, **62**, 776–800.

Winkler, R.L. (1967b) The quantification of judgement: some methodological suggestions, *J. Am. Stat. Assoc.*, **62** 1105–1120.

Winkler, R.L. (1971) Probabilistic prediction: some experimental results, *J. Am. Stat. Assoc.*, **66**, 675–685.

Winkler, R.L. (1972) *Introduction to Bayesian Inference and Decision*, New York, Holt, Rinehart and Winston.

Winkler, R.L. (1980) Prior information, predictive distributions, and Bayesian model-building: In Zellner, 1980.

Winkler, R.L. (1981) Combining probability distributions from dependent information sources, *Manage. Sci.* **27**, 479–488.

Woodward, M.W. and Lesley, M.A.F. (1988) *Statistics for Health Management and Research*, London, Edward Arnold.

Yates, F. (1981) *Sampling Methods for Censuses and Surveys (4th edn.)*, London, Griffin.

Zeleny, M. (1982) *Multiple Criteria Decision Making*, New York, McGraw-Hill.

Zellner, A. (1971) *An Introduction to Bayesian Inference in Econometrics*, New York, Wiley.

Zellner, A. (ed.) (1980) *Bayesian Analysis in Econometrics and Statistics: Essays in Honor of Harold Jeffreys*, Amsterdam, North-Holland.

Zernike, J. (1972) *Entropy: The Devil on the Pillion*, Deventer, Kluwer.

Zionts, S. (ed.) (1978) *Multiple Criteria Problem Solving*, Berlin, Springer-Verlag.

Zurek, W.H. (ed.) (1990) *Complexity, Entropy and the Physics of Information*, Redwood City, Calif., Addison-Wesley.

Answers to Numerical Exercises

NOTE: Solutions are given for exercises with numerical results. Where answers have been found iteratively or have necessarily involved sensible yet arbitrary decisions your solutions and those given here may differ slightly. If the solutions involve tests of significance a level of 5% has been used unless otherwise stated. The abbreviation df is used for degrees of freedom.

CHAPTER 1

1.1   mean = 37.08, variance = 18760951.7
1.2   for the raw data mean = 58.33 and standard deviation = 12.09
      Your answers will depend on how you grouped the data.
1.3   mean = 423.75, standard deviation = 257.63, probability = 0.288
1.4   prob(zero failures) = 0.912576
      prob(one failure) = 0.084872
      prob(two failures) = 0.002528
      prob(three failures) = 0.000024
1.5   0.89
1.6   0.94
1.7   0.31
1.8   4
1.9   0.206
1.10 15/16 = 0.9375

CHAPTER 2

2.2   2.270, 0.207
2.3   prob(blonde | blue eyes) = 3/8, so surprisal = 0.981
2.5   (a)   ln(2) = 0.693
      (b)   events are independent so no change
      (c)   ln(26) = 3.258 if you think I am equally likely to choose any letter or 2.897 using Table 2.4 (why are these two values different?)
      (d)   your answer will depend on what you think the next letter will be (p to give export, for instance) and how likely you think the alternatives to be. In fact, I choose both randomly by selecting the numbers 5 and 24 from a table of random numbers (yes, honestly!) so I have no idea what the next letter would have been. (See the remarks early in the next chapter for comments on our abilities to recognise patterns and the like.)
2.6   (a)   The cat sat on the mat.
      (b)   I do not much like puzzles.
      (c)   Nsd sdclfpc dfuw we vdpoasc brghjkwef c asj.
            (Don't ask me, I just closed my eyes and hit the keyboard.)

(d) Lying is indeed an accursed vice.
(Montaigne, *Essays*, Book 1 Chapter 9, "On Liars")
(e) My wife is called Sylvia and my daughter is Kate.
(f) En verité le mentir est un maudit vice. (As (d) but in the original French.)

CHAPTER 3

3.1 0.37, 0.28, 0.20, 0.15
3.2 0.511, 0.109
3.3 35.9%
3.4 0.239
3.5 0.063. Entropy reduced from 1.885 to 1.366.
3.6 0.536

CHAPTER 4

4.1 [0.231, 0.355, 0.183, 0.146, 0.054, 0.031]
4.2 0.648
4.3 0.651
4.4 3.5%
4.5 0.992 kg
4.6 (a) 0.337   (b) 0.581   (c) 3.188
4.7 (a) 0.403   (b) 0.455   (c) 4.300
4.8 0.115
4.9 9.7%
4.10 8.7%

CHAPTER 5

5.1 (0.124, 0.624, 0.252)
5.2 0.371, 0.044, −8%
5.3 0.5, 0.95
5.4 (0.094, 0.159, 0.747)
5.5 0.333
5.6 (a) 0.377   (b) 92
5.7 0.367
5.8 0.030
5.9 0.015, 0, 0.531
5.10 The three cups are A, B and C and I choose A. Here are the likelihoods of B and C being lifted.

		lift cup		
		B	C	
prize	A	0.5	0.5	1.0
is	B	0	1	1.0
under	C	1	0	1.0

Assume uniform prior and that B is lifted.

	prior	l'hood	prior.l'hood	revised
A	1/3	1/2	1/6	1/3
B	1/3	0	0	0
C	1/3	1	1/3	2/3
	1.0		1/2	1.00

By symmetry it is always twice as likely that the prize is under the cup that was not initially chosen. So always switch and by so doing double your chances of winning.

CHAPTER 6

6.1 Large; expected profit = 0.76 > 0.66
6.2 Small; expected utility = 0.81 > 0.78
6.3 No, 3602 psi
6.4 No, still choose 2 with expected profit = 15.85
6.5 7, which gives minimum expected loss = £25.23
6.6 £100000
6.8 95p, £241891
6.10

		prediction of friend		
		wooden	both	
your	wooden	100	0	table of
choice	both	110	10	gains

(a) By the principle of dominance take both since whatever the wooden box contains you will get £10 more by taking both.

(b) Let p be the probability that your friend will correctly predict your behaviour. The table shows the probabilities and the expected gains.

		prediction of friend		
		wooden	both	
your	wooden	p	$1-p$	$100p$
choice	both	$1-p$	p	$110(1-p) + 10p$

You will take the wooden box if $100p > (110 - 100p)$, i.e. $p > 0.55$. Since you think your friend predicts with uncanny accuracy then certainly you think $p > 0.55$ and so you will take only the wooden box.

CHAPTER 7

7.1 (1462.10, 1589.15)
7.2 prob(storm) $\geq 1/51 = 0.02$
7.3 (55.84, 60.82), 150
7.4 (0, 2.495), (0, 4.991) (2.763, 4.437)
7.5 Probabilities of the observed number of unacceptable jobs or worse are 0.223, 0.034, 0.520 and 0.1095. Assuming a 5% risk of wrongly getting rid of a good contractor Acme should not be used any more.
7.6 $G^2 = 2.42$, df = 5, $\chi^2_{CRIT} = 11.071$ so hypothesis of Poisson behaviour not rejected.
7.7 Poisson. $G^2 = 0.61$, df = 3, $\chi^2_{CRIT} = 7.815$ so hypothesis of Poisson behaviour not rejected.
7.8 0.164
7.9 0.068
7.10 No. $G^2 = 1.74$; df = 1, $\chi^2_{CRIT} = 3.84$, 3 or less.

CHAPTER 8

8.1 $G^2 = 0.513$ or $Y^2 = 0.506$; df = 1, $\chi^2_{CRIT} = 3.84$ so do not reject the hypothesised independence
8.2 $G^2 = 32.059$; df = 6, $\chi^2_{CRIT} = 12.59$ so accept that there has been a change in the pattern of usage.
8.3 $G^2 = 237.52$; df = 6, $\chi^2_{CRIT} = 12.59$ so accept that there is a difference in the proportions of ships of different size in collision at the different ports.
8.4 $G^2 = 9.338$; df = 2, $\chi^2_{CRIT} = 5.991$ so accept that there is a difference in comprehension between immigrant groups.
8.5 Using a discrete approximation to the prior and revised probability distributions (in steps of 0.001) the successive values of H are 5.602, 5.672, 5.308. Compare with Figure 5.10.
8.6 Equivocation = 0.503. Transmission = 1.576.
       prob($X = 2$) = 0.826, prob($X = 3$) = prob($X = 6$) = 0.083, prob($X = 7$) = 0.008.
8.7 $G^2 = 52.13$; df = 6, $\chi^2_{CRIT} = 12.59$ so the distributions were different.
8.8 No. $G^2 = 0.414$; df = 2, $\chi^2_{CRIT} = 5.99$. Also, taking student groups in pairs gives no evidence to refute independence.

CHAPTER 9

9.1

394575	27592	18333
22863	161583	7054
21562	11225	37214

9.2 64%

9.3

20.80	29.20	20.48	29.52	21.04	28.96
29.20	20.80	29.52	20.48	28.96	21.04

9.4 12.47, 5.63

9.5 [24.68, 8.34, 30.32, 14.68, 5.32, 12.66, 0.68, 2.32], $G^2 = 0.23$

9.6 With variables G = gender, C = convict, Q = qualified:

model	G^2
[G][C][Q]	473.50
[GC][Q]	8.09
[GQ][C]	5.77
[CQ][G]	5.96
[GC][GQ]	7.18
[GC][CQ]	3.65
[GQ][CQ]	5.05
[GC][CQ][GQ]	2.58

9.7

model	G^2	df	χ^2_{CRIT}
uniform	33.59	15	25.00
rows effect only	27.88	12	21.03
col. effect only	21.19	12	21.03
rows and cols.	15.49	9	16.92

9.8

model	G^2	df	χ^2_{CRIT}
uniform	128.313	3	7.81
rows effect only	37.907	2	5.99
col. effect only	119.937	2	5.99
rows and cols.	29.532	1	3.84

9.9

12.06	2.23
6.77	4.72
10.25	4.04
6.49	7.79
5.13	9.16
0.01	14.27
6.49	7.79

9.10 (a) 34.1% (b) 56.4%

CHAPTER 10

10.1 cov = 7141.4, $r = 0.853$, $T = 0.650$, y_{est} = (5219,5633)
10.2 $y = 34.00 + 8.35x$, slope = (6.43,10.27), intercept = (31.89,36.11)
10.3 $T = 0.622$, (63.032,83.122), (70.662,75.492)
10.4 1.88
10.5 $r^2 = 0.848$ and so $T = 0.942$. With a 3 × 3 table with uniform marginal distributions (almost), $T = 0.595$.
10.6 (a) $a = 6.225$, $b = -0.126$, $r = 0.660$
 (b) (−0.150,−0.101)
 (c) (−0.144,−0.100)
 (d) (−0.140,−0.110)
10.7 Using all the data gives $r^2 = 0.559$, $y = 2997.732 + 3.480x$ and a predicted value of 3154.3. But look at the trajectory. Surely better to begin at, say, the 17th value to give $r^2 = 0.903$, $y = 2886.865 + 6.955x$ and a prediction of 3201.6. Might this second model be overoptimistic (think about turning points)? The first *Footsie* value in November was 3164.4.
10.8 No; all prices lie within the 95% credible interval of their predicted price, given their OG's.
10.9 [0.269, 0.372, 0.173, 0.124, 0.042, 0.020]; mean = 2.37

Appendices

APPENDIX A1
THE MATHEMATICS OF ENTROPY MAXIMISATION

To proceed we need to know some calculus. You will find the details in just about any mathematics book; I shall just sketch the basic results, without proofs.

FINDING A SIMPLE MAXIMUM

If we have some function, $f(x)$, we can think of the rate of change of the value of the function as being measured by the slope of the tangent at any given point (Figure A1.1). In particular, notice that at the maximum value of the function the tangent is horizontal and so the value of the slope (the gradient of the tangent) is zero.

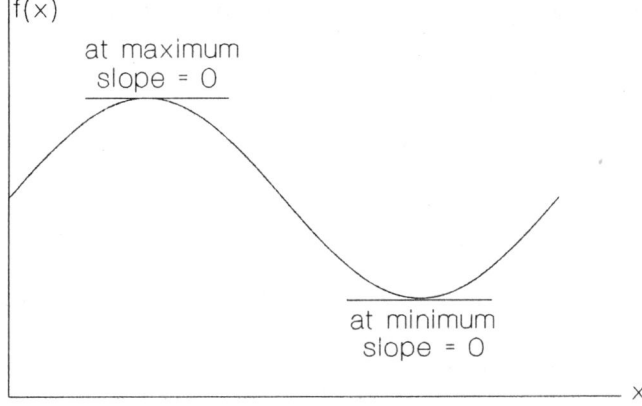

Figure A1.1

So, if we have an equation that defines the function $f(x)$ and can obtain from this another equation for the rate of change or slope, $f'(x)$, all we have to do is find the value of x for which $f'(x) = 0$.

The expression $f'(x)$ is called the *differential* of $f(x)$. Calculus is used to find $f'(x)$ given $f(x)$ and we shall be using some standard results (Table A1.1) in what follows.

function $f(x)$	rate of change $f'(x)$
c	0
cx	c
x	1
$-x\ln(x)$	$-\ln(x) - 1$
$-x\ln(x/c)$	$-\ln(x/c) - 1$

c is a constant

Table A1.1

CONSTRAINED OPTIMISATION

Suppose now that we wish to impose certain constraints on our problem so that instead of finding a simple unconstrained maximum for $f(x)$ we wish to

maximise $\quad f(x)$

subject to $\quad g(x) = k_1$
$\qquad\qquad\quad h(x) = k_2$
$\qquad\qquad\quad$ etc.

the k's being constants.

The theory due to Lagrange says that this can be solved by maximising

$$L = f(x) + \alpha g(x) + \beta h(x) + \ldots \qquad (A1.1)$$

L is called the Lagrangian and α, β the Lagrange multipliers, whose values are to be determined as part of the analysis.

If a function is the sum of a number of other functions then the differential of the sum is the sum of the differentials:

if $\quad f(x) = a(x) + b(x) + \ldots$

then $\quad f'(x) = a'(x) + b'(x) + \ldots$

Applying this to (A1.1) gives

$$L' = f'(x) + \alpha g'(x) + \beta h'(x) + \ldots$$

and the solution to the optimisation problem is found by solving

$$L' = f'(x) + \alpha g'(x) + \beta h'(x) + \ldots = 0 \tag{A1.2}$$

MAXIMISING ENTROPY

Having established these basic results we can now find expressions for maximum entropy probability estimates.

We wish to maximise

$$H = -\Sigma\, p_i \ln(p_i)$$

subject to some constraints.

We shall do this by finding an expression for a typical p_i, rather than look explicitly at all the different p_i's.

We are therefore only interested in how much H changes as p_i changes so other probabilities, p_1, p_2, etc., are held constant and the rate of change of H with respect to any of these other p's is zero.

So $\quad H = -p_1\ln(p_1) - p_2\ln(p_2) - \ldots - p_i\ln(p_i) - \ldots$

and $\quad H' = -0 - 0 - \ldots - \ln(p_i) - 1 - \ldots$
$\quad\quad\quad = -\ln(p_i) - 1$

Our general problem is to

$$\text{maximise} \quad H = -\Sigma\, p_i \ln(p_i)$$

$$\text{subject to} \quad \begin{aligned} g(p_i) &= k_1 \\ h(p_i) &= k_2 \\ &\text{etc.} \end{aligned}$$

and this is solved by forming

$$L = H + \alpha g(p_i) + \beta h(p_i) + \ldots \tag{A1.3}$$

and then solving

$$L' = H' + \alpha g'(p_i) + \beta' h(p_i) + \ldots$$

$$= -\ln(p_i) - 1 + \alpha g'(p_i) + \beta h'(p_i) + \ldots = 0$$

Rearranging gives

$$\ln(p_i) = -1 + \alpha g'(p_i) + \beta h'(p_i) + ...$$

and so

$$p_i = \exp[-1 + \alpha g'(p_i) + \beta h'(p_i) + ...] \qquad (A1.4)$$

There is one constraint that will always be present:

$$\Sigma p_i = 1.0$$

is needed to ensure that a complete probability distribution is obtained.

Since this constraint is always present let us put it in now with multiplier τ to give

$$L = H + \alpha g(p_i) + \beta h(p_i) + ... + \tau \Sigma p_i$$

Remember that we are only considering one particular p_i so the differential of Σp_i is just the differential of p_i which is 1.0. Equation (A1.4) is therefore

$$p_i = \exp[-1 + \alpha g'(p_i) + \beta h'(p_i) + ... + \tau]$$

i.e. $\qquad p_i = k \exp[\alpha g'(p_i) + \beta h'(p_i) + ...] \qquad (A1.5)$

where $k = \exp(-1 + \tau)$

Equation (A1.5) is the basic equation that we will use. The constant k is associated with the constraint $\Sigma p_i = 1.0$ and so is a scaling factor to ensure that that constraint is met.

MINIMISING RELATIVE ENTROPY

Suppose that we have some prior estimate of a probability distribution, typically q_i, and wish to change or update it in the light of some new data or requirement concerning the mean, variance or whatever. Let the new probabilities be p_i. The difference in surprisal between the existing and updated distributions is

$$[-\ln(q_i)] - [-\ln(p_i)] = \ln(p_i/q_i)$$

and its expected value is

$$I = \Sigma p_i \ln(p_i/q_i)$$

Appendices

This measures *information gain* or *relative entropy*.

Using the methodology of the previous section we can minimise information gain (i.e. keep as close as possible to the original distribution) by minimising

$$L = I + \alpha g(p_i) + \beta h(p_i) + ... \qquad (A1.6)$$

A minimum, like a maximum, is found when $L' = 0$ (Figure A1.1):

$$L' = I' + \alpha g'(p_i) + \beta h'(p_i) + ...$$

$$= \ln(p_i/q_i) + 1 + \alpha g'(p_i) + \beta h'(p_i) + ... = 0$$

to get (with a change of sign of the parameters)

$$p_i = q_i \exp[-1 + \alpha g'(p_i) + \beta h'(p_i) + ...] \qquad (A1.7)$$

or, ensuring always that $\Sigma p_i = 1.0$,

$$p_i = k q_i \exp[\alpha g'(p_i) + \beta h'(p_i) + ...] \qquad (A1.8)$$

JOINT DISTRIBUTIONS

Suppose now that we wish to estimate a joint probability distribution, p_{ij}, subject only to the marginal constraints

$$\sum_j p_{ij} = r_i \qquad (A1.9)$$

and $\quad \sum_i p_{ij} = c_j \qquad (A1.10)$

where r_i and c_j are the row and column marginal probabilities.

As before, we maximise

$$H = -\sum_i \sum_j p_{ij} \ln(p_{ij}) \qquad (A1.11)$$

In forming the Lagrangian we use multipliers α_i and β_j for the row and column constraints giving

$$L = -\sum_i \sum_j p_{ij} \ln(p_{ij}) + \alpha_1 \sum_j p_{1j} + \alpha_2 \sum_j p_{2j} + \alpha_3 \sum_j p_{3j} + \ldots + \beta_1 \sum_i p_{i1} + \beta_2 \sum_i p_{i2} + \beta_3 \sum_i p_{i3} + \ldots$$

$$L = -\sum_i \sum_j p_{ij} \ln(p_{ij}) + \sum_i \alpha_i \sum_j p_{ij} + \sum_j \beta_j \sum_i p_{ij} \qquad (A1.12)$$

Note that we do not need a constraint that the probabilities sum to 1.0 since this is ensured by the correct specification of the marginal probabilities. Differentiating L to find the maximum:

$$L' = -\ln(p_{ij}) - 1 + \alpha_i + \beta_j = 0$$

so $\quad p_{ij} = \exp(-1 + \alpha_i + \beta_j)$

$$p_{ij} = a_i b_j \qquad (A1.13)$$

with

$$a_i = \exp(-1 + \alpha_i) \quad \text{and} \quad b_j = \exp(\beta_j)$$

To find out what values to give a_i substitute from (A1.13) into the appropriate constraint equation (A1.9):

$$\sum_j p_{ij} = r_i$$

$$\sum_j a_i b_j = a_i \sum_j b_j = r_i \qquad (A1.14)$$

and similarly

$$b_j \sum_i a_i = c_j \qquad (A1.15)$$

Now

$$\sum_i \sum_j p_{ij} = \sum_i \sum_j a_i b_j = \sum_i a_i \sum_j b_j = 1$$

Substituting from (A1.14) and (A1.15) gives

$$(c_j / b_j)(r_i / a_i) = 1$$

so $\quad a_i b_j = r_i c_j = p_{ij} \qquad (A1.16)$

Appendices

which means that if the only *a priori* constraints are the marginal distributions the unbiased estimate of the joint distribution is obtained by assuming that the two variables are independent.

Other estimates of joint distributions may be found by applying the same ideas that were used above for the distribution of a single variable.

If, as well as the marginal totals, we have other constraints

$$g(p_{ij}) = k_1$$
$$h(p_{ij}) = k_2$$
etc.

then, as with (A1.5),

$$p_{ij} = a_i b_j \exp[\alpha g'(p_{ij}) + \beta h'(p_{ij}) + ...] \quad (A1.17)$$

If, in addition, we have some earlier estimate, q_{ij}, which may be treated as an *a priori* distribution, we minimise the expected information gain just as we did to get (A1.8) and obtain

$$p_{ij} = a_i b_j q_{ij} \exp[\alpha g'(p_{ij}) + \beta h'(p_{ij}) + ...] \quad (A1.18)$$

THREE-DIMENSIONAL TABLES AND THE LOG-LINEAR MODEL

We wish to estimate the probabilities p_{ijk} subject to the sum

$$\sum_i \sum_j \sum_k p_{ijk} = 1$$

and also the marginal constraints

$$\sum_i \sum_j p_{ijk} = x_k$$

$$\sum_i \sum_k p_{ijk} = y_j$$

$$\sum_j \sum_k p_{ijk} = z_i$$

and also the two-dimensional interaction effects

$$\sum_k p_{ijk} = X_{ij}$$

$$\sum_j p_{ijk} = Y_{ik}$$

$$\sum_i p_{ijk} = Z_{jk}$$

The Lagrangian is

$$L = -p_{ijk}\ln(p_{ijk})$$
$$+ \tau \sum_i \sum_j \sum_k p_{ijk} + \sum_k \beta_k \sum_i \sum_j p_{ijk} + \sum_j \gamma_j \sum_i \sum_k p_{ijk} + \sum_i \delta_i \sum_j \sum_k p_{ijk}$$
$$+ \sum_i \sum_j \varepsilon_{ij} \sum_k p_{ijk} + \sum_i \sum_k \xi_{ik} \sum_j p_{ijk} + \sum_j \sum_k \eta_{jk} \sum_i p_{ijk}$$

Setting $L' = 0$ gives

$$\ln(p_{ijk}) = \alpha + \beta_k + \gamma_j + \delta_i + \varepsilon_{ij} + \xi_{ik} + \eta_{jk} \qquad (A1.19)$$

where $\alpha = \tau - 1$. The model in this form is commonly known as the *log-linear model* showing the logarithm of the probability as the sum of marginal effects, interaction effects and a constant which could be seen as a uniform effect ensuring the correct sum. This model is easily extended to more dimensions and is much used in the study of contingency tables. We shall adopt our usual procedure and exponentiate both sides to give

$$p_{ijk} = a b_k c_j d_i e_{ij} f_{ik} g_{jk} \qquad (A1.20)$$

where $a = \exp(\alpha)$ and so on. The same effects are here as in (A1.19) but now in multiplicative form.

APPENDIX A2
LOGARITHMS

Any positive number can be written as any other positive number raised to a power. For instance, nine is just three squared:

$$9 = 3^2$$

Another way of making this statement is to say that the *logarithm* of nine to *base* 3 is two:

$$\log_3(9) = 2$$

Logarithms are simply powers.
Since any base raised to the power zero is 1.0 we have

$$\log_k(1.0) = 0.0$$

whatever the value of k. Also remember that negative powers correspond to reciprocals so that the logarithms of numbers between zero and 1.0 are negative. For example,

$$0.111 = 1/9 = 3^{-2}$$

so $\log_3(0.111) = -2$

The situation is shown in Figure A2.1.

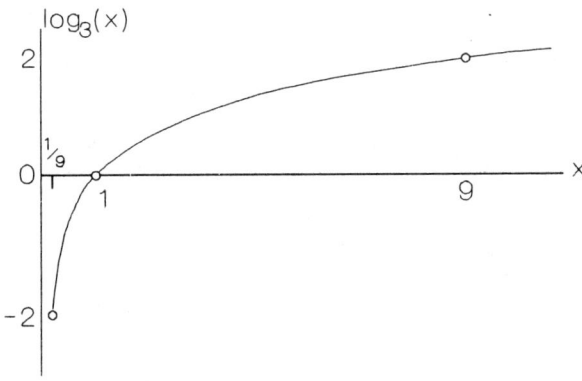

Figure A2.1 Logarithms to the base 3

The value of the base for the logarithms is usually not that significant, though it is important not to alter the base within any application.

There are three common bases. First, using the base 2 gives a quantity measured in *bits* (short for *binary digits*) which is of some significance in communication theory. Second, the base 10 is popular. When it is used the subscript showing the value of the base is often dropped. Finally, when the base is e the logarithms are called *natural logarithms* and are written as ln. (e is a mathematical constant that has certain useful properties. Its value is 2.718281... .)

Here are the different notations:

base	notation
k	$\log_k(x)$
2	$\log_2(x)$
10	$\log(x)$
e	$\ln(x)$

Changing the base is equivalent to introducing a scaling factor. In general

$$\log_a(x) = \log_a(b)\log_b(x)$$

e.g. $\log_2(x) = \log_2(e)\ln(x) = 1.4427\ln(x)$

and similarly

$$\ln(x) = 0.6931\log_2(x)$$

The situation is shown in Figure A2.2.

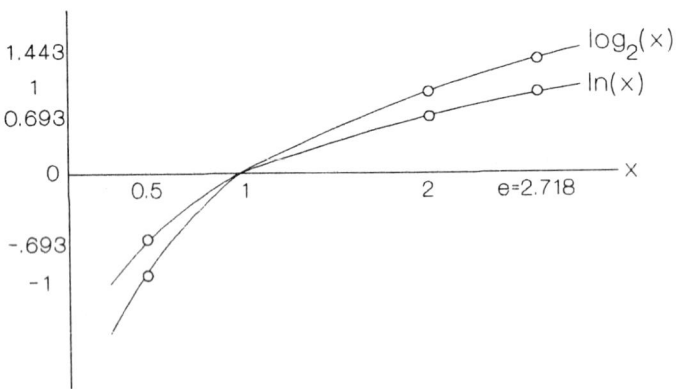

Figure A2.2 Logarithms to different bases

We mainly use natural logarithms in this book so, in summary, if we have

$$y = \log_e(x) = \ln(x)$$

then $x = e^y = \exp(y)$

Note the two equivalent forms of notation in this last equation.

Because logarithms are just powers, relationships that are multiplicative in their natural form become additive in their logarithmic form. Think of three variables

$$a = \ln(A)$$
$$b = \ln(B)$$
$$c = \ln(C)$$

We may have the relationship

$$A = BC$$

so $e^a = e^b e^c = e^{b+c}$

Taking logarithms gives, since $\ln(e^a) = a$ etc.,

$$a = b + c$$

or $\ln(A) = \ln(BC) = \ln(B) + \ln(C)$

Thus multiplication in natural units is, equivalently, the addition of logarithms. Similarly,

$$\ln(B/C) = \ln(B) - \ln(C)$$

and $\ln(B^C) = \ln(BBB...) = C\ln(B)$

APPENDIX A3
NUMERICAL INTEGRATION

It is not always possible to find areas under curves analytically. In such cases we can adopt some simple approximations by dividing the area to be found into a series of rectangles or other simple shapes. We shall give two examples here.

THE MID-POINT RULE

The simplest approximation is to divide the area into n rectangles each with base h and, typically, height y_i as shown in Figure A3.1. The area under the curve is then just the sum of the areas under the rectangles:

$$\text{area} = \Sigma\, hy_i = h \Sigma\, y_i \tag{A3.1}$$

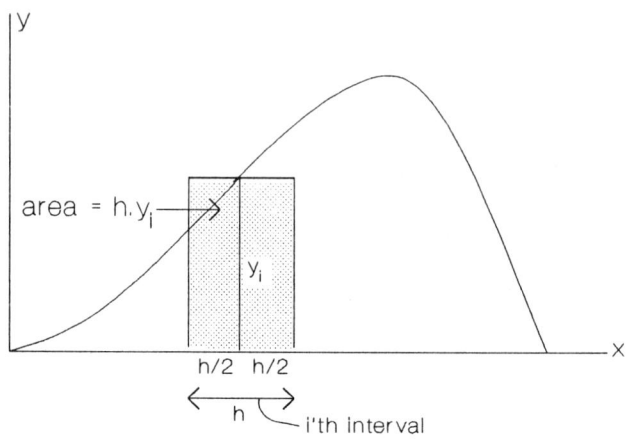

Figure A3.1 Mid-point rule

The area is the sum of the values of the function, y, at the mid-points multiplied by the interval, h.

SIMPSON'S RULE

Greater accuracy is given by more complex approximations to the shape of the curve. Fitting a quadratic curve of the form $y = a + bx + cx^2$ to triplets of points (Figure A3.2) gives an approximation to the area in a pair of intervals as

$$(y_i + 4y_{i+1} + y_{i+2})h/3$$

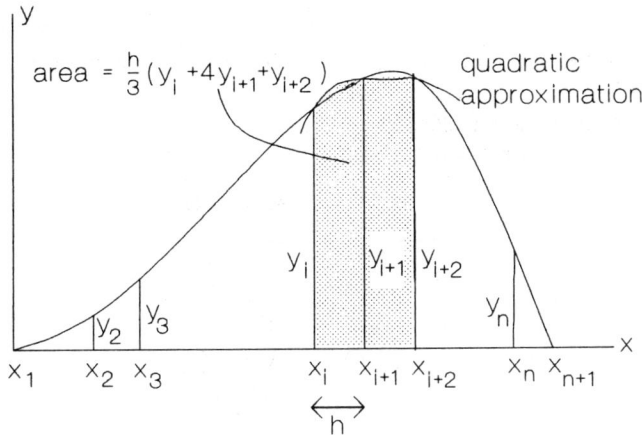

Figure A3.2   Simpson's Rule

For a proof see, for instance, Jessop (1990, App. 4).

Applying this result to consecutive pairs of intervals gives the required approximation:

$$\text{area} = (y_1 + 4y_2 + y_3)h/3 + (y_3 + 4y_4 + y_5)h/3$$

$$+ \ldots + (y_{n-1} + 4y_n + y_{n+1})h/3$$

$$= [(y_1 + y_{n+1}) + 4(y_2 + y_4 + \ldots + y_n)$$

$$+ 2(y_1 + y_3 + \ldots + y_{n-1})]h/3 \qquad (A3.2)$$

The number of intervals, n, must be even.

EXAMPLE

We want the area under the curve $y = 3x^4 + 5x^3 - 10x$ between $x = 1$ and $x = 3$.

The correct answer is 205.20.

Using (A3.1) with $h = 0.5$ gives

x	y
1.25	4.59
1.75	37.43
2.25	111.34
2.75	248.06
	401.42

and so

$$\text{area} = 0.5 \times 401.42 = 200.71$$

A better approximation can be found by increasing the number of intervals. With $h = 0.1$:

x	y
1.05	−1.07
1.15	1.35
⋮	⋮
2.85	285.17
2.95	326.06
	2050.20

The estimate is now improved to

$$\text{area} = 0.1 \times 2050.20 = 205.02$$

To decide how many intervals to use make a series of estimates increasing the number of intervals each time. When two consecutive estimates differ by an acceptably small amount take the last estimate.

Appendices

Now, applying Simpson's Rule (A3.2) with ten intervals gives

x	y		
1.0	−2.0		
1.2		2.86	
1.4			11.24
1.6		24.14	
1.8			42.65
2.0		68.00	
2.2			101.52
2.4		144.65	
2.6			198.97
2.8		266.16	
3.0	348.00		
	346.00	505.81	354.39

The estimate is

$$\text{area} = [346.00 + (4 \times 505.81) + (2 \times 354.39)]0.2/3$$
$$= 205.20$$

which is the correct answer. As expected, Simpson's Rule is more accurate.

APPENDIX A4
THE BINOMIAL, POISSON AND NORMAL DISTRIBUTIONS

This appendix is for the mathematically curious. The steps in the derivations are set out in full but there will be some standard results used that you will have to accept without explanation or else consult a mathematics textbook.

For concreteness a particular example will be used but the results apply to any population of objects that either do or do not possess a particular property of interest.

THE BINOMIAL DISTRIBUTION

We ask five people, A, B, C, D, and E, whether they saw a certain television programme. Three of them, B, C, and E say that they did. What is the probability of this result?

Assume that each person has an *a priori* probability p of having seen the programme so that

prob(A did not see programme) $= 1 - p$
prob(B did see programme) $= p$
prob(C did see programme) $= p$
prob(D did not see programme) $= 1 - p$
prob(E did see programme) $= p$

Now *assume* that the probability of one person having seen the programme is independent of whether anyone else did so that the probability of the observed result is

prob[(A didn't) AND (B did) AND (C did) AND (D didn't) AND (E did)]

$$= (1-p)pp(1-p)p$$
$$= p^3(1-p)^2$$

Note that this would be the result whichever three watched the programme.

We now extend our question and ask what is the probability that any three people out of five watched the programme? Each pattern or *combination* of three people from five has the same probability of occurrence $p^3(1-p)^2$. Let the different combinations be $C1$, $C2$, etc., so that the probability of *any* of these combinations is

prob($C1$ or $C2$ or $C3$ or ...) = prob($C1$) + prob($C2$) + ...

$$= C_{5,3} p^3 (1-p)^2 \qquad (A4.1)$$

where $C_{5,3}$ is the number of combinations of three from five.

To get an expression for $C_{5,3}$ we have also to think about the number of *permutations* $P_{5,3}$. The relationship between the two is straightforward. The three interviewees (B, C, E) represent one combination of three from five. If the order in which we spoke to them was important we would have six possible permutations

$$\begin{array}{ccc} B & C & E \\ B & E & C \\ C & B & E \\ C & E & B \\ E & B & C \\ E & C & B \end{array}$$

To see how the six permutations arise imagine a process of writing down one of the triplets above. You may choose any one of three letters to write first and then, having done so, you choose from the two remaining letters to write next and, finally, you write the remaining letter. So the total number of triplets that may be constructed is

$$P_{3,3} = 3 \times 2 \times 1 = 3!$$

This gives the relationship between combinations and permutations as

$$P_{5,3} = 3! \times C_{5,3}$$

so $\quad C_{5,3} = P_{5,3}/3!$ \hfill (A4.2)

Consider again the process of writing a triplet, but this time we shall choose from any of the five letters representing the interviewees. You now select from five for the first character in the triplet, from four for the second and from three for the third so that the total number of possibilities is

$$P_{5,3} = 5 \times 4 \times 3$$

$$= \frac{5 \times 4 \times 3 \times 2 \times 1}{2 \times 1} = \frac{5!}{2!}$$

From (A4.2)

$$C_{5,3} = \frac{5!}{3! \times 2!}$$

Substituting into (A4.1) gives the probability of any three having seen the programme given an *a priori* probability p and a sample of five as

$$\text{prob}(3|5, p) = \frac{5!}{3! \times 2!} p^3 (1-p)^2$$

In general the probability of x objects in a sample of n having a given property is

$$\text{prob}(x|n, p) = \frac{n!}{x!(n-x)!} p^x (1-p)^{n-x} \qquad (A4.3)$$

which is the *Binomial distribution*.

THE POISSON DISTRIBUTION

Let our sample of size n be drawn from a population of size N of which Q have the property of having seen the television programme so that $p = Q/N$. Applying the Binomial distribution (A4.3) gives

$$\text{prob}(x|n, p) = \frac{n!}{x!(n-x)!} \left[\frac{Q}{N}\right]^x \left[1 - \frac{Q}{N}\right]^{n-x}$$

$$= \frac{n(n-1)(n-2)\ldots(n-x+1)}{x!} \frac{1}{n^x} \left[\frac{nQ}{N}\right]^x \left[1 - \frac{Q}{N}\right]^{n-x}$$

Let the sampling proportion be $d = n/N$ and simplify to give

$$\text{prob}(x|n, p) = \frac{1(1-1/n)(1-2/n)\ldots(1-(x-1)/n)}{x!} (Qd)^x (1-Qd/n)^{n-x}$$

As n gets very large terms such as $1/n$ and $2/n$ get very small and may, at the limit, be ignored so that $(1 - 1/n)$, $(1 - 2/n)$, etc., all become just 1 and

$$\text{prob}(x|n, p) = \frac{1}{x!} (Qd)^x (1-Qd/n)^{n-x} \qquad (A4.4)$$

The last term may be expanded using the Binomial expansion which states

$$(1+x)^a = \sum_{k=0}^{a} C_{a,k} x^k$$

so

$$(1-Qd/n)^{n-x} = \sum_{k=0}^{n-x} C_{n-x,k} (-Qd/n)^k$$

$$= 1 - (n-x)\frac{(Qd)}{n} + \frac{(n-x)(n-x-1)}{2!}\frac{(Qd)^2}{n^2} + \ldots$$

$$= 1 - (1-x/n)(Qd) + \frac{(1-x/n)(1-x/n-1/n)}{2!}(Qd)^2 + \ldots$$

Again, as n gets very large terms such as x/n may be ignored to give

$$(1-Qd/n)^{n-x} = 1 - (Qd) + \frac{(Qd)^2}{2!} + \ldots$$

$$= \exp(-Qd)$$

using the standard definition

$$\exp(x) = 1 + x + x^2/2! + x^3/3! + \ldots$$

So we get, from (A4.4),

$$\text{prob}(x|n, p) = \frac{(Qd)^x \exp(-Qd)}{x!}$$

Now the expected value of x is $np = Qd$. Call this value m to give

$$\text{prob}(x|m) = \frac{m^x \exp(-m)}{x!}$$

which is the *Poisson* distribution and has been obtained as an approximation to the Binomial distribution when n is large.

THE NORMAL DISTRIBUTION

The derivation of the Normal distribution as a generalisation of the Binomial/Poisson distribution is beyond the scope of this book but is covered elsewhere. See, for instance, Chung (1974, pp210–217).

APPENDIX A5
STATISTICAL TABLES

The following tables give values of the distribution function, prob$(X \leq x)$, for selected parameter values.

If the parameter values in which you are interested fall between those given in the table you will have to *interpolate*. This just means assuming that the function $F(x)$ is linear over the short range between tabulated values. If you want $F(X = x)$ but the distribution function is only tabulated at $X = a$ and $X = b$, values which lie either side of $X = x$, then

$$F(x) = F(a) + [F(b) - F(a)] \frac{(x-a)}{(b-a)}, \qquad a < x < b$$

For example, suppose that we wanted $F(z = 0.653)$ for the standard Normal distribution function in A5.3. The nearest values we have are

$$F(0.65) = 0.74215$$

and $\quad F(0.66) = 0.74537$

so

$$F(0.653) = F(0.65) + [F(0.66) - F(0.65)] \frac{(0.653 - 0.65)}{(0.66 - 0.65)}$$

$$= 0.74215 + [0.74537 - 0.74215] \frac{0.003}{0.010}$$

$$= 0.74312$$

The proof of the formulae given for the Beta and Gamma distributions in Appendices A5.4 and A5.5 may be found in Appendix 5 of Jessop (1990), as well as in many other texts. As you would expect, there is a close relationship between these convenience priors and the corresponding likelihood distributions, the Binomial and Poisson. The formulae are only valid for integer parameter values. For non-integer values you will have either to interpolate, which will probably be good enough, or use one of the numerical methods given in Appendix A3.

Increasingly, spreadsheets contain formulae which estimate a number of distribution functions.

A5.1 BINOMIAL DISTRIBUTION

This table shows prob$(X \leq x)$ where

$$\text{prob}(x) = \frac{n!}{x!(n-x)!} p^x (1-p)^{n-x}, \qquad 0 \leq p \leq 1 \quad \text{and} \quad x = 0,1,\ldots,n$$

e.g. if $n = 7$ and $p = 0.2$ then prob$(X \leq 4) = 0.99533$.

The tabulations have only been made for $p \leq 0.5$. For greater values use the symmetry of the Binomial distribution; you can think either of x "successes" each with probability p or, equivalently, of $n - x$ "failures" each with probability $1 - p$. This gives

$$F(x \mid p, n) = \text{prob}(X \leq x \mid p, n) = \text{prob}(X \geq n - x \mid 1 - p, n)$$

$$= 1 - F(n - x - 1 \mid 1 - p, n)$$

e.g. $\text{prob}(X \leq 4 \mid p = 0.7, n = 6) = F(X = 4 \mid p = 0.7, n = 6)$

$$= 1 - F(X = 1 \mid p = 0.3, n = 6)$$

$$= 1 - 0.42018 = 0.57982$$

When n is large and p is small (less than 0.1) use the Poisson distribution with $m = np$. For example, if we want $F(3)$ with $p = 0.04$ and $n = 50$ then use the Poisson tables in Appendix A5.2 with $m = 0.04 \times 50 = 2.0$ and find $F(3) = 0.85712$.

Appendices

BINOMIAL DISTRIBUTION – distribution function $F(X) = \text{prob}(X \leq x)$

	p=	0.01	0.02	0.03	0.04	0.05	0.06	0.07	0.08	0.09
n=2	x=0	0.98010	0.96040	0.94090	0.92160	0.90250	0.88360	0.86490	0.84640	0.82810
	1	0.99990	0.99960	0.99910	0.99840	0.99750	0.99640	0.99510	0.99360	0.99190
	2	1.00000	1.00000	1.00000	1.00000	1.00000	1.00000	1.00000	1.00000	1.00000
	p=	0.01	0.02	0.03	0.04	0.05	0.06	0.07	0.08	0.09
n=3	x=0	0.97030	0.94119	0.91267	0.88474	0.85738	0.83058	0.80436	0.77869	0.75357
	1	0.99970	0.99882	0.99735	0.99533	0.99275	0.98963	0.98599	0.98182	0.97716
	2	1.00000	0.99999	0.99997	0.99994	0.99988	0.99978	0.99966	0.99949	0.99927
	3	1.00000	1.00000	1.00000	1.00000	1.00000	1.00000	1.00000	1.00000	
	p=	0.01	0.02	0.03	0.04	0.05	0.06	0.07	0.08	0.09
n=4	x=0	0.96060	0.92237	0.88529	0.84935	0.81451	0.78075	0.74805	0.71639	0.68575
	1	0.99941	0.99766	0.99481	0.99090	0.98598	0.98009	0.97327	0.96557	0.95704
	2	1.00000	0.99997	0.99989	0.99975	0.99952	0.99917	0.99870	0.99807	0.99728
	3		1.00000	1.00000	1.00000	0.99999	0.99999	0.99998	0.99996	0.99993
	4					1.00000	1.00000	1.00000	1.00000	1.00000
	p=	0.01	0.02	0.03	0.04	0.05	0.06	0.07	0.08	0.09
n=5	x=0	0.95099	0.90392	0.85873	0.81537	0.77378	0.73390	0.69569	0.65908	0.62403
	1	0.99902	0.99616	0.99153	0.98524	0.97741	0.96813	0.95751	0.94564	0.93262
	2	0.99999	0.99992	0.99974	0.99940	0.99884	0.99803	0.99692	0.99547	0.99366
	3	1.00000	1.00000	1.00000	0.99999	0.99997	0.99994	0.99989	0.99981	0.99970
	4				1.00000	1.00000	1.00000	1.00000	1.00000	0.99999
	5									1.00000
	p=	0.01	0.02	0.03	0.04	0.05	0.06	0.07	0.08	0.09
n=6	x=0	0.94148	0.88584	0.83297	0.78276	0.73509	0.68987	0.64699	0.60636	0.56787
	1	0.99854	0.99431	0.98754	0.97845	0.96723	0.95408	0.93918	0.92271	0.90485
	2	0.99998	0.99985	0.99950	0.99883	0.99777	0.99624	0.99416	0.99149	0.98817
	3	1.00000	1.00000	0.99999	0.99996	0.99991	0.99982	0.99968	0.99946	0.99915
	4			1.00000	1.00000	1.00000	1.00000	0.99999	0.99998	0.99997
	5							1.00000	1.00000	1.00000
	p=	0.01	0.02	0.03	0.04	0.05	0.06	0.07	0.08	0.09
n=7	x=0	0.93207	0.86813	0.80798	0.75145	0.69834	0.64848	0.60170	0.55785	0.51676
	1	0.99797	0.99214	0.98291	0.97062	0.95562	0.93822	0.91873	0.89741	0.87452
	2	0.99997	0.99974	0.99914	0.99802	0.99624	0.99371	0.99031	0.98599	0.98067
	3	1.00000	0.99999	0.99997	0.99992	0.99981	0.99961	0.99929	0.99882	0.99816
	4		1.00000	1.00000	1.00000	0.99999	0.99999	0.99997	0.99994	0.99989
	5					1.00000	1.00000	1.00000	1.00000	1.00000
	p=	0.01	0.02	0.03	0.04	0.05	0.06	0.07	0.08	0.09
n=8	x=0	0.92274	0.85076	0.78374	0.72139	0.66342	0.60957	0.55958	0.51322	0.47025
	1	0.99731	0.98966	0.97766	0.96185	0.94276	0.92084	0.89653	0.87024	0.84232
	2	0.99995	0.99958	0.99865	0.99692	0.99421	0.99038	0.98530	0.97890	0.97111
	3	1.00000	0.99999	0.99995	0.99984	0.99963	0.99925	0.99866	0.99780	0.99659
	4		1.00000	1.00000	0.99999	0.99998	0.99996	0.99992	0.99985	0.99974
	5				1.00000	1.00000	1.00000	1.00000	0.99999	0.99999
	6								1.00000	1.00000

p=	0.01	0.02	0.03	0.04	0.05	0.06	0.07	0.08	0.09
n=9 x=0	0.91352	0.83375	0.76023	0.69253	0.63025	0.57299	0.52041	0.47216	0.42793
1	0.99656	0.98689	0.97184	0.95223	0.92879	0.90216	0.87295	0.84168	0.80883
2	0.99992	0.99939	0.99802	0.99552	0.99164	0.98620	0.97909	0.97021	0.95952
3	1.00000	0.99998	0.99991	0.99973	0.99936	0.99872	0.99773	0.99628	0.99430
4		1.00000	1.00000	0.99999	0.99997	0.99992	0.99983	0.99969	0.99945
5				1.00000	1.00000	1.00000	0.99999	0.99998	0.99996
6							1.00000	1.00000	1.00000
p=	0.01	0.02	0.03	0.04	0.05	0.06	0.07	0.08	0.09
n=10 x=0	0.90438	0.81707	0.73742	0.66483	0.59874	0.53862	0.48398	0.43439	0.38942
1	0.99573	0.98382	0.96549	0.94185	0.91386	0.88241	0.84827	0.81212	0.77455
2	0.99989	0.99914	0.99724	0.99379	0.98850	0.98116	0.97166	0.95992	0.94596
3	1.00000	0.99997	0.99985	0.99956	0.99897	0.99797	0.99642	0.99420	0.99117
4		1.00000	0.99999	0.99998	0.99994	0.99985	0.99969	0.99941	0.99899
5			1.00000	1.00000	1.00000	0.99999	0.99998	0.99996	0.99992
6						1.00000	1.00000	1.00000	1.00000
p=	0.01	0.02	0.03	0.04	0.05	0.06	0.07	0.08	0.09
n=20 x=0	0.81791	0.66761	0.54379	0.44200	0.35849	0.29011	0.23424	0.18869	0.15164
1	0.98314	0.94010	0.88016	0.81034	0.73584	0.66045	0.58686	0.51686	0.45160
2	0.99900	0.99293	0.97899	0.95614	0.92452	0.88503	0.83900	0.78795	0.73343
3	0.99996	0.99940	0.99733	0.99259	0.98410	0.97103	0.95287	0.92938	0.90067
4	1.00000	0.99996	0.99974	0.99904	0.99743	0.99437	0.98929	0.98166	0.97096
5		1.00000	0.99998	0.99990	0.99967	0.99913	0.99807	0.99620	0.99321
6			1.00000	0.99999	0.99997	0.99989	0.99972	0.99936	0.99871
7				1.00000	1.00000	0.99999	0.99997	0.99991	0.99980
8						1.00000	1.00000	0.99999	0.99997
9								1.00000	1.00000
p=	0.01	0.02	0.03	0.04	0.05	0.06	0.07	0.08	0.09
n=50 x=0	0.60501	0.36417	0.21807	0.12989	0.07694	0.04533	0.02656	0.01547	0.00896
1	0.91056	0.73577	0.55528	0.40048	0.27943	0.19000	0.12649	0.08271	0.05324
2	0.98618	0.92157	0.81080	0.67671	0.54053	0.41625	0.31079	0.22597	0.16054
3	0.99840	0.98224	0.93724	0.86087	0.76041	0.64730	0.53274	0.42530	0.33034
4	0.99985	0.99679	0.98319	0.95103	0.89638	0.82060	0.72903	0.62895	0.52766
5	0.99999	0.99952	0.99626	0.98559	0.96222	0.92236	0.86495	0.79187	0.70719
6	1.00000	0.99994	0.99930	0.99639	0.98821	0.97108	0.94169	0.89813	0.84037
7		0.99999	0.99989	0.99922	0.99681	0.99062	0.97799	0.95621	0.92316
8		1.00000	0.99998	0.99985	0.99924	0.99733	0.99268	0.98335	0.96717
9			1.00000	0.99998	0.99984	0.99933	0.99784	0.99437	0.98748
10				1.00000	0.99997	0.99985	0.99943	0.99829	0.99572
11					1.00000	0.99997	0.99986	0.99953	0.99868
12						0.99999	0.99997	0.99989	0.99963
13						1.00000	0.99999	0.99997	0.99991
14							1.00000	0.99999	0.99998
15								1.00000	1.00000

	p=	0.01	0.02	0.03	0.04	0.05	0.06	0.07	0.08	0.09
n=100	x=0	0.36603	0.13262	0.04755	0.01687	0.00592	0.00205	0.00071	0.00024	0.00008
	1	0.73576	0.40327	0.19462	0.08716	0.03708	0.01517	0.00601	0.00232	0.00087
	2	0.92063	0.67669	0.41978	0.23214	0.11826	0.05661	0.02579	0.01127	0.00476
	3	0.98163	0.85896	0.64725	0.42948	0.25784	0.14302	0.07441	0.03671	0.01730
	4	0.99657	0.94917	0.81785	0.62886	0.43598	0.27678	0.16316	0.09034	0.04739
	5	0.99947	0.98452	0.91916	0.78837	0.61600	0.44069	0.29142	0.17988	0.10452
	6	0.99993	0.99594	0.96877	0.89361	0.76601	0.60635	0.44428	0.30316	0.19398
	7	0.99999	0.99907	0.98938	0.95249	0.87204	0.74835	0.59878	0.44711	0.31279
	8	1.00000	0.99981	0.99678	0.98101	0.93691	0.85371	0.73397	0.59263	0.44940
	9		0.99997	0.99913	0.99316	0.97181	0.92246	0.83798	0.72198	0.58751
	10		0.99999	0.99979	0.99776	0.98853	0.96239	0.90922	0.82433	0.71180
	11		1.00000	0.99995	0.99933	0.99573	0.98325	0.95310	0.89715	0.81238
	12			0.99999	0.99982	0.99854	0.99312	0.97759	0.94412	0.88616
	13			1.00000	0.99995	0.99954	0.99739	0.99007	0.97176	0.93555
	14				0.99999	0.99986	0.99908	0.99591	0.98670	0.96590
	15				1.00000	0.99996	0.99970	0.99843	0.99415	0.98312
	16					0.99999	0.99991	0.99944	0.99759	0.99216
	17					1.00000	0.99997	0.99981	0.99907	0.99658
	18						0.99999	0.99994	0.99966	0.99860
	19						1.00000	0.99998	0.99988	0.99946
	20							0.99999	0.99996	0.99980
	21							1.00000	0.99999	0.99993
	22								1.00000	0.99998
	23									0.99999
	24									1.00000

	p=	0.10	0.15	0.20	0.25	0.30	0.35	0.40	0.45	0.50
n=2	x=0	0.81000	0.72250	0.64000	0.56250	0.49000	0.42250	0.36000	0.30250	0.25000
	1	0.99000	0.97750	0.96000	0.93750	0.91000	0.87750	0.84000	0.79750	0.75000
	2	1.00000	1.00000	1.00000	1.00000	1.00000	1.00000	1.00000	1.00000	1.00000
	p=	0.10	0.15	0.20	0.25	0.30	0.35	0.40	0.45	0.50
n=3	x=0	0.72900	0.61413	0.51200	0.42188	0.34300	0.27463	0.21600	0.16638	0.12500
	1	0.97200	0.93925	0.89600	0.84375	0.78400	0.71825	0.64800	0.57475	0.50000
	2	0.99900	0.99663	0.99200	0.98438	0.97300	0.95713	0.93600	0.90888	0.87500
	3	1.00000	1.00000	1.00000	1.00000	1.00000	1.00000	1.00000	1.00000	1.00000
	p=	0.10	0.15	0.20	0.25	0.30	0.35	0.40	0.45	0.50
n=4	x=0	0.65610	0.52201	0.40960	0.31641	0.24010	0.17851	0.12960	0.09151	0.06250
	1	0.94770	0.89048	0.81920	0.73828	0.65170	0.56298	0.47520	0.39098	0.31250
	2	0.99630	0.98802	0.97280	0.94922	0.91630	0.87352	0.82080	0.75852	0.68750
	3	0.99990	0.99949	0.99840	0.99609	0.99190	0.98499	0.97440	0.95899	0.93750
	4	1.00000	1.00000	1.00000	1.00000	1.00000	1.00000	1.00000	1.00000	1.00000
	p=	0.10	0.15	0.20	0.25	0.30	0.35	0.40	0.45	0.50
n=5	x=0	0.59049	0.44371	0.32768	0.23730	0.16807	0.11603	0.07776	0.05033	0.03125
	1	0.91854	0.83521	0.73728	0.63281	0.52822	0.42842	0.33696	0.25622	0.18750
	2	0.99144	0.97339	0.94208	0.89648	0.83692	0.76483	0.68256	0.59313	0.50000
	3	0.99954	0.99777	0.99328	0.98438	0.96922	0.94598	0.91296	0.86878	0.81250
	4	0.99999	0.99992	0.99968	0.99902	0.99757	0.99475	0.98976	0.98155	0.96875
	5	1.00000	1.00000	1.00000	1.00000	1.00000	1.00000	1.00000	1.00000	1.00000
	p=	0.10	0.15	0.20	0.25	0.30	0.35	0.40	0.45	0.50
n=6	x=0	0.53144	0.37715	0.26214	0.17798	0.11765	0.07542	0.04666	0.02768	0.01563
	1	0.88574	0.77648	0.65536	0.53394	0.42018	0.31908	0.23328	0.16357	0.10938
	2	0.98415	0.95266	0.90112	0.83057	0.74431	0.64709	0.54432	0.44152	0.34375
	3	0.99873	0.99411	0.98304	0.96240	0.92953	0.88258	0.82080	0.74474	0.65625
	4	0.99995	0.99960	0.99840	0.99536	0.98907	0.97768	0.95904	0.93080	0.89063
	5	1.00000	0.99999	0.99994	0.99976	0.99927	0.99816	0.99590	0.99170	0.98438
	6		1.00000	1.00000	1.00000	1.00000	1.00000	1.00000	1.00000	1.00000
	p=	0.10	0.15	0.20	0.25	0.30	0.35	0.40	0.45	0.50
n=7	x=0	0.47830	0.32058	0.20972	0.13348	0.08235	0.04902	0.02799	0.01522	0.00781
	1	0.85031	0.71658	0.57672	0.44495	0.32942	0.23380	0.15863	0.10242	0.06250
	2	0.97431	0.92623	0.85197	0.75641	0.64707	0.53228	0.41990	0.31644	0.22656
	3	0.99727	0.98790	0.96666	0.92944	0.87396	0.80015	0.71021	0.60829	0.50000
	4	0.99982	0.99878	0.99533	0.98712	0.97120	0.94439	0.90374	0.84707	0.77344
	5	0.99999	0.99993	0.99963	0.99866	0.99621	0.99099	0.98116	0.96429	0.93750
	6	1.00000	1.00000	0.99999	0.99994	0.99978	0.99936	0.99836	0.99626	0.99219
	7			1.00000	1.00000	1.00000	1.00000	1.00000	1.00000	1.00000
	p=	0.10	0.15	0.20	0.25	0.30	0.35	0.40	0.45	0.50
n=8	x=0	0.43047	0.27249	0.16777	0.10011	0.05765	0.03186	0.01680	0.00837	0.00391
	1	0.81310	0.65718	0.50332	0.36708	0.25530	0.16913	0.10638	0.06318	0.03516
	2	0.96191	0.89479	0.79692	0.67854	0.55177	0.42781	0.31539	0.22013	0.14453
	3	0.99498	0.97865	0.94372	0.88618	0.80590	0.70640	0.59409	0.47696	0.36328
	4	0.99957	0.99715	0.98959	0.97270	0.94203	0.89391	0.82633	0.73962	0.63672
	5	0.99998	0.99976	0.99877	0.99577	0.98871	0.97468	0.95019	0.91154	0.85547
	6	1.00000	0.99999	0.99992	0.99962	0.99871	0.99643	0.99148	0.98188	0.96484
	7		1.00000	1.00000	0.99998	0.99993	0.99977	0.99934	0.99832	0.99609
	8				1.00000	1.00000	1.00000	1.00000	1.00000	1.00000

p=	0.10	0.15	0.20	0.25	0.30	0.35	0.40	0.45	0.50
n=9 x=0	0.38742	0.23162	0.13422	0.07508	0.04035	0.02071	0.01008	0.00461	0.00195
1	0.77484	0.59948	0.43621	0 30034	0.19600	0.12109	0.07054	0.03852	0.01953
2	0.94703	0.85915	0.73820	0.60068	0.46283	0.33727	0.23179	0.14950	0.08984
3	0.99167	0.96607	0.91436	0.83427	0.72966	0.60889	0.48261	0.36138	0.25391
4	0.99911	0.99437	0.98042	0.95107	0.90119	0.82828	0.73343	0.62142	0.50000
5	0.99994	0.99937	0.99693	0.99001	0.97471	0.94641	0.90065	0.83418	0.74609
6	1.00000	0.99995	0.99969	0.99866	0.99571	0.98882	0.97497	0.95023	0.91016
7		1.00000	0.99998	0.99989	0.99957	0.99860	0.99620	0.99092	0.98047
8			1.00000	1.00000	0.99998	0.99992	0.99974	0.99924	0.99805
9					1.00000	1.00000	1.00000	1.00000	1.00000
p=	0.10	0.15	0.20	0.25	0.30	0.35	0.40	0.45	0.50
n=10 x=0	0.34868	0.19687	0.10737	0.05631	0.02825	0.01346	0.00605	0.00253	0.00098
1	0.73610	0.54430	0.37581	0.24403	0.14931	0.08595	0.04636	0.02326	0.01074
2	0.92981	0.82020	0.67780	0.52559	0.38278	0.26161	0.16729	0.09956	0.05469
3	0.98720	0.95003	0.87913	0.77588	0.64961	0.51383	0.38228	0.26604	0.17188
4	0.99837	0.99013	0.96721	0.92187	0.84973	0.75150	0.63310	0.50440	0.37695
5	0.99985	0.99862	0.99363	0.98027	0.95265	0.90507	0.83376	0.73844	0.62305
6	0.99999	0.99987	0.99914	0.99649	0.98941	0.97398	0.94524	0.89801	0.82813
7	1.00000	0.99999	0.99992	0.99958	0.99841	0.99518	0.98771	0.97261	0.94531
8		1.00000	1.00000	0.99997	0.99986	0.99946	0.99832	0.99550	0.98926
9				1.00000	0.99999	0.99997	0.99990	0.99966	0.99902
10					1.00000	1.00000	1.00000	1.00000	1.00000
p=	0.10	0.15	0.20	0.25	0.30	0.35	0.40	0.45	0.50
n=20 x=0	0.12158	0.03876	0.01153	0.00317	0.00080	0.00018	0.00004	0.00001	0.00000
1	0.39175	0.17556	0.06918	0.02431	0.00764	0.00213	0.00052	0.00011	0.00002
2	0.67693	0.40490	0.20608	0.09126	0.03548	0.01212	0.00361	0.00093	0.00020
3	0.86705	0.64773	0.41145	0.22516	0.10709	0.04438	0.01596	0.00493	0.00129
4	0.95683	0.82985	0.62965	0.41484	0.23751	0.11820	0.05095	0.01886	0.00591
5	0.98875	0.93269	0.80421	0.61717	0.41637	0.24540	0.12560	0.05533	0.02069
6	0.99761	0.97806	0.91331	0.78578	0.60801	0.41663	0.25001	0.12993	0.05766
7	0.99958	0.99408	0.96786	0.89819	0.77227	0.60103	0.41589	0.25201	0.13159
8	0.99994	0.99867	0.99002	0.95907	0.88667	0.76238	0.59560	0.41431	0.25172
9	0.99999	0.99975	0.99741	0.98614	0.95204	0.87822	0.75534	0.59136	0.41190
10	1.00000	0.99996	0.99944	0.99606	0.98286	0.94683	0.87248	0.75071	0.58810
11		1.00000	0.99990	0.99906	0.99486	0.98042	0.94347	0.86924	0.74828
12			0.99998	0.99982	0.99872	0.99398	0.97897	0.94197	0.86841
13			1.00000	0.99997	0.99974	0.99848	0.99353	0.97859	0.94234
14				1.00000	0.99996	0.99969	0.99839	0.99357	0.97931
15					0.99999	0.99995	0.99968	0.99847	0.99409
16					1.00000	0.99999	0.99995	0.99972	0.99871
17						1.00000	0.99999	0.99996	0.99980
18							1.00000	1.00000	0.99998
19									1.00000

p=	0.10	0.15	0.20	0.25	0.30	0.35	0.40	0.45	0.50
n=50 x=0	0.00515	0.00030	0.00001	0.00000	0.00000	0.00000	0.00000	0.00000	0.00000
1	0.03379	0.00291	0.00019	0.00001	0.00000	0.00000	0.00000	0.00000	0.00000
2	0.11173	0.01419	0.00129	0.00009	0.00000	0.00000	0.00000	0.00000	0.00000
3	0.25029	0.04605	0.00566	0.00050	0.00003	0.00000	0.00000	0.00000	0.00000
4	0.43120	0.11211	0.01850	0.00211	0.00017	0.00001	0.00000	0.00000	0.00000
5	0.61612	0.21935	0.04803	0.00705	0.00072	0.00005	0.00000	0.00000	0.00000
6	0.77023	0.36130	0.10340	0.01939	0.00249	0.00022	0.00001	0.00000	0.00000
7	0.87785	0.51875	0.19041	0.04526	0.00726	0.00080	0.00006	0.00000	0.00000
8	0.94213	0.66810	0.30733	0.09160	0.01825	0.00248	0.00023	0.00001	0.00000
9	0.97546	0.79109	0.44374	0.16368	0.04023	0.00670	0.00076	0.00006	0.00000
10	0.99065	0.88008	0.58356	0.26220	0.07885	0.01601	0.00220	0.00020	0.00001
11	0.99678	0.93719	0.71067	0.38162	0.13904	0.03423	0.00569	0.00063	0.00005
12	0.99900	0.96994	0.81394	0.51099	0.22287	0.06613	0.01325	0.00177	0.00015
13	0.99971	0.98683	0.88941	0.63704	0.32788	0.11633	0.02799	0.00449	0.00047
14	0.99993	0.99471	0.93928	0.74808	0.44683	0.18778	0.05396	0.01038	0.00130
15	0.99998	0.99805	0.96920	0.83692	0.56918	0.28010	0.09550	0.02195	0.00330
16	1.00000	0.99934	0.98556	0.90169	0.68388	0.38886	0.15609	0.04265	0.00767
17		0.99979	0.99374	0.94488	0.78219	0.50597	0.23688	0.07653	0.01642
18		0.99994	0.99749	0.97127	0.85944	0.62159	0.33561	0.12735	0.03245
19		0.99998	0.99907	0.98608	0.91520	0.72644	0.44648	0.19737	0.05946
20		1.00000	0.99968	0.99374	0.95224	0.81395	0.56103	0.28617	0.10132
21			0.99990	0.99738	0.97491	0.88126	0.67014	0.38996	0.16112
22			0.99997	0.99898	0.98772	0.92904	0.76602	0.50191	0.23994
23			0.99999	0.99963	0.99441	0.96036	0.84383	0.61341	0.33591
24			1.00000	0.99988	0.99763	0.97933	0.90219	0.71604	0.44386
25				0.99996	0.99907	0.98996	0.94266	0.80337	0.55614
26				0.99999	0.99966	0.99546	0.96859	0.87207	0.66409
27				1.00000	0.99988	0.99809	0.98397	0.92204	0.76006
28					0.99996	0.99925	0.99238	0.95562	0.83888
29					0.99999	0.99973	0.99664	0.97646	0.89868
30					1.00000	0.99991	0.99863	0.98840	0.94054
31						0.99997	0.99948	0.99470	0.96755
32						0.99999	0.99982	0.99776	0.98358
33						1.00000	0.99994	0.99913	0.99233
34							0.99998	0.99969	0.99670
35							1.00000	0.99990	0.99870
36								0.99997	0.99953
37								0.99999	0.99985
38								1.00000	0.99995
39									0.99999
40									1.00000

p=	0.10	0.15	0.20	0.25	0.30	0.35	0.40	0.45	0.50
n=100 x=0	0.00003	0.00000	0.00000	0.00000	0.00000	0.00000	0.00000	0.00000	0.00000
1	0.00032	0.00000	0.00000	0.00000	0.00000	0.00000	0.00000	0.00000	0.00000
2	0.00194	0.00002	0.00000	0.00000	0.00000	0.00000	0.00000	0.00000	0.00000
3	0.00784	0.00009	0.00000	0.00000	0.00000	0.00000	0.00000	0.00000	0.00000
4	0.02371	0.00043	0.00000	0.00000	0.00000	0.00000	0.00000	0.00000	0.00000
5	0.05758	0.00155	0.00002	0.00000	0.00000	0.00000	0.00000	0.00000	0.00000
6	0.11716	0.00470	0.00008	0.00000	0.00000	0.00000	0.00000	0.00000	0.00000
7	0.20605	0.01217	0.00028	0.00000	0.00000	0.00000	0.00000	0.00000	0.00000
8	0.32087	0.02748	0.00086	0.00001	0.00000	0.00000	0.00000	0.00000	0.00000
9	0.45129	0.05509	0.00233	0.00004	0.00000	0.00000	0.00000	0.00000	0.00000
10	0.58316	0.09945	0.00570	0.00014	0.00000	0.00000	0.00000	0.00000	0.00000
11	0.70303	0.16349	0.01257	0.00039	0.00001	0.00000	0.00000	0.00000	0.00000
12	0.80182	0.24730	0.02533	0.00103	0.00002	0.00000	0.00000	0.00000	0.00000
13	0.87612	0.34743	0.04691	0.00246	0.00006	0.00000	0.00000	0.00000	0.00000
14	0.92743	0.45722	0.08044	0.00542	0.00016	0.00000	0.00000	0.00000	0.00000
15	0.96011	0.56832	0.12851	0.01108	0.00040	0.00001	0.00000	0.00000	0.00000
16	0.97940	0.67246	0.19234	0.02111	0.00097	0.00002	0.00000	0.00000	0.00000
17	0.98999	0.76328	0.27119	0.03763	0.00216	0.00005	0.00000	0.00000	0.00000
18	0.99542	0.83717	0.36209	0.06301	0.00452	0.00014	0.00000	0.00000	0.00000
19	0.99802	0.89346	0.46016	0.09953	0.00889	0.00034	0.00001	0.00000	0.00000
20	0.99919	0.93368	0.55946	0.14883	0.01646	0.00078	0.00002	0.00000	0.00000
21	0.99969	0.96072	0.65403	0.21144	0.02883	0.00169	0.00004	0.00000	0.00000
22	0.99989	0.97786	0.73893	0.28637	0.04787	0.00343	0.00011	0.00000	0.00000
23	0.99996	0.98811	0.81091	0.37108	0.07553	0.00662	0.00025	0.00000	0.00000
24	0.99999	0.99392	0.86865	0.46167	0.11357	0.01213	0.00056	0.00001	0.00000
25	1.00000	0.99703	0.91252	0.55347	0.16313	0.02114	0.00119	0.00003	0.00000
26		0.99862	0.94417	0.64174	0.22440	0.03514	0.00240	0.00007	0.00000
27		0.99939	0.96585	0.72238	0.29637	0.05581	0.00460	0.00016	0.00000
28		0.99974	0.97998	0.79246	0.37678	0.08482	0.00843	0.00036	0.00001
29		0.99989	0.98875	0.85046	0.46234	0.12360	0.01478	0.00076	0.00002
30		0.99996	0.99394	0.89621	0.54912	0.17302	0.02478	0.00154	0.00004
31		0.99998	0.99687	0.93065	0.63311	0.23311	0.03985	0.00297	0.00009
32		0.99999	0.99845	0.95540	0.71072	0.30288	0.06150	0.00550	0.00020
33			1.00000	0.99926	0.97241	0.77926	0.38029	0.09125	0.00976	0.00044
34				0.99966	0.98357	0.83714	0.46243	0.13034	0.01663	0.00089
35				0.99985	0.99059	0.88392	0.54584	0.17947	0.02724	0.00176
36				0.99994	0.99482	0.92012	0.62692	0.23861	0.04290	0.00332
37				0.99998	0.99725	0.94695	0.70245	0.30681	0.06507	0.00602
38				0.99999	0.99860	0.96602	0.76987	0.38219	0.09514	0.01049
39				1.00000	0.99931	0.97901	0.82758	0.46208	0.13425	0.01760
40					0.99968	0.98750	0.87498	0.54329	0.18306	0.02844
41					0.99985	0.99283	0.91232	0.62253	0.24149	0.04431
42					0.99994	0.99603	0.94057	0.69674	0.30865	0.06661
43					0.99997	0.99789	0.96109	0.76347	0.38277	0.09667
44					0.99999	0.99891	0.97540	0.82110	0.46133	0.13563
45					1.00000	0.99946	0.98499	0.86891	0.54132	0.18410
46						0.99974	0.99116	0.90702	0.61956	0.24206
47						0.99988	0.99498	0.93621	0.69312	0.30865
48						0.99995	0.99725	0.95770	0.75957	0.38218

p=	0.10	0.15	0.20	0.25	0.30	0.35	0.40	0.45	0.50
n=100 x=49					0.99998	0.99855	0.97290	0.81727	0.46021
50					0.99999	0.99926	0.98324	0.86542	0.53979
51					1.00000	0.99964	0.98999	0.90405	0.61782
52						0.99983	0.99424	0.93383	0.69135
53						0.99992	0.99680	0.95589	0.75794
54						0.99997	0.99829	0.97161	0.81590
55						0.99999	0.99912	0.98236	0.86437
56						0.99999	0.99956	0.98943	0.90333
57						1.00000	0.99979	0.99389	0.93339
58							0.99990	0.99660	0.95569
59							0.99996	0.99818	0.97156
60							0.99998	0.99906	0.98240
61							0.99999	0.99953	0.98951
62							1.00000	0.99978	0.99398
63								0.99990	0.99668
64								0.99996	0.99824
65								0.99998	0.99911
66								0.99999	0.99956
67								1.00000	0.99980
68									0.99991
69									0.99996
70									0.99998
71									0.99999
72									1.00000

A5.2 POISSON DISTRIBUTION

This table shows prob$(X \leq x)$ where

$$\text{prob}(x) = \frac{m^x \exp(-m)}{x!}, \qquad m > 0 \text{ and } x = 0, 1, \ldots, n$$

When $m > 20$ use the Normal distribution with $\mu = \sigma^2 = m$.

For example, suppose we want $F(32)$ with $m = 25$. Remember that we are using a continuous variable in the Normal distribution to approximate an integer value. This means, for instance, that the integer value 32 becomes the continuous range $31.5 < x < 32.5$ and so prob(32) becomes $F(32.5) - F(31.5)$.

The required Poisson probability, $F(32)$, is approximated by a Normal probability $F(32.5)$ with $\mu = \sigma^2 = 25$, and so $\sigma = 5$. Using the tables in Appendix A5.3 find

$$\begin{aligned} F(X = 32.5) &= F(Z = (32.5 - 25)/5) = F(Z = 1.5) \\ &= 0.93319 \end{aligned}$$

POISSON DISTRIBUTION – distribution function F(X) = prob(X ≤ x)

m=	0.10	0.20	0.30	0.40	0.50	0.60	0.70	0.80	0.90	1.00
x=0	0.90484	0.81873	0.74082	0.67032	0.60653	0.54881	0.49659	0.4493	0.40657	0.36788
1	0.99532	0.98248	0.96306	0.93845	0.90980	0.87810	0.84420	0.80879	0.77248	0.73576
2	0.99985	0.99885	0.99640	0.99207	0.98561	0.97688	0.96586	0.95258	0.93714	0.91970
3	1.00000	0.99994	0.99973	0.99922	0.99825	0.99664	0.99425	0.99092	0.98654	0.98101
4		1.00000	0.99998	0.99994	0.99983	0.99961	0.99921	0.99859	0.99766	0.99634
5			1.00000	1.00000	0.99999	0.99996	0.99991	0.99982	0.99966	0.99941
6					1.00000	1.00000	0.99999	0.99998	0.99996	0.99992
7							1.00000	1.00000	1.00000	0.99999
8										1.00000

m=	1.10	1.20	1.30	1.40	1.50	1.60	1.70	1.80	1.90	2.00
x=0	0.33287	0.30119	0.27253	0.24660	0.22313	0.20190	0.18268	0.16530	0.14957	0.13534
1	0.69903	0.66263	0.62682	0.59183	0.55783	0.52493	0.49325	0.46284	0.43375	0.40601
2	0.90042	0.87949	0.85711	0.83350	0.80885	0.78336	0.75722	0.73062	0.70372	0.67668
3	0.97426	0.96623	0.95690	0.94627	0.93436	0.92119	0.90681	0.89129	0.87470	0.85712
4	0.99456	0.99225	0.98934	0.98575	0.98142	0.97632	0.97039	0.96359	0.95592	0.94735
5	0.99903	0.99850	0.99777	0.99680	0.99554	0.99396	0.99200	0.98962	0.98678	0.98344
6	0.99985	0.99975	0.99960	0.99938	0.99907	0.99866	0.99812	0.99743	0.99655	0.99547
7	0.99998	0.99996	0.99994	0.99989	0.99983	0.99974	0.99961	0.99944	0.99921	0.99890
8	1.00000	1.00000	0.99999	0.99998	0.99997	0.99995	0.99993	0.99989	0.99984	0.99976
9			1.00000	1.00000	1.00000	0.99999	0.99999	0.99998	0.99997	0.99995
10						1.00000	1.00000	1.00000	0.99999	0.99999
11									1.00000	1.00000

m=	2.10	2.20	2.30	2.40	2.50	2.60	2.70	2.80	2.90	3.00
x=0	0.12246	0.11080	0.10026	0.09072	0.08208	0.07427	0.06721	0.06081	0.05502	0.04979
1	0.37961	0.35457	0.33085	0.30844	0.28730	0.26738	0.24866	0.23108	0.21459	0.19915
2	0.64963	0.62271	0.59604	0.56971	0.54381	0.51843	0.49362	0.46945	0.44596	0.42319
3	0.83864	0.81935	0.79935	0.77872	0.75758	0.73600	0.71409	0.69194	0.66962	0.64723
4	0.93787	0.92750	0.91625	0.90413	0.89118	0.87742	0.86291	0.84768	0.83178	0.81526
5	0.97955	0.97509	0.97002	0.96433	0.95798	0.95096	0.94327	0.93489	0.92583	0.91608
6	0.99414	0.99254	0.99064	0.98841	0.98581	0.98283	0.97943	0.97559	0.97128	0.96649
7	0.99851	0.99802	0.99741	0.99666	0.99575	0.99467	0.99338	0.99187	0.99012	0.98810
8	0.99966	0.99953	0.99936	0.99914	0.99886	0.99851	0.99809	0.99757	0.99694	0.99620
9	0.99993	0.99990	0.99986	0.99980	0.99972	0.99962	0.99950	0.99934	0.99914	0.99890
10	0.99999	0.99998	0.99997	0.99996	0.99994	0.99991	0.99988	0.99984	0.99978	0.99971
11	1.00000	1.00000	0.99999	0.99999	0.99999	0.99998	0.99997	0.99996	0.99995	0.99993
12			1.00000	1.00000	1.00000	1.00000	0.99999	0.99999	0.99999	0.99998
13							1.00000	1.00000	1.00000	1.00000

m=	3.10	3.20	3.30	3.40	3.50	3.60	3.70	3.80	3.90	4.00
x=0	0.04505	0.04076	0.03688	0.03337	0.03020	0.02732	0.02472	0.02237	0.02024	0.01832
1	0.18470	0.17120	0.15860	0.14684	0.13589	0.12569	0.11620	0.10738	0.09919	0.09158
2	0.40116	0.37990	0.35943	0.33974	0.32085	0.30275	0.28543	0.26890	0.25313	0.23810
3	0.62484	0.60252	0.58034	0.55836	0.53663	0.51522	0.49415	0.47348	0.45325	0.43347
4	0.79819	0.78061	0.76259	0.74418	0.72544	0.70644	0.68722	0.66784	0.64837	0.62884
5	0.90567	0.89459	0.88288	0.87054	0.85761	0.84412	0.83009	0.81556	0.80056	0.78513
6	0.96120	0.95538	0.94903	0.94215	0.93471	0.92673	0.91819	0.90911	0.89948	0.88933
7	0.98579	0.98317	0.98022	0.97693	0.97326	0.96921	0.96476	0.95989	0.95460	0.94887
8	0.99532	0.99429	0.99309	0.99171	0.99013	0.98833	0.98630	0.98402	0.98147	0.97864
9	0.99860	0.99824	0.99781	0.99729	0.99669	0.99598	0.99515	0.99420	0.99311	0.99187
10	0.99962	0.99950	0.99936	0.99919	0.99898	0.99873	0.99843	0.99807	0.99765	0.99716
11	0.99990	0.99987	0.99983	0.99978	0.99971	0.99963	0.99953	0.99941	0.99926	0.99908
12	0.99998	0.99997	0.99996	0.99994	0.99992	0.99990	0.99987	0.99983	0.99978	0.99973
13	1.00000	0.99999	0.99999	0.99999	0.99998	0.99997	0.99997	0.99996	0.99994	0.99992
14		1.00000	1.00000	1.00000	1.00000	0.99999	0.99999	0.99999	0.99999	0.99998
15						1.00000	1.00000	1.00000	1.00000	1.00000

Appendices

m=	4.10	4.20	4.30	4.40	4.50	4.60	4.70	4.80	4.90	5.00
x=0	0.01657	0.01500	0.01357	0.01228	0.01111	0.01005	0.00910	0.00823	0.00745	0.00674
1	0.08452	0.07798	0.07191	0.06630	0.06110	0.05629	0.05184	0.04773	0.04393	0.04043
2	0.22381	0.21024	0.19735	0.18514	0.17358	0.16264	0.15230	0.14254	0.13333	0.12465
3	0.41418	0.39540	0.37715	0.35945	0.34230	0.32571	0.30968	0.29423	0.27934	0.26503
4	0.60931	0.58983	0.57044	0.55118	0.53210	0.51323	0.49461	0.47626	0.45821	0.44049
5	0.76931	0.75314	0.73666	0.71991	0.70293	0.68576	0.66844	0.65101	0.63350	0.61596
6	0.87865	0.86746	0.85579	0.84365	0.83105	0.81803	0.80461	0.79080	0.77665	0.76218
7	0.94269	0.93606	0.92897	0.92142	0.91341	0.90495	0.89603	0.88667	0.87686	0.86663
8	0.97551	0.97207	0.96830	0.96420	0.95974	0.95493	0.94974	0.94418	0.93824	0.93191
9	0.99046	0.98887	0.98709	0.98511	0.98291	0.98047	0.97779	0.97486	0.97166	0.96817
10	0.99659	0.99593	0.99518	0.99431	0.99333	0.99222	0.99098	0.98958	0.98803	0.98630
11	0.99887	0.99863	0.99833	0.99799	0.99760	0.99714	0.99661	0.99601	0.99532	0.99455
12	0.99966	0.99957	0.99947	0.99934	0.99919	0.99902	0.99882	0.99858	0.99830	0.99798
13	0.99990	0.99987	0.99984	0.99980	0.99975	0.99969	0.99961	0.99953	0.99942	0.99930
14	0.99997	0.99997	0.99996	0.99994	0.99993	0.99991	0.99988	0.99985	0.99982	0.99977
15	0.99999	0.99999	0.99999	0.99998	0.99998	0.99997	0.99997	0.99996	0.99995	0.99993
16	1.00000	1.00000	1.00000	1.00000	0.99999	0.99999	0.99999	0.99999	0.99998	0.99998
17					1.00000	1.00000	1.00000	1.00000	1.00000	0.99999
18										1.00000

m=	5.20	5.40	5.60	5.80	6.00	6.20	6.40	6.60	6.80	7.00
x=0	0.00552	0.00452	0.00370	0.00303	0.00248	0.00203	0.00166	0.00136	0.00111	0.00091
1	0.03420	0.02891	0.02441	0.02059	0.01735	0.01461	0.01230	0.01034	0.00869	0.00730
2	0.10879	0.09476	0.08239	0.07151	0.06197	0.05362	0.04632	0.03997	0.03444	0.02964
3	0.23807	0.21329	0.19062	0.16996	0.15120	0.13423	0.11892	0.10515	0.09281	0.08177
4	0.40613	0.37331	0.34215	0.31272	0.28506	0.25918	0.23507	0.21270	0.19203	0.17299
5	0.58091	0.54613	0.51186	0.47831	0.44568	0.41411	0.38374	0.35467	0.32698	0.30071
6	0.73239	0.70167	0.67026	0.63839	0.60630	0.57421	0.54233	0.51084	0.47992	0.44971
7	0.84492	0.82166	0.79698	0.77103	0.74398	0.71602	0.68732	0.65808	0.62849	0.59871
8	0.91806	0.90265	0.88568	0.86719	0.84724	0.82591	0.80331	0.77956	0.75477	0.72909
9	0.96033	0.95125	0.94087	0.92916	0.91608	0.90162	0.88580	0.86864	0.85018	0.83050
10	0.98230	0.97749	0.97178	0.96510	0.95738	0.94856	0.93859	0.92743	0.91507	0.90148
11	0.99269	0.99037	0.98751	0.98405	0.97991	0.97502	0.96930	0.96271	0.95517	0.94665
12	0.99719	0.99617	0.99486	0.99321	0.99117	0.98868	0.98568	0.98211	0.97790	0.97300
13	0.99899	0.99857	0.99802	0.99730	0.99637	0.99520	0.99375	0.99196	0.98979	0.98719
14	0.99966	0.99950	0.99928	0.99899	0.99860	0.99809	0.99744	0.99661	0.99557	0.99428
15	0.99989	0.99984	0.99976	0.99964	0.99949	0.99928	0.99901	0.99865	0.99818	0.99759
16	0.99997	0.99995	0.99992	0.99988	0.99983	0.99975	0.99964	0.99949	0.99930	0.99904
17	0.99999	0.99999	0.99998	0.99996	0.99994	0.99991	0.99987	0.99982	0.99974	0.99964
18	1.00000	1.00000	0.99999	0.99999	0.99998	0.99997	0.99996	0.99994	0.99991	0.99987
19			1.00000	1.00000	0.99999	0.99999	0.99999	0.99998	0.99997	0.99996
20					1.00000	1.00000	1.00000	0.99999	0.99999	0.99999
21								1.00000	1.00000	1.00000

m=	7.20	7.40	7.60	7.80	8.00	8.20	8.40	8.60	8.80	9.00
x=0	0.00075	0.00061	0.00050	0.00041	0.00034	0.00027	0.00022	0.00018	0.00015	0.00012
1	0.00612	0.00513	0.00430	0.00361	0.00302	0.00253	0.00211	0.00177	0.00148	0.00123
2	0.02547	0.02187	0.01876	0.01607	0.01375	0.01176	0.01005	0.00858	0.00731	0.00623
3	0.07192	0.06315	0.05537	0.04848	0.04238	0.03700	0.03226	0.02809	0.02443	0.02123
4	0.15552	0.13953	0.12494	0.11167	0.09963	0.08874	0.07891	0.07005	0.06210	0.05496
5	0.27590	0.25256	0.23068	0.21025	0.19124	0.17359	0.15728	0.14223	0.12839	0.11569
6	0.42036	0.39196	0.36462	0.33841	0.31337	0.28956	0.26699	0.24568	0.22561	0.20678
7	0.56894	0.53933	0.51004	0.48121	0.45296	0.42541	0.39865	0.37277	0.34783	0.32390
8	0.70267	0.67565	0.64819	0.62044	0.59255	0.56465	0.53689	0.50940	0.48228	0.45565
9	0.80965	0.78773	0.76485	0.74111	0.71662	0.69152	0.66592	0.63995	0.61374	0.58741
10	0.88668	0.87068	0.85351	0.83523	0.81589	0.79555	0.77430	0.75223	0.72942	0.70599
11	0.93709	0.92647	0.91477	0.90197	0.88808	0.87310	0.85707	0.84001	0.82197	0.80301
12	0.96734	0.96088	0.95357	0.94535	0.93620	0.92609	0.91500	0.90292	0.88984	0.87577
13	0.98410	0.98047	0.97625	0.97138	0.96582	0.95952	0.95244	0.94453	0.93578	0.92615
14	0.99272	0.99082	0.98856	0.98588	0.98274	0.97910	0.97490	0.97010	0.96466	0.95853
15	0.99685	0.99593	0.99480	0.99342	0.99177	0.98980	0.98747	0.98475	0.98160	0.97796
16	0.99871	0.99829	0.99776	0.99710	0.99628	0.99528	0.99408	0.99263	0.99092	0.98889
17	0.99950	0.99932	0.99909	0.99879	0.99841	0.99793	0.99734	0.99662	0.99574	0.99468
18	0.99982	0.99974	0.99964	0.99952	0.99935	0.99914	0.99886	0.99852	0.99810	0.99757
19	0.99994	0.99991	0.99987	0.99982	0.99975	0.99966	0.99954	0.99938	0.99919	0.99894
20	0.99998	0.99997	0.99995	0.99993	0.99991	0.99987	0.99982	0.99975	0.99967	0.99956
21	0.99999	0.99999	0.99998	0.99998	0.99997	0.99995	0.99993	0.99991	0.99987	0.99983
22	1.00000	1.00000	0.99999	0.99999	0.99999	0.99998	0.99998	0.99997	0.99995	0.99993
23			1.00000	1.00000	1.00000	0.99999	0.99999	0.99999	0.99999	0.99998
24						1.00000	1.00000	1.00000	0.99999	0.99999
25									1.00000	1.00000

Appendices

m=	9.20	9.40	9.60	9.80	10.00	12.00	14.00	16.00	18.00	20.00
x=0	0.00010	0.00008	0.00007	0.00006	0.00005	0.00001	0.00000	0.00000	0.00000	0.00000
1	0.00103	0.00086	0.00072	0.00060	0.00050	0.00008	0.00001	0.00000	0.00000	0.00000
2	0.00531	0.00452	0.00384	0.00326	0.00277	0.00052	0.00009	0.00002	0.00000	0.00000
3	0.01842	0.01597	0.01383	0.01196	0.01034	0.00229	0.00047	0.00009	0.00002	0.00000
4	0.04858	0.04288	0.03779	0.03327	0.02925	0.00760	0.00181	0.00040	0.00008	0.00002
5	0.10407	0.09347	0.08381	0.07504	0.06709	0.02034	0.00553	0.00138	0.00032	0.00007
6	0.18917	0.17273	0.15745	0.14327	0.13014	0.04582	0.01423	0.00401	0.00104	0.00026
7	0.30100	0.27917	0.25843	0.23878	0.22022	0.08950	0.03162	0.01000	0.00289	0.00078
8	0.42961	0.40424	0.37961	0.35578	0.33282	0.15503	0.06206	0.02199	0.00706	0.00209
9	0.56108	0.53486	0.50886	0.48319	0.45793	0.24239	0.10940	0.04330	0.01538	0.00500
10	0.68203	0.65764	0.63295	0.60804	0.58304	0.34723	0.17568	0.07740	0.03037	0.01081
11	0.78318	0.76257	0.74124	0.71928	0.69678	0.46160	0.26004	0.12699	0.05489	0.02139
12	0.86074	0.84476	0.82788	0.81012	0.79156	0.57597	0.35846	0.19312	0.09167	0.03901
13	0.91562	0.90419	0.89185	0.87860	0.86446	0.68154	0.46445	0.27451	0.14260	0.06613
14	0.95169	0.94410	0.93572	0.92654	0.91654	0.77202	0.57044	0.36753	0.20808	0.10486
15	0.97381	0.96910	0.96380	0.95786	0.95126	0.84442	0.66936	0.46674	0.28665	0.15651
16	0.98653	0.98379	0.98064	0.97704	0.97296	0.89871	0.75592	0.56596	0.37505	0.22107
17	0.99342	0.99192	0.99016	0.98810	0.98572	0.93703	0.82720	0.65934	0.46865	0.29703
18	0.99693	0.99616	0.99523	0.99412	0.99281	0.96258	0.88264	0.74235	0.56224	0.38142
19	0.99864	0.99826	0.99779	0.99723	0.99655	0.97872	0.92350	0.81225	0.65092	0.47026
20	0.99942	0.99924	0.99902	0.99875	0.99841	0.98840	0.95209	0.86817	0.73072	0.55909
21	0.99976	0.99969	0.99959	0.99946	0.99930	0.99393	0.97116	0.91077	0.79912	0.64370
22	0.99991	0.99988	0.99983	0.99978	0.99970	0.99695	0.98329	0.94176	0.85509	0.72061
23	0.99997	0.99995	0.99993	0.99991	0.99988	0.99853	0.99067	0.96331	0.89889	0.78749
24	0.99999	0.99998	0.99998	0.99997	0.99995	0.99931	0.99498	0.97768	0.93174	0.84323
25	1.00000	0.99999	0.99999	0.99999	0.99998	0.99969	0.99739	0.98688	0.95539	0.88782
26		1.00000	1.00000	1.00000	0.99999	0.99987	0.99869	0.99254	0.97177	0.92211
27					1.00000	0.99994	0.99936	0.99589	0.98268	0.94752
28						0.99998	0.99970	0.99781	0.98970	0.96567
29						0.99999	0.99986	0.99887	0.99406	0.97818
30						1.00000	0.99994	0.99943	0.99667	0.98653
31							0.99997	0.99972	0.99819	0.99191
32							0.99999	0.99987	0.99904	0.99527
33							1.00000	0.99994	0.99951	0.99731
34								0.99997	0.99975	0.99851
35								0.99999	0.99988	0.99920
36								1.00000	0.99994	0.99958
37									0.99997	0.99978
38									0.99999	0.99989
39									0.99999	0.99995
40									1.00000	0.99997
41										0.99999
42										0.99999
43										1.00000

A5.3 NORMAL DISTRIBUTION

This table shows prob$(Z \leq z)$ where

$$f(z) = \frac{\exp(-0.5z^2)}{(2\pi)^{0.5}}$$

Z is a standardised Normal variable with $\mu = 0$ and $\sigma = 1$.

If X is a Normal variable with arbitrary parameters μ and σ then

$$F(X = x) = F(Z = (x - \mu)/\sigma)$$

For example, if X is Normally distributed with $\mu = 3.6$ and $\sigma = 0.4$ then

$$\begin{aligned}\text{prob}(X \leq 4.22) = F(X = 4.22) &= F(Z = (4.22 - 3.6)/0.4) \\ &= F(Z = 1.55) \\ &= 0.93943\end{aligned}$$

The distribution function is only tabulated for $z \geq 0$ but since the density function is symmetrical we have

$$F(z) = 1 - F(-z), \qquad z < 0$$

e.g. $\quad\begin{aligned}F(Z = -0.35) &= 1 - F(Z = 0.35) \\ &= 1 - 0.63683 = 0.36317\end{aligned}$

Appendices

STANDARD NORMAL DISTRIBUTION FUNCTION $F(z) = \text{prob}(Z \leq z)$

z	0.00	0.01	0.02	0.03	0.04	0.05	0.06	0.07	0.08	0.09
0.0	0.50000	0.50399	0.50798	0.51197	0.51595	0.51994	0.52392	0.52790	0.53188	0.53586
0.1	0.53983	0.54380	0.54776	0.55172	0.55567	0.55962	0.56356	0.56750	0.57142	0.57535
0.2	0.57926	0.58317	0.58706	0.59095	0.59483	0.59871	0.60257	0.60642	0.61026	0.61409
0.3	0.61791	0.62172	0.62552	0.62930	0.63307	0.63683	0.64058	0.64431	0.64803	0.65173
0.4	0.65542	0.65910	0.66276	0.66640	0.67003	0.67364	0.67724	0.68082	0.68439	0.68793
0.5	0.69146	0.69497	0.69847	0.70194	0.70540	0.70884	0.71226	0.71566	0.71931	0.72240
0.6	0.72575	0.72907	0.73237	0.73565	0.73891	0.74215	0.74537	0.74857	0.75175	0.75490
0.7	0.75804	0.76115	0.76424	0.76731	0.77035	0.77337	0.77637	0.77935	0.78230	0.78524
0.8	0.78814	0.79103	0.79389	0.79673	0.79955	0.80234	0.80511	0.80785	0.81057	0.81327
0.9	0.81594	0.81859	0.82121	0.82381	0.82639	0.82894	0.83147	0.83398	0.83646	0.83891
1.0	0.84134	0.84375	0.84614	0.84850	0.85083	0.85314	0.85543	0.85769	0.85993	0.86214
1.1	0.86433	0.86650	0.86864	0.87076	0.87286	0.87493	0.87698	0.87900	0.88100	0.88298
1.2	0.88493	0.88686	0.88877	0.89065	0.89251	0.89435	0.89617	0.89796	0.89973	0.90147
1.3	0.90320	0.90490	0.90658	0.90824	0.90988	0.91149	0.91308	0.91466	0.91621	0.91774
1.4	0.91924	0.92073	0.92220	0.92364	0.92507	0.92647	0.92785	0.92922	0.93056	0.93189
1.5	0.93319	0.93448	0.93574	0.93699	0.93822	0.93943	0.94062	0.94179	0.94295	0.94408
1.6	0.94520	0.94630	0.94738	0.94845	0.94950	0.95053	0.95154	0.95254	0.95352	0.95449
1.7	0.95543	0.95637	0.95728	0.95818	0.95907	0.95994	0.96080	0.96164	0.96246	0.96327
1.8	0.96407	0.96485	0.96562	0.96638	0.96712	0.96784	0.96856	0.96926	0.96995	0.97062
1.9	0.97128	0.97193	0.97257	0.97320	0.97381	0.97441	0.97500	0.97558	0.97615	0.97670
2.0	0.97725	0.97778	0.97831	0.97882	0.97932	0.97982	0.98030	0.98077	0.98124	0.98169
2.1	0.98214	0.98257	0.98300	0.98341	0.98382	0.98422	0.98461	0.98500	0.98537	0.98574
2.2	0.98610	0.98645	0.98679	0.98713	0.98745	0.98778	0.98809	0.98840	0.98870	0.98899
2.3	0.98928	0.98956	0.98983	0.99010	0.99036	0.99061	0.99086	0.99111	0.99134	0.99158
2.4	0.99180	0.99202	0.99224	0.99245	0.99266	0.99286	0.99305	0.99324	0.99343	0.99361
2.5	0.99379	0.99396	0.99413	0.99430	0.99446	0.99461	0.99477	0.99492	0.99506	0.99520
2.6	0.99534	0.99547	0.99560	0.99573	0.99585	0.99598	0.99609	0.99621	0.99632	0.99643
2.7	0.99653	0.99664	0.99674	0.99683	0.99693	0.99702	0.99711	0.99720	0.99728	0.99736
2.8	0.99744	0.99752	0.99760	0.99767	0.99774	0.99781	0.99788	0.99795	0.99801	0.99807
2.9	0.99813	0.99819	0.99825	0.99831	0.99836	0.99841	0.99846	0.99851	0.99856	0.99861
3.0	0.99865	0.99869	0.99874	0.99878	0.99882	0.99886	0.99889	0.99893	0.99896	0.99900
3.1	0.99903	0.99906	0.99910	0.99913	0.99916	0.99918	0.99921	0.99924	0.99926	0.99929
3.2	0.99931	0.99934	0.99936	0.99938	0.99940	0.99942	0.99944	0.99946	0.99948	0.99950
3.3	0.99952	0.99953	0.99955	0.99957	0.99958	0.99960	0.99961	0.99962	0.99964	0.99965
3.4	0.99966	0.99968	0.99969	0.99970	0.99971	0.99972	0.99973	0.99974	0.99975	0.99976
3.5	0.99977	0.99978	0.99978	0.99979	0.99980	0.99981	0.99981	0.99982	0.99983	0.99983
3.6	0.99984	0.99985	0.99985	0.99986	0.99986	0.99987	0.99987	0.99988	0.99988	0.99989
3.7	0.99989	0.99990	0.99990	0.99990	0.99991	0.99991	0.99992	0.99992	0.99992	0.99992
3.8	0.99993	0.99993	0.99993	0.99994	0.99994	0.99994	0.99994	0.99995	0.99995	0.99995
3.9	0.99995	0.99995	0.99996	0.99996	0.99996	0.99996	0.99996	0.99996	0.99997	0.99997

A5.4 BETA DISTRIBUTION

The Beta density function is

$$f(x) = kx^a(1-x)^b, \qquad 0 \le x \le 1$$

which is in form similar to the Binomial distribution. For integer values of a and b the two distribution functions are related:

$$F_{BETA}(x|a,b) = 1 - F_{BINOMIAL}(y|n,p)$$

where

$y = a$
$n = a + b + 1$
$p = x$

If, for instance, we have a Beta distribution with parameters $a = 2$ and $b = 3$ and we wish to find $F(0.3)$ then the corresponding Binomial distribution has parameters

$y = 2$
$n = 2 + 3 + 1 = 6$
$p = 0.3$

From the tables of the Binomial distribution function in A5.1 we get a value of 0.74431 and so

$$F(0.3) = 1 - 0.74431 = 0.25569$$

Some tabulations of the Beta distribution are given in Winkler (1972) while Philips (1973) gives tables of 95% and 99% credible intervals.

A5.5 GAMMA DISTRIBUTION

The Gamma density function is

$$f(x) = kx^a \exp(-bx), \qquad x \geq 0$$

which is in form similar to the Poisson distribution. As with the Beta and Binomial distributions there is a correspondence between the distribution functions for integer values of a and b

$$F_{\text{GAMMA}}(x|a,b) = 1 - F_{\text{POISSON}}(y|m)$$

where

$y = a$
$m = bx$

If for a Gamma distribution with parameters $a = 3$ and $b = 4$ we wish to find $F(1.4)$ then the corresponding Poisson distribution has parameters

$y = 3$
$m = 4 \times 1.4 = 5.6$

From the tables of the Poisson distribution function in Appendix A5.2 we get a value of 0.19062 and so

$$F(1.4) = 1 - 0.19062 = 0.80938$$

A5.6 χ-SQUARED DISTRIBUTION

The χ^2 distribution is the sum of squares of m independent standard normal distributions, $X_1^2, X_2^2, \ldots, X_m^2$.

$$\chi^2 = X_1^2 + X_2^2 + \ldots + X_m^2$$

is said to be a χ^2 distribution with m degrees of freedom and density function

$$f(x) = kx^{(m/2)-1}\exp(-x/2), \qquad x > 0$$

This function leads to the tabulations below.

Comparing this formula with the Gamma density function in the previous section shows that the two are equivalent. The χ^2 density is equivalent to a Gamma density with parameters $a = (m/2) - 1$ and $b = 1/2$.

The χ^2 distribution has a mean of m and a variance of $2m$.

χ^2 distributions are additive. If A and B are both χ^2 distributed with degrees of freedom a and b then $A + B$ is also χ^2 with $a + b$ degrees of freedom.

The distribution has many applications. Here are two.

Sample variances

If a random sample of size n is taken from a Normal population with variance σ^2 then ns^2/σ^2 has a χ^2 distribution with $n - 1$ degrees of freedom. s^2 is the variance of the sample.

Discrimination

Given expected and observed frequencies e_i and o_i then

$$G^2 = 2\Sigma\, o_i\ln(o_i/e_i)$$

has a sampling distribution well approximated by χ^2 with an error of the magnitude of $n^{-0.5}$, where n is the sample size (Jaynes, 1983, pp262–263). Agresti (1990, p246) quotes suggestions that the χ^2 approximation is poor if $n/N < 5$ (N is the number of classes) and also if a large number of the expected frequencies are less than 0.5.

The number of degrees of freedom is determined by the method by which the expected frequencies were generated. In general

degrees of freedom = no. of classes − no. of parameters

G^2 is derived from information theoretic considerations by Kullback (1959) but is also known via another derivation as the likelihood-ratio statistic or Wilks' statistic.

Pearson provided another measure of dissimilarity whose sampling distribution is$_2$ also well approximated by χ^2 and which is often used instead of G^2: $X^2 = \Sigma(o_i - e_i)^2/e_i$. See Agresti (1990, secs. 3.3.2 and 7.7.3), Delucchi (1983), Chapman (1976) and also Cochran (1952) for a discussion of the use of both measures. The χ^2 approximations are good for large samples and less so as n gets smaller.

POINTS FROM THE χ^2 DISTRIBUTION

degrees of freedom	probability that χ^2 exceeds the critical value given in the table				
	0.250	0.100	0.050	0.025	0.010
1	1.3233	2.7055	3.8415	5.0239	6.6349
2	2.7726	4.6052	5.9915	7.3778	9.2103
3	4.1083	6.2514	7.8147	9.3484	11.3449
4	5.3853	7.7794	9.4877	11.1433	13.2767
5	6.6527	9.2364	11.0705	12.8325	15.0863
6	7.8408	10.6446	12.5916	14.4494	16.8119
7	9.0372	12.0170	14.0671	16.0128	18.4753
8	10.2189	13.3616	15.5073	17.5345	20.0902
9	11.3888	14.6837	16.9190	19.0228	21.6660
10	12.5489	15.9872	18.3070	20.4832	23.2093
11	13.7007	17.2750	19.6751	21.9200	24.7250
12	14.8454	18.5493	21.0261	23.3367	26.2170
13	15.9839	19.8119	22.3620	24.7356	27.6882
14	17.1169	21.0641	23.6848	26.1189	29.1412
15	18.2451	22.3071	24.9958	27.4884	30.5779
16	19.3689	23.5418	26.2962	28.8454	31.9999
17	20.4887	24.7690	27.2871	30.1910	33.4087
18	21.6049	25.9894	28.8693	31.5264	34.8053
19	22.7178	27.2036	30.1435	32.8523	36.1909
20	23.8277	28.4120	31.4104	34.1696	37.5662
21	24.9348	29.6151	32.6706	35.4789	38.9322
22	26.0393	30.8133	33.9244	36.7807	40.2894
23	27.1413	32.0069	35.1725	38.0756	41.6384
24	28.2412	33.1962	36.4150	39.3641	42.9798
25	29.3389	34.3816	37.6525	40.6465	44.3141
26	30.4346	35.5632	38.8851	41.9232	45.6417
27	31.5284	36.7412	40.1133	43.1945	46.9629
28	32.6205	37.9159	41.3371	44.4608	48.2782
29	33.7109	39.0875	42.5570	45.7223	49.5879
30	34.7997	40.2560	43.7730	46.9792	50.8922
40	45.6160	51.8051	55.7585	59.3417	63.6907
50	56.3336	63.1671	67.5048	71.4202	76.1539
60	66.9815	74.3970	79.0819	83.2977	88.3794
70	77.5767	85.5270	90.5312	95.0232	100.4250
80	88.1303	96.5782	101.8790	106.6290	112.3290
90	98.6499	107.5650	113.1450	118.1360	124.1160
100	109.1410	118.4980	124.3420	129.5610	135.8070

A5.7 THE t DISTRIBUTION

The t distribution gives the distribution of sample means from a population whose variance is unknown and estimated using (7.4). It is therefore appropriate as both a likelihood and posterior inferential distribution in such cases.

The table gives t values such that $1 - F(t) = \alpha$, where the probability α is shown at the head of each column.

Note that the values given in the last line of the table are exactly those that are obtained from the standardised Normal distribution of Appendix A5.3.

When making an inference about a population mean from a sample of size n use the t distribution with $(n - 1)$ degrees of freedom.

POINTS FROM THE t DISTRIBUTION

degrees of freedom	probability, α, that t exceeds the value in the table					
	0.250	0.100	0.050	0.025	0.010	0.005
1	1.000	3.078	6.314	12.706	31.821	63.657
2	0.816	1.886	2.920	4.303	6.965	9.925
3	0.765	1.638	2.353	3.182	4.541	5.841
4	0.741	1.533	2.132	2.776	3.747	4.604
5	0.727	1.476	2.015	2.571	3.365	4.032
6	0.718	1.440	1.943	2.447	3.143	3.707
7	0.711	1.415	1.895	2.365	2.998	3.499
8	0.706	1.397	1.860	2.306	2.896	3.355
9	0.703	1.383	1.833	2.262	2.821	3.250
10	0.700	1.372	1.812	2.228	2.764	3.169
11	0.697	1.363	1.796	2.201	2.718	3.106
12	0.695	1.356	1.782	2.179	2.681	3.055
13	0.694	1.350	1.771	2.160	2.650	3.012
14	0.692	1.345	1.761	2.145	2.624	2.977
15	0.691	1.341	1.753	2.131	2.602	2.947
16	0.690	1.337	1.746	2.120	2.583	2.921
17	0.689	1.333	1.740	2.110	2.567	2.898
18	0.688	1.330	1.734	2.101	2.552	2.878
19	0.688	1.328	1.729	2.093	2.539	2.861
20	0.687	1.325	1.725	2.086	2.528	2.845
21	0.686	1.323	1.721	2.080	2.518	2.831
22	0.686	1.321	1.717	2.074	2.508	2.819
23	0.685	1.319	1.714	2.069	2.500	2.807
24	0.685	1.318	1.711	2.064	2.492	2.797
25	0.684	1.316	1.708	2.060	2.485	2.787
26	0.684	1.315	1.706	2.056	2.479	2.779
27	0.684	1.314	1.703	2.052	2.473	2.771
28	0.683	1.313	1.701	2.048	2.467	2.763
29	0.683	1.311	1.699	2.045	2.462	2.756
30	0.683	1.310	1.697	2.042	2.457	2.750
40	0.681	1.303	1.684	2.021	2.423	2.704
60	0.679	1.296	1.671	2.000	2.390	2.660
120	0.677	1.289	1.658	1.980	2.358	2.617
infinity	0.674	1.282	1.645	1.960	2.326	2.576

APPENDIX A6
MEASURES OF LOCATION AND LOSS FUNCTIONS

A variable X is described by a probability density function $f(x)$ and distribution function $F(x)$. We have to choose a particular value, A, as our best guess at a value $X = x$ to be chosen at random from $f(x)$. The penalty incurred by a wrong guess is described by the loss function $g(x|A)$. We choose A to minimise the expected loss

$$L = \int_{-\infty}^{\infty} g(x|A) f(x) dx$$

LINEAR LOSS FUNCTION (Figure 7.1)

If the loss is of the form

$$g(x|A) = |x - A|$$

then L is evaluated in two parts, for $x \leq A$ and $x > A$.

$$L = \int_{-\infty}^{A} (A - x) f(x) dx + \int_{A}^{\infty} (x - A) f(x) dx$$

$$= A \int_{-\infty}^{A} f(x) dx - \int_{-\infty}^{A} x f(x) dx + \int_{A}^{\infty} x f(x) dx - A \int_{A}^{\infty} f(x) dx$$

$$= AF(A) - \int_{-\infty}^{A} x f(x) dx + \int_{A}^{\infty} x f(x) dx - A(1 - F(A))$$

$$L = 2AF(A) - A - \int_{-\infty}^{A} x f(x) dx + \int_{A}^{\infty} x f(x) dx$$

To minimise L

$$\frac{\partial L}{\partial A} = 2Af(A) + 2F(A) - 1 - Af(A) - Af(A) = 2F(A) - 1 = 0$$

so $\quad F(A) = 0.5$

and A is the *median*.

QUADRATIC LOSS FUNCTION (Figure 7.2)

Given

$$g(x|A) = (x - A)^2 = x^2 - 2Ax + A^2$$

then $L = \int_{-\infty}^{\infty} x^2 f(x)dx - 2A\int_{-\infty}^{\infty} xf(x)dx + A^2 \int_{-\infty}^{\infty} f(x)dx$

$ = E[x^2] - 2A\mu + A^2$

which has a minimum at

$$\frac{\partial L}{\partial A} = -2\mu + 2A = 2(A - \mu) = 0$$

so A is the *mean*.

Note that at this optimum value of $A = \mu$, the minimum loss is $L = E[x^2] - \mu^2 = \sigma^2$, the variance of X.

CONSTANT LOSS FUNCTION WITH GAP (Figure 7.3)

We have

$g(x|A) = 0 \qquad\qquad A - d \leq x \leq A + d$
$ = k \qquad\qquad$ elsewhere

so

$L = k\int_{-\infty}^{A-d} f(x)dx + k\int_{A+d}^{\infty} f(x)dx$

$ = kF(A-d) + k(1 - F(A+d))$

so

$$\frac{\partial L}{\partial A} = kf(A-d) - kf(A+d)$$

$\phantom{\frac{\partial L}{\partial A}} = k(f(A-d) - f(A+d)) = 0$

gives $f(A - d) = f(A + d)$.

If the width of the gap (= $2d$) is very small compared to the standard deviation of the distribution it seems sensible to take the mid-point of the gap as the best estimate of A and so A is the *mode*. You can see that as d tends towards a value of zero the gap will shrink and, in the limit, will isolate the mode.

APPENDIX A7
CONDITIONS FOR A CREDIBLE INTERVAL

A variable X has density function $f(x)$ and distribution function $F(x)$.

We require an interval between $x = L$ and $x = H$ such that the width of the interval is minimised and $\text{prob}(L \leq x \leq H) = c$.

So we wish to minimise

$$R = H - L$$

subject to

$$F(H) - F(L) = c$$

Form Lagrangian M with multiplier a:

$$M = H - L + a(F(H) - F(L))$$

Now

$$\frac{\partial M}{\partial H} = 1 + af(H) = 0$$

$$f(H) = -1/a$$

and

$$\frac{\partial M}{\partial L} = -1 - af(L) = 0$$

$$f(L) = -1/a = f(H)$$

So for a credible interval $x = (L, H)$, $f(L) = f(H)$.

APPENDIX A8
THE REGRESSION MODEL

Given data points (x_i, y_i), $i = 1,\ldots,n$, we wish to find the linear model

$$y = a + bx \tag{A8.1}$$

that minimises

$$S = \Sigma(y_i - a - bx_i)^2 \tag{A8.2}$$

Differentiating with respect to the parameters gives, first,

$$\frac{\partial S}{\partial a} = -2\Sigma(y_i - a - bx_i) = 0$$

so $\quad \Sigma y_i - na - b\Sigma x_i = 0 \tag{A8.3}$

Dividing by n and rearranging gives

$$a = \bar{y} - b\bar{x} \tag{A8.4}$$

Differentiating with respect to b gives

$$\frac{\partial S}{\partial b} = -2\Sigma x_i (y_i - a - bx_i) = 0$$

so $\quad \Sigma x_i y_i - a\Sigma x_i - b\Sigma x_i^2 = 0$

Substituting for a from (A8.4) and dividing by n

$$b[\Sigma x_i^2 / n - (\bar{x})^2] = \Sigma x_i y_i / n - \bar{y}\bar{x} \tag{A8.5}$$

Now,

$$\text{cov}(x, y) = \Sigma(x_i - \bar{x})(y_i - \bar{y})/n$$

$$= \Sigma x_i y_i / n - \bar{y}\Sigma x_i / n - \bar{x}\Sigma y_i / n + n\bar{x}\bar{y}/n$$

$$= \Sigma x_i y_i / n - \bar{x}\bar{y}$$

and, as a special case,

$$\text{var}(x) = \text{cov}(x,x) = \Sigma x_i^2 / n - (\bar{x})^2$$

Making these substitutions in (A8.5) gives

$$b = \text{cov}(x,y)/\text{var}(x) \qquad (A8.6)$$

Substituting from (A8.4) into (A8.1) gives

$$y = \bar{y} + b(x - \bar{x}) \qquad (A8.7)$$

which is an alternative statement of the model. Note that when $x = \bar{x}$ then $y = \bar{y}$, so the regression line passes through the point $(\bar{x}, \bar{y})$, the centroid of the data.

The mean of the residuals is

$$\Sigma(y_i - a - bx_i)/n$$

But from (A8.3) the sum is zero and so, therefore, is the mean. This shows that the model is unbiased.

Now, consider the sum of squares of the residuals using the model in (A8.7). This gives

$$S = \Sigma[(y_i - \bar{y}) - b(x_i - \bar{x})]^2$$

$$= \Sigma(y_i - \bar{y})^2 + b^2 \Sigma(x_i - \bar{x})^2 - 2b\Sigma(x_i - \bar{x})(y_i - \bar{y})$$

But from (A8.6)

$$b = \Sigma(x_i - \bar{x})(y_i - \bar{y}) / \Sigma(x_i - \bar{x})^2$$

so $\quad S = \Sigma(y_i - \bar{y})^2 - \Sigma[(x_i - \bar{x})(y_i - \bar{y})]^2 / \Sigma(x_i - \bar{x})^2$

From (10.12)

$$R^2 = 1 - S / \Sigma(y_i - \bar{y})^2$$

which, on substituting for S, gives

$$R^2 = [\Sigma(x_i - \bar{x})(y_i - \bar{y})]^2 / [\Sigma(x_i - \bar{x})^2 \Sigma(y_i - \bar{y})^2] = r^2$$

from (10.2).

Suppose now that we consider the data to constitute a sample drawn from some population. The model permits estimates of y conditional upon the known value of x. The conditional variance of y is

$$s_y^2 = S/(n-2) \tag{A8.8}$$

where $(n-2)$ is the number of degrees of freedom.

The product $\Sigma(x_i - \bar{x})(y_i - \bar{y})$ may be simplified:

$$\Sigma(x_i - \bar{x})(y_i - \bar{y}) = \Sigma y_i(x_i - \bar{x}) - \bar{y}\Sigma(x_i - \bar{x})$$

$$= \Sigma y_i(x_i - \bar{x}) - \bar{y}\Sigma x_i - n\bar{x}\bar{y}$$

$$= \Sigma y_i(x_i - \bar{x})$$

Substituting into (A8.6) gives

$$b = \Sigma y_i(x_i - \bar{x}) / \Sigma(x_i - \bar{x})^2$$

Using (7.1) and remembering that all x's are assumed known so that the only source of variation is in y

$$\text{var}(b) = \text{var}(y)\Sigma(x_i - \bar{x})^2 / [\Sigma(x_i - \bar{x})^2]^2$$

so $\quad \text{var}(b) = s_y^2 / \Sigma(x_i - \bar{x})^2 \tag{A8.9}$

and this forms the basis of inferences about the population value of b, usually denoted by β.

The estimated value of y is the value predicted by the model plus the error

$$e = y - (a + bx)$$

so from (A8.7) the estimated value is

$$y = \bar{y} + b(x - \bar{x}) + e$$

Using (7.1) again and considering only variation due to y

$$\text{var}(y) = \text{var}(\bar{y}) + (x - \bar{x})^2 \text{var}(b) + \text{var}(e)$$

From (7.2) $\text{var}(\bar{y}) = \text{var}(y)/n = s_y^2/n$, so

$$\text{var}(y) = s_y^2/n + (x-\bar{x})^2 s_y^2 / \Sigma(x_i - \bar{x})^2 + s_y^2$$

$$= s_y^2 [1 + 1/n + (x-\bar{x})^2 / \Sigma(x_i - \bar{x})^2]$$

(A8.10)

This gives a basis for constructing confidence intervals for any forecast (interpolation) made using the model.

Index

α, 175
χ^2, see chi-squared
μ, 18, 164
σ, 23, 164
σ^2, 22, 24
σ^2_{EST}, 166
Σ, 18
ψ, 173-4, 195

addition of probabilities, 4-6
association between variables, 191
authorship, determination of, 129-31
average, 18, 158-9

Bayes' Rule, 107-9
    as information transmission, 198-200
Beta distribution, 113-14, 123, 135, 331
    Normal approximation to, 122
    tables of, 348
bias, 58-9, 258
binary numbers, 210
Binomial expansion, 328-9
Binomial distribution, 109-10, 326-8
    Normal approximation to, 122, 132-3
    Poisson approximation to, 117, 332
    tables of, 332-40
bits, 40, 320
box plot, 23-4

Central Limit Theorem, 165
chi-squared, 174-6, 194, 350-1
    tables of, 352
coding, 54-6
coefficient of determination, 259-60
    adjusted for degrees of freedom, 282
coefficient of variation, 23
combinations, 326-7
complementary events, 4
computer graphics, 24
conditional probabilities, 8
    entropy of, 188-9

confidence interval, 161
conjugate prior, 112
contingency tables, 186
    entropies of, 187-91
    maximum entropy estimates of, 212-18, 225-36, 315-18
    structural comparison of, 220-4
    three dimensional, 228-36, 317-18
continuity correction, 180
convenience prior, 112
correlation, 192, 255, 260
correlation coefficient, 254-5, 260, 276
covariance, 254
credibility level, 163
credible interval, 161, 357
cumulative probability distribution, 11

decile, 21
degrees of freedom, 175-6, 194, 196, 261, 282
density function, 14
    rectangular, 83-4, 92
    triangular, 16-17, 29-32
differential, 60-1, 311-12
discrimination, between distributions, 173
disorder, 49
dissimilarity, between distributions, 174
    testing for, 175-6
distribution function, 14
dominated action, 142, 151

e, see $\exp(x)$
elasticity of demand, 156
efficient market hypothesis, 207-8
English language, 46-7, 51
entropy, 49-51
    for contingency tables, 187-91
    and correlation, 260
    maximisation, 59, 311-18
    of probability distribution
        Normal, 259

rectangular, 84
uniform, 84
relative, 76, 314-15
equivocation, 197
errors with hypothesis, 169-70
estimated population variance, 166
events
complementary, 4
independent, 10
mutually exclusive, 5
exp(x), 320-1, 329
expected loss, 149-50
of linear model, 259
expected utility, 148
expected value, 24
of conditional entropies, 189
of perfect information, 152
exponential change, 271, 275

forecast, 261-3, 271-2, 276, 361

G^2, 174-5, 195-6, 350
Gamma distribution, 117-18, 123, 135, 331, 350
Normal approximation to, 122
tables of, 349
God, belief in, 150
grouped data, 33-4, 95

H, 40, 45, 47-8
for contingency tables, 187-91
hypothesis, 169

I, 76-7
as a measure of dissimilarity, 173
ignorance, 40
induction, 124
independent events, 10, 192, 317
testing for, 193-6
inference, 164
with linear model, 261, 360
information, 46, 50-1, 196
gain, 76, 199

perfect, value of, 152
transmission of, 196-8
information change, 76
information potential, 40, 46, 51
integration, 322-5
inter-decile range, 21, 23
interpolation, 162, 331
inter-quartile range, 20, 23
interval
confidence, 161
credible, 161
reporting, 160
intropy, 50
iterative proportional fitting procedure, 215, 236

jazz, 52
joint probability distribution, 8
entropy of, 187
maximum entropy estimates of, 212-18, 315-18

Lagrange multipliers, 312-13
least squares regression, 257
learning, 124-7
likelihood distribution, 106
for sample mean, 165
linear regression, 255-63, 358-61
computer solution, 280-2
linear transformation, 145
logarithms, 319-21
logarithmic search, 43, 86-7
logarithmic transformation, 272
log-linear model, 236, 317-18
loss, 149
loss functions for averages, 158-9, 355-6

MAD, 22, 24
marginal probability distribution, 7
as constraint, 213-17, 226-36, 315-18
entropy of, 187
from joint distribution, 219-20
maximum entropy, 311-18
principle, 59, 78

procedure, 62, 93-4
maximum entropy distribution
    given mean, 64, 78
    given mean and variance, 66-9, 83
maximum entropy density function
    given mean, 84-8
    given mean and variance, 90, 92-3
maximum entropy estimate for joint
        probabilities, 213, 225-36, 315-18
    with marginal constraints, 212-17
    with mean constraint, 216-18
Maxwell's demon, 49-50
mean, 18, 24, 159, 356
    of grouped data, 34
mean absolute deviation, 22, 24
median, 18, 23, 32, 158, 355
mid-point rule, 322, 324
minimum information discrimination
        statistic, 173-4
mode, 18, 23, 159, 356
multiplication of probabilities, 7-10
mutually exclusive events, 5

nats, 40
negative exponential distribution, 88-90, 97, 118, 120
Newcomb's paradox, 157
noise, 46, 164, 197-8
    white, 121, 259
noiseless transmission, 196
nominal variable, 10
non-informative distribution, 41
Normal distribution, 90-2, 97-8, 120-1, 330
    approximation to other distributions, 122
    entropy of, 259
    standardised, 91
    tables of, 346-7
numerical integration, 322-5

opportunity loss, 150
organisation, 52

pdf, 14
perfect information, value of, 152
permutations, 327
Poisson distribution, 116-17, 120, 328-9
    Normal approximation to, 122, 341
    tables of, 341-5
population variance, estimate of, 166
posterior distribution, 106
precision, 121
predictive distribution, 108
prior distribution, 106
probability
    addition, 4-6
    conditional, 8
    density function, 14
    distribution, 10
        cumulative, 11
    joint, 8
    marginal, 7
    multiplication, 7-10

quartiles, 20-1

r, see correlation coefficient
R^2, see coefficient of determination
random number generators, 202-4
range, 20
real variable, 13, 250
rectangular density function, 83-4
    entropy of, 84
    variance of, 92
redundancy, 49, 51-2
regression, 257, 358-61
    computer solution, 280-2
relative entropy, 76, 314-15
    as a measure of dissimilarity, 173
relative frequency, 32-3
reporting interval, 160
residual error, 257, 262
    variance of, 261, 276
risk, 143
    aversion, 143, 145
    premium, 143

root mean square error, 282

s^2, 166
s_y^2, 261-2
scattergram, 251-2
sd(X), 23
signal, 46
Simpson's rule, 322-3, 325
skew, 19-20
standard deviation, 23
    of grouped data, 34
standard error, 166
standardised Normal distribution, 92, 346
surprisal, 39, 52
surprise, 39

t distribution, 166-7, 182
    tables of, 353-4
T, 190
    and correlation, 260
    testing for, 193-4
testable requirement, 93

transmission of information, 196-8
triangular density function, 16-17, 29-32
type I and type II errors, 169-70
    probability of, 175

uniform distribution, 41, 62
utility, 143-7
    function, 145-7

validation, of a model, 171
variable
    nominal, 10
    real, 13, 250
var(X), 23
variance, 22, 24, 259, 356
    of a sum, 165

white noise, 121, 259

$\bar{x}$, 165-6

Y^2, 194, 261